AF389212

Electronics, Close-Range Sensors and Artificial Intelligence in Forestry

Electronics, Close-Range Sensors and Artificial Intelligence in Forestry

Editors

Stelian Alexandru Borz
Andrea R. Proto
Robert Keefe
Mihai Nita

MDPI • Basel • Beijing • Wuhan • Barcelona • Belgrade • Manchester • Tokyo • Cluj • Tianjin

Editors

Stelian Alexandru Borz
Department of Forest
Engineering, Forest
Management Planning and
Terrestrial Measurements
Transilvania University of
Brasov
Brasov
Romania

Andrea R. Proto
Department of AGRARIA
Mediterranean University of
Reggio Calabria
Reggio Calabria
Italy

Robert Keefe
College of Natural Resources
University of Idaho
Experimental Forest
Moscow
United States

Mihai Nita
Department of Forest
Engineering, Forest
Management Planning and
Terrestrial Measurements,
Faculty of Silviculture and
Forest Engineering
Transilvania University of
Brasov
Brasov
Romania

Editorial Office
MDPI
St. Alban-Anlage 66
4052 Basel, Switzerland

This is a reprint of articles from the Special Issue published online in the open access journal *Forests* (ISSN 1999-4907) (available at: www.mdpi.com/journal/forests/special_issues/forest_engineering).

For citation purposes, cite each article independently as indicated on the article page online and as indicated below:

LastName, A.A.; LastName, B.B.; LastName, C.C. Article Title. *Journal Name* **Year**, *Volume Number*, Page Range.

ISBN 978-3-0365-6172-1 (Hbk)
ISBN 978-3-0365-6171-4 (PDF)

Contents

forests

Editorial

Electronics, Close-Range Sensors and Artificial Intelligence in Forestry

Stelian Alexandru Borz [1,*], Andrea Rosario Proto [2], Robert Keefe [3] and Mihai Daniel Niță [1]

1   Department of Forest Engineering, Forest Management Planning and Terrestrial Measurements,
    Faculty of Silviculture and Forest Engineering, Transilvania University of Brasov, 500123 Brasov, Romania
2   Department of AGRARIA, Mediterranean University of Reggio Calabria, 89122 Reggio Calabria, Italy
3   College of Natural Resources, University of Idaho Experimental Forest, Moscow, ID 83844, USA
*   Correspondence: stelian.borz@unitbv.ro; Tel.: +40-742-042-455

Citation: Borz, S.A.; Proto, A.R.; Keefe, R.; Niță, M.D. Electronics, Close-Range Sensors and Artificial Intelligence in Forestry. *Forests* **2022**, *13*, 1669. https://doi.org/10.3390/f13101669

Received: 5 October 2022
Accepted: 8 October 2022
Published: 11 October 2022

Publisher's Note: MDPI stays neutral with regard to jurisdictional claims in published maps and institutional affiliations.

The use of electronics, close-range sensing and artificial intelligence has changed the management paradigm in many of the current industries in which big data analytics by automated processes has become the backbone of decision making and improvement. Acknowledging the integration of electronics, devices, sensors and intelligent algorithms in much of the equipment used in forest operations, as well as their use in various forestry-related applications, we are still seeing that many disciplines within forestry and forest science still rely on data collected traditionally, which is resource-intensive. In turn, this brings limitations in characterizing the specific behaviors of the forest product systems and wood supply chains, and often prevents the development of solutions for improvement or inferring the laws behind the operation and management of such systems.

Undoubtedly, many solutions still need to be developed in the future to provide the technology required for the effective management of forests. In this regard, the Special Issue "Electronics, Close-Range Sensors and Artificial Intelligence in Forestry" highlights many examples of how technological improvements can be brought to forestry and to other related fields of science and practice.

For instance, the work of [1] has shown a new approach on how to improve tree ring identification technology which, in turn, supports the science in many scientific topics, including forest growth and dendrochronology, and the effect that climate changes have on forests. The work of [2] describes a solution for the long-term monitoring of sawmilling operations by developing a highly accurate machine learning framework which works on limited amounts of data and enables the use of inexpensive sensors for extended periods of time. Changing the existing modalities of accounting for quantitative estimates in the wood supply chain has been found to be one of the drivers of automation in forestry which will support a more effective management. The comparative study of [3] concluded that there is a lot of potential in using affordable digital solutions in wood measurement applications, which could be a feasible alternative when balancing the running costs and the ergonomics of wood measurement activities. The management of future forests would have to rely on high amounts of data collected in real time. In turn, this would have to use proper protocols to extract useful information. To support such needs, the work of [4] describes and operationalizes a concept to support data curation for Tree-Talker-based applications. Prototyping technologies that meet practical and scientific sampling purposes is one of the challenges in many scientific disciplines. The work of [5] describes the design of an Unoccupied Aircraft System with a lot of potential in physical sampling, thereby enhancing our ability to obtain samples from rather inaccessible parts of the trees. Remote sensing coupled with statistical learning may support large-scale spatial forest management. The work of [6] successfully tests these solutions for dense forests with the aim of removing the bottlenecks brought by traditional field sampling in estimating the aboveground biomass of trees. Understanding the drivers behind the land use change, including forest loss, may help in designing better policies and practices. The work of [7] evaluates the drivers of forest loss using a machine learning approach in a spatially heterogeneous space. LiDAR technology has been proved to support many applications and decisions in forestry. LiDAR-based data twinning may even be a prerequisite in our attempt to create virtual copies of forests in the future. The work of [8] describes a non-biased highly accurate tree-segmentation platform which supports the extraction of tree-level attributes such as diameter at breast height (DBH), tree height and estimates of tree volume. At our level of development, forest management is not possible without a well-designed road infrastructure, which should provide the connectivity to resources, and should enable mobility for various purposes. The management of forest roads by monitoring provides the possibility to take early action and requires advanced solutions to support management's responsivity. The work of [9] proposes an ultrasound-based solution which has strong potential for road monitoring, with a road geometry interpretation rate of up to 91.2%. Forest disturbances, including forest fires, are shaping our forests in ways that can hinder their sustainability. For this reason, the early detection of fire can contribute to taking responsive measures in order to prevent losses due to damage. In this regard, and based on convolutional neural networks, the work of [10] provides a competitive solution for fire detection. The work of [11] provides a solution for monitoring motor-manual operations, in order to remove the effects brought by the variability in placement of acceleration sensors; although the solution was tested on a specific equipment, the approach has the potential of being adapted to many other applications in which motor-manual tools are used in operations. Last, but not least, the technique of ensemble learning may help discover patterns that are rather inaccessible to conventional machine learning. Guided by this, the work of [12] describes a novel ensemble learning method to detect forest fires in various scenarios, which improves detection performance by 2.5%–10.9%.

In summary, to promote a better understanding of the usefulness of advanced solutions in forest management, the Special Issue "Electronics, Close-Range Sensors and Artificial Intelligence in Forestry" compiled advanced knowledge, techniques and solutions specific to several disciplines, starting with the monitoring of forest resources and infrastructure, and ending with the tools needed in disturbance management, sampling, and operational forestry.

Funding: This research received no external funding.

Conflicts of Interest: The authors declare no conflict of interest.

References

1. Hu, X.; Zheng, Y.; Xing, D.; Sun, Q. Research on Tree Ring Micro-Destructive Detection Technology Based on Digital Micro-Drilling Resistance Method. *Forests* **2022**, *13*, 1139. [CrossRef]
2. Borz, S.A.; Forkuo, G.O.; Oprea-Sorescu, O.; Proto, A.R. Development of a Robust Machine Learning Model to Monitor the Operational Performance of Fixed-Post Multi-Blade Vertical Sawing Machines. *Forests* **2022**, *13*, 1115. [CrossRef]
3. Borz, S.A.; Morocho Toaza, J.M.; Forkuo, G.O.; Marcu, M.V. Potential of Measure App in Estimating Log Biometrics: A Comparison with Conventional Log Measurement. *Forests* **2022**, *13*, 1028. [CrossRef]
4. Tomelleri, E.; Belelli Marchesini, L.; Yaroslavtsev, A.; Asgharinia, S.; Valentini, R. Toward a Unified TreeTalker Data Curation Process. *Forests* **2022**, *13*, 855. [CrossRef]
5. Krisanski, S.; Taskhiri, M.S.; Montgomery, J.; Turner, P. Design and Testing of a Novel Unoccupied Aircraft System for the Collection of Forest Canopy Samples. *Forests* **2022**, *13*, 153. [CrossRef]
6. Moradi, F.; Darvishsefat, A.A.; Pourrahmati, M.R.; Deljouei, A.; Borz, S.A. Estimating Aboveground Biomass in Dense Hyrcanian Forests by the Use of Sentinel-2 Data. *Forests* **2022**, *13*, 104. [CrossRef]
7. Park, J.; Lim, B.; Lee, J. Analysis of Factors Influencing Forest Loss in South Korea: Statistical Models and Machine-Learning Model. *Forests* **2021**, *12*, 1636. [CrossRef]
8. Niță, M.D. Testing Forestry Digital Twinning Workflow Based on Mobile LiDAR Scanner and AI Platform. *Forests* **2021**, *12*, 1576. [CrossRef]
9. Starke, M.; Kunneke, A.; Ziesak, M. Monitoring of Carriageway Cross Section Profiles on Forest Roads: Assessment of an Ultrasound Data Based Road Scanner with TLS Data Reference. *Forests* **2021**, *12*, 1191. [CrossRef]
10. Pan, J.; Ou, X.; Xu, L. A Collaborative Region Detection and Grading Framework for Forest Fire Smoke Using Weakly Supervised Fine Segmentation and Lightweight Faster-RCNN. *Forests* **2021**, *12*, 768. [CrossRef]

11. Borz, S.A. Development of a Modality-Invariant Multi-Layer Perceptron to Predict Operational Events in Motor-Manual Willow Felling Operations. *Forests* **2021**, *12*, 406. [CrossRef]
12. Xu, R.; Lin, H.; Lu, K.; Cao, L.; Liu, Y. A Forest Fire Detection System Based on Ensemble Learning. *Forests* **2021**, *12*, 217. [CrossRef]

Article

Research on Tree Ring Micro-Destructive Detection Technology Based on Digital Micro-Drilling Resistance Method

Xueyang Hu [1,2,3], Yili Zheng [1,2,3,*], Da Xing [1,2,3] and Qingfeng Sun [1,2,3]

1 School of Technology, Beijing Forestry University, Beijing 100083, China; huxueyang@bjfu.edu.cn (X.H.); xingda@bjfu.edu.cn (D.X.); qf3210261@bjfu.edu.cn (Q.S.)
2 Beijing Laboratory of Urban and Rural Ecological Environment, Beijing 100083, China
3 Key Lab of State Forestry Administration for Forestry Equipment and Automation, Beijing 100083, China
* Correspondence: zhengyili@bjfu.edu.cn

Abstract: Micro-drilling resistance method is a widely used tree ring micro-destructive detection technology. To solve the problem that the detection signal of the analog micro-drilling resistance method has excessive noise interference and cannot intuitively identify tree ring information, this research proposes a digital micro-drilling resistance method and provides a recommended hardware implementation. The digital micro-drilling resistance method adopts the photoelectric encoder instead of ADC as the signal sampling module. Through the theoretical analysis of the DC motor characteristic, the PWM closed-loop speed control, the detection principle of the digital method is given. Additionally, the experimental equipment that can complete the detection of the digital method and the analog method simultaneously is designed to carry out comparative experiments. The experimental results show that: (1) The detection results of the digital method have a better-quality signal which can intuitively identify the tree rings. (2) The average correlation coefficient reaches 0.9365 between the detection results of the digital method and the analog method. (3) The average Signal-to-Noise Ratio (SNR) of the digital method is 39.0145 dB, which is 19.2590 dB higher than that of the analog method. The average noise interference energy in the detection result of the digital method is only 1.27% of the analog method. In summary, hardware implementation of the digital micro-drilling resistance method can correctly reflect the tree ring information and significantly improve the signal quality of the micro-drilling resistance technology. This research is helping to improve the identification accuracy of micro-drilling resistance technology, and to develop the application of tree ring micro-destructive detection technology in the high-precision field.

Keywords: tree ring; forestry detection; resistance sensor; micro-drilling resistance method; signal processing; Signal-to-Noise Ratio (SNR)

Citation: Hu, X.; Zheng, Y.; Xing, D.; Sun, Q. Research on Tree Ring Micro-Destructive Detection Technology Based on Digital Micro-Drilling Resistance Method. *Forests* **2022**, *13*, 1139. https://doi.org/10.3390/f13071139

Academic Editors: Stelian Alexandru Borz, Andrea R. Proto, Robert Keefe and Mihai Nita

Received: 17 June 2022
Accepted: 15 July 2022
Published: 19 July 2022

1. Introduction

Tree rings are the chronology of tree growth. The detection of tree rings can reveal the growth of trees, judge their age, and provide an important basis for the cultivation, utilization, and protection of trees [1]. In addition, tree rings also record the impact of external factors such as environment and climate on tree growth [2]. Tree ring detection has become an important way of obtaining forest growth and ecological environment information [3–6]. Dendrochronology which is widely used in archaeology, climatology, ecology, and geomorphology, has also become an interdisciplinary subject [7].

Facing the goal of developing better-quality forestry, it is imperative to develop modern and smart forestry and improve the level of digitalization and intelligence [8,9]. Electronic and intelligent tools are constantly being applied to tree ring detection. In the traditional tree ring detection method of tree disc sampling, the STD4800 scanner is introduced to obtain high-definition tree disk images [10]. Then digital image recognition is carried out through a special tree ring analysis system such as WinDendro [11–14]. It also can use the LINTAB CNC measuring platform and TSAP standard annual ring analysis software to observe the tree disk by the high-resolution microscope [15–18]. With the support of electronic tools, the detection accuracy of the tree disc sampling method is very satisfactory. However, the great damage caused by felling trees is unavoidable with the tree disk sampling method. The increment borer sampling method is an improved method [19,20]. Taking the widely used HAGLOF increment borer as an example, it drills a hollow cone tube with a diameter of 5~12 mm into the tree trunk to obtain tree core samples [21–23]. However, the penetrating wound left on the trunk after the increment borer sampling greatly increases the risk of disease infection of the tree, which causes great damage [24].

With the strengthening of forest protection, the tree-ring detection technology is developing in the direction of reducing the detection damage, and the non-destructive or micro-destructive detection method of tree rings has attracted more and more attention. From the traditional tree disk sampling method and the increment borer sampling method to the computer tomography technology and the micro-drilling resistance technology, the detection of tree rings is constantly trying new methods and technologies in forestry operations [25,26]. However, the non-destructive tree ring detection equipment represented by computer tomography has the disadvantages of high cost and large equipment size, so it is difficult to be widely used in wild forestry practice [27–29]. Therefore, the micro-drilling resistance tree ring detection technology has high expectations and has become a widely used technology in the micro-destructive detection of tree rings.

The micro-drilling resistance tree ring detection technology refers to using a slender drill needle to drill into the tree's interior using a motor drive, detecting the tree ring by sensing the resistance change during the drilling process [30]. The essence of micro-drilling resistance technology is to build a sensor system that measures the change in density and resistance caused by tree ring distribution [31]. The diameter of the micro-drill drilled into the tree is generally less than 3 mm [32], so the damage to the tree's phloem will be significantly reduced, and the sieve tube that transports nutrients will not affect the growth of the entire tree due to individual damage. Therefore, the micro-drilling resistance technology can be regarded as a method of tree ring micro-destructive detection. With the application of electrical recording in micro-drilling resistance equipment, research on tree rings, internal structure, density, elastic modulus, etc., has been gradually carried out by analyzing resistance signal waveforms [33–36]. The device, capable of acquiring resistance waveforms, was named Resistograph by Rinntech.

Concerning the requirements of micro-destructive detection, obtaining a higher quality signal and higher detection accuracy has become a research hotspot of the micro-drilling resistance method. Rinn, F gives the recommended micro-drill bit shape and mechanical structure to improve the tree rings sensitivity of the drill pin and reduce the interference of mechanical vibration in the detection process [37]. Cao, Y et al. attempt to improve detection accuracy by selecting the best detection path concerning the tree pith [38]. Oh, J et al. showed the most proper feed speed to better evaluate the number of tree rings for each tree species [15].

For the widely used and fully disclosed analog micro-drilling resistance method, the main factor that affects the detection and identification accuracy is the excessive noise interference in the output signal [39,40]. The signal in the analog method transmits in analog quantities form, which results in poor anti-interference ability. And the complex signal sampling process inevitably introduces noise interference, which causes poor signal quality. These shortcomings, determined by the principle of the analog method, limit the detection accuracy and mean the original waveform cannot be visually identified, so the original waveform has to be processed by a filtering algorithm to identify the tree rings. The researchers use various filtering algorithms to improve the original detection signal from the analog method. For example, Pan H's research uses Kalman filtering to process the detection signal, and the processed signal is used to evaluate the tree age [41]. The research of Yao, J et al. use an adaptive filtering algorithm to improve the accuracy of tree ring identification [42]. The research of Hu, X et al. uses the FIR filtering algorithm and IIR filtering algorithm to process the output signals and evaluate the filtering results [43]. However, the filtering algorithms can only reduce the influence of noise interference but cannot eliminate it. At the same time, the filtering process also brings side effects. On the one hand, the complex filtering algorithm reduces the real-time performance of the result. On the other hand, the threshold setting of filtering parameters is still quite difficult. The strict filtering parameter settings will lead to the lack of effective signal, while the loose filtering parameter settings will lead to difficulties in tree ring identification, which will greatly reduce detection accuracy.

To solve the problem that the detection signal of the analog micro-drilling resistance method has excessive noise interference and cannot intuitively identify the tree ring information, this research proposes a digital micro-drilling resistance method. The digital micro-drilling resistance method attempts to design a new detection principle and hardware implementation to realize the transmission of the signals in the digital quantities form, fundamentally eliminate part of the noise interference, and improve the signal quality. Compared with the analog method, the digital method has the advantages of less noise interference, a simple detection circuit, and easy identification of tree rings.

This research aims to improve the signal quality and the detection accuracy of tree ring micro-drilling detection technology, and to develop the application of tree ring micro-destructive detection technology in the high-precision field. By reading this article, readers will acquire the sensor principle and hardware implementation of the digital micro-drilling resistance method, which has better signal quality and more easily identifiable waveforms than the widely used analog micro-drilling resistance method.

2. Background

2.1. Dendrochronological Basis for Micro-Drilling Resistance Technology

Tree rings refer to the concentric rings on the cross-section of the tree trunk, and their density and resistance characteristics are the basic principles of micro-drilling resistance technology. Each round consists of earlywood and latewood, generally representing secondary wood formed within a year.

The climate in spring and summer is warm and humid, which is suitable for tree growth. The cambium cells grow and divide rapidly, and the formed xylem cells have a larger size, thinner cell wall, less fiber content, and more ducts for transporting water. Therefore, this part of the tree ring is loose in texture with a lighter color, and it is used to be called earlywood or spring wood. On the contrary, the activity of cambium cells is significantly weakened in autumn and winter, and the formed xylem cells become narrow, thick, and fiber-rich. Therefore, the part of the tree ring in autumn and winter is dense in texture with a darker color, habitually called latewood or autumn wood [39].

The microscopic structure of conifers is simple and regular, and mainly composed of tracheids and xylem rays. The xylem rays of conifers are very thin and invisible to the naked eye. In general, the tree ring circle of a coniferous tree is obvious, and the difference between early and late wood is obvious. The earlywood has a thin wall with a large cavity and a lighter color, while the latewood has a thick wall with a small cavity and a darker color.

The broad-leaved wood is mainly composed of vessel, wood fiber, axial parenchyma, and xylem ray. The structure is complicated. The size and distribution of the vessel holes are divided into ring-porous trees, diffuse-porous trees, and semi-ring-porous trees. In the ring-porous trees, the diameter of the vessels in the earlywood is significantly larger, while the vessels in the latewood are quite small, so the density differences are very obvious and the tree rings are easy to identify. The vessel size and distribution of diffuse-porous trees reflect consistency or slight graduality. Therefore, there is no significant boundary from earlywood to latewood; only a thin boundary exists between the latewood of the previous growing season and the earlywood of the next growing season, so it is not easy to distinguish. The semi-ring-porous trees are an intermediate type between the ring-porous and the diffuse-porous. In the earlywood part, there will be bands formed by large vessels, or rings formed by many small vessels, making the earlywood more obvious [44,45].

The micro-drilling resistance technology can reflect the density during the drilling process and distinguish the earlywood and the latewood through the obvious change in the wood density. Then the operator can carry out tree ring identification and analysis to infer the tree's age.

2.2. Classification of Micro-Drilling Resistance Method

According to the different sensor principles for determining resistance changes, this research divides the micro-drilling resistance technology into three types: mechanical method, analog method, and digital method.

2.2.1. Mechanical Micro-Drilling Resistance Method

The mechanical micro-drilling resistance method refers to driving the micro-drill using a DC motor on a constant voltage and relying on the mechanical vibration to perceive the measurement results and identify the change in the tree's internal resistance. The sensor principle of the method is based on the mechanical characteristic of the DC motor. And the motor running at a constant voltage can be called the open-loop control mode. When the wood density contacted by the micro-drill increases or decreases, the output torque of the DC motor will increase or decrease accordingly, and the drilling needle driven by the DC motor will suddenly decelerate or accelerate. This sudden speed change can be perceived by the user through the instrument's mechanical vibration or be recorded by the spring-loader [46]. The mechanical method does not set up a signal detection circuit, so the measurement results cannot be quantified. Therefore, the mechanical method is difficult to identify the subtle resistance changes caused by the tree rings and gradually withdraws from the application of tree ring detection. As the first micro-drilling resistance detection method, the mechanical method realizes the micro-destructive detection of the tree's internal material by simple equipment and sensor principles. IML-RESI MD300 is a representative micro-destructive detection equipment based on the mechanical micro-drilling resistance method [47]. MD300 is a drilling instrument working purely mechanically that does not electronically record a measurement curve. The user feels the result and can read the penetration depth on the 300 mm scale. Often an abrupt and fast penetration is perceptible when the instrument detects a rot zone. It is currently used in the detection of hollows, rots, and cracks in trees, and other fields which do not require high precision. Since the mechanical method is replaced by the analog method in tree ring detection, the mechanical method will not be involved in the following experiments and comparative analysis.

2.2.2. Analog Micro-Drilling Resistance Method

To solve the problem acknowledging that the measurement results of the mechanical method cannot be quantified and the automation degree of the detection process is low, the analog micro-drilling resistance method came into being. The analog micro-drilling resistance method refers to using a closed-loop control motor to drive the micro-drill rotated at a constant speed, converting the resistance amplitude into an analog signal by connecting a sampling resistor in the motor armature circuit, and finally transferring the signal to SoC (System-on-a-Chip) by the ADC (Analog-to-Digital Converter) sampling module. The analog method solves the disadvantage that the mechanical method cannot quantify the results and has become a widely used method for micro-drilling resistance technology. The design principle of the tree ring micro-destructive detection equipment published by Hu, X and Chen, X can be considered the analog method [39,40]. Currently, the tree ring micro-destructive detection system designed by Beijing Forestry University (BJFU) and the tree-ring acupuncture instrument designed by the Chinese Academy of Forestry (CAF) are both designed by the analog micro-drilling resistance method. Using the equipment based on the analog method, the operator can automatically complete the drilling and simply obtain the result data by sending control commands through the button. And in the process of detection, the results can be printed on wax paper in real-time waveforms or stored on an SD card. The distribution of earlywood and latewood in tree rings was analyzed by the peaks and troughs of the waveform. The analog method that can quantify the measurement results ensures the feasibility of the micro-drilling resistance method for tree ring detection [48,49].

2.2.3. Digital Micro-Drilling Resistance Method

The digital micro-drilling resistance method proposed in this research is a new micro-drilling resistance tree ring detection method, which is based on the principle of the DC motor output characteristic and the PWM closed-loop speed control. The digital micro-drilling resistance method uses a photoelectric encoder to obtain digital signals for transmission, which is different from the widely used analog method and does not need to use the ADC conversion module, which avoids noise interference during signal sampling and improves the anti-interference ability of the signal. Compared with the analog method, the digital method has the advantages of less noise interference, a simple detection circuit, and easy identification of tree rings.

Table 1 shows the comparison of the mechanical method, the analog method, and the digital method proposed in this research.

Table 1. Comparison of the mechanical method, the analog method, and the digital method.

Characteristic	Mechanical	Analog	Digital
Quantification of results	No	Yes	Yes
Circuit complexity	Simple	Complex	Simple
Motor control	Open-loop	Closed-loop	Closed-loop
Sampling module	Mechanical vibration	ADC	Photoelectric encoder
Signal type	Mechanical	Analog	Digital
Anti-interference ability	Weak	Weak	Strong
Signal quality	Weak	Poor	Good

3. Materials and Methods

In this section, the principle of the digital micro-drilling resistance method is firstly deduced by formula. The recommended hardware implementation of the digital micro-drilling resistance method is also given. The experimental equipment which can complete the detection of the digital method and the analog method is designed to simultaneously carry out comparative experiments. Finally, the experimental sample is introduced.

3.1. Principle of Digital Micro-Drilling Resistance Method

The digital micro-destructive resistance method proposed in this research is based on the output torque characteristic of the DC motor and the PWM closed-loop speed control. It uses the photoelectric encoder to obtain the detection data. The formula is derived as follows.

The digital micro-drilling resistance method detects the resistance torque given by the tree rings on the micro-drill bit to obtain the density change of the tree rings.

Equation (1) is the theorem of rigid body rotation with a fixed axis:

$$\vec{M} = J\vec{\beta} = J\frac{d\vec{\omega}}{dt} \tag{1}$$

M: combined external torque acting on the drill; J: moment of Inertia of the drill; β: angular acceleration of the drill; ω: angular velocity of the drill; t: time.

The moment of inertia J of the drill needle is determined by three factors: the quality, the mass distribution, and the position of the rotating axis. During the detection, the moment of inertia J of the drill is a constant value.

Equation (2) is derived from the force analysis of the micro-drill:

$$\vec{M} = \vec{T} + \vec{T_L} + \vec{T_0} \tag{2}$$

$\vec{T}$: electromagnetic torque; $\vec{T_L}$: load torque or resistance torque; $\vec{T_0}$: no-load torque.

The direction of electromagnetic torque $\vec{T}$ is opposite to the direction of resistance torque $\vec{T_L}$ and no-load torque $\vec{T_0}$. Therefore, the scalarized Equation (2) combines with Equation (1) to form Equation (3):

$$T - T_L - T_0 = J\beta = \frac{d\omega}{dt} \tag{3}$$

Shift the terms of Equation (3) to get the value of load torque T_L:

$$T_L = T - T_0 - \frac{d\omega}{dt} \tag{4}$$

To obtain a linear relationship between the T_L and the T, the angular acceleration of the micro-drill must always be equal to zero—that is, the derivative of the angular velocity with time is always equal to zero. The drill needle must move at a constant angular velocity during the drilling process. In this case, Equation (4) can be simplified to Equation (5):

$$T_L = T - T_0 \tag{5}$$

The no-load torque T_0 of the motor is much smaller than the resistance torque T_L during drilling, so T_0 can be ignored. Under the condition that the drill needle is kept moving at a constant angular velocity and the load torque T_0 is ignored, the Equation (5) can be approximated as:

$$T_L \approx T \tag{6}$$

The above derivation converts the resistance torque T_L into the measurement of the electromagnetic torque T of the motor and provides an electrical way to measure the resistance torque T_L.

Equation (7) is the torque characteristic of the DC brush motor:

$$T = C_t \phi I_a \tag{7}$$

C_t: torque constant; ϕ: flux per pole, determined by the characteristic of the motor; I_a: armature current.

Equation (8) can be obtained by combining Equations (6) and (7):

$$T_L = C_t \phi I_a \tag{8}$$

Equation (8) shows that the armature current I_a is linearly proportional to the resistance torque T_L. Through the value of the armature current I_a, the resistance change in the drilling process can be obtained, and the tree ring detection can be realized.

Different from the analog method in which the armature current I_a is obtained by the sampling resistance method, the digital method converts I_a into a digital signal and transmits it to the SoC.

The DC motor voltage balance equation is shown in Equation (9):

$$U = E + I_a R_a \tag{9}$$

U: armature voltage; E: armature electromotive force; R_a armature resistance.

The principle of electromagnetics shows that the relationship between the motor speed n and the armature electromotive force E is shown in Equation (10):

$$E = C_e \phi n \tag{10}$$

C_e: potential coefficient.

The relationship between the motor speed n and the rotational angular velocity ω is shown in Equation (11):

$$\omega = 2\pi n \tag{11}$$

The prerequisite of Equation (8) is that the angular velocity ω is a constant value, so during the detection process, the motor speed n is a constant value, and the armature electromotive force E is also a constant value.

The PID algorithm and PWM technology control a constant angular velocity ω of the micro-drill during the detection process. PWM is a method for the SoC to control analog circuits by outputting digital signals [50,51]. And PWM control technology obtains the required waveform or equivalent amplitude according to the principle of area equivalence by modulating the width of the pulse. The rectangular wave voltage PWM is shown in Equation (12):

$$U = D_{PWM} U_m \tag{12}$$

D_{PWM}: PWM duty ratio; U_m: maximum voltage.

Equation (13) can be derived from Equations (8)–(10):

$$T_L = C_t \phi \frac{D_{PWM} U_m - C_e \phi n}{R_a} \tag{13}$$

The PID algorithm is a method of control by deviation, where P means proportional, I means integral, and D means differential. The algorithm is simple, robust, reliable, and widely used in various control fields [52–54]. And Figure 1 shows the structure diagram of the PID speed closed-loop control algorithm.

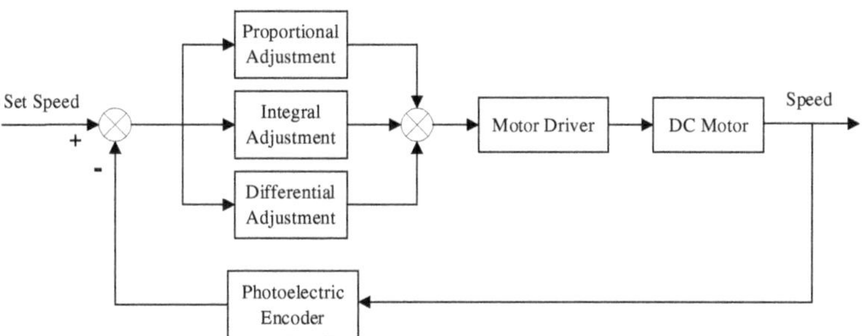

Figure 1. The structure diagram of the PID speed closed-loop control algorithm.

D_{PWM} is calculated by SoC according to PID speed closed-loop control algorithm and speed error, as shown in Equation (14):

$$D_{PWM} = \frac{T_{on}}{T} = \frac{K_p e(k) + K_i \sum_{n=0}^{k} e(k) + K_d (e(k) - e(k-1))}{T} \tag{14}$$

T_{on}: pulse width time; T: cycle time of PWM; $e(k)$: speed error; K_p: proportional adjustment coefficient; K_i: integral adjustment coefficient: K_d: differential adjustment coefficient.

The PWM cycle time T is a constant value during the control process, and the rotational speed error $e(k)$ is calculated by the photoelectric encoder.

The above formula derivation proves the relationship shown in Equation (15):

$$T_L \propto D_{PWM} \propto T_{on} \tag{15}$$

D_{PWM} and T_{on} are proportional to the resistance torque T_L, and the resistance change can be reflected by the value change of D_{PWM} or T_{on}. The detection signal is sampled and transmitted as a digital signal, and ADC is not used in the whole signal flow process. Therefore, this research named the method "digital micro-drilling resistance method".

3.2. Hardware Implementation of Digital Micro-Drilling Resistance Method

According to the principle of the digital micro-drilling resistance method, we give the recommended hardware system architecture and hardware implementation scheme. The hardware system architecture of the digital micro-drilling resistance method is shown in Figure 2, which consists of a SoC module, a DC motor drive module, a digital signal sampling module, and a data transmission module.

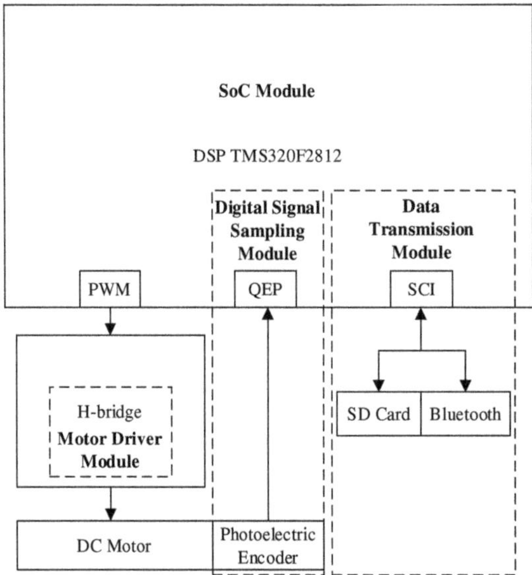

Figure 2. The hardware system architecture.

3.2.1. SoC Module

The SoC module is composed of DSP chip and peripheral circuits. As the core of the detection and control circuit, the DSP chip adopts a TMS320F2812 high-speed real-time digital signal processing chip. The chip is a high-performance 32-bit data processor with excellent digital signal processing and motion control capabilities. Abundant peripheral functions and interfaces can meet the needs of digital micro-drilling resistance methods. The PWM function realizes the control of the DC motor, the Serial Communication Interface (SCI) realizes the transmission of detection data, and the Quadrature Encoder Pulse (QEP) module is used for digital encoder signal sampling [55,56].

3.2.2. H-Bridge Motor Driver Module

The RE35 DC motor is selected as the drive motor for the high-speed rotation of the micro-drill. The motor has the characteristics of low-speed fluctuation, high conversion efficiency, high operation stability, and easy control.

In the process of drilling into the tree and exiting the tree, the micro-drill needs to rotate in the opposite direction, so the first control requirement for the motor is to realize forward and reverse control. At the same time, the principle of the micro-drilling resistance method requires that the motor must rotate at a constant angular velocity to ensure the validity of tree ring identification. Therefore, the second control requirement for the motor is to realize speed control. Further, the principle of the digital micro-drilling resistance method requires the use of a PWM voltage modulation signal to control the speed of the motor. For this purpose, the H-bridge motor driver circuit, as shown in Figure 3, is designed.

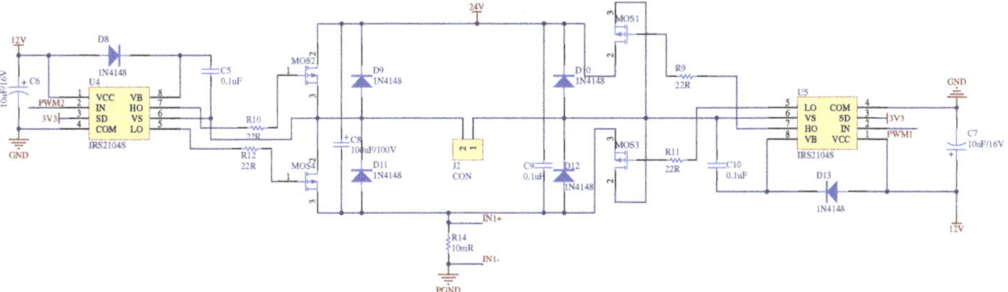

Figure 3. The H-bridge motor driver circuit.

The H-bridge motor driver circuit is composed of 4 MOSFETs distributed on 4 bridge arms, and the on and off of the MOSFET is controlled by the PWM signal [57]. The PWM control signals of MOS1 and MOS3, MOS2 and MOS4, are complementary channels. PWM1 and PWM2 are a pair of control signals with opposite polarities but the same period and duty cycle so that the two MOSFETs on the diagonal can be turned on and off simultaneously. The H-bridge motor driver circuit can control the forward and reverse rotation of the motor and has the advantages of small speed regulation static difference, large range, and fast dynamic response, which meets the control requirements of the micro-drill drive motor.

The maximum output voltage of the DSP pin is 3.3 V, which cannot meet the turn-on requirements of driving the upper bridge arm. It is necessary to use a bootstrap circuit to boost the PWM signal to control the MOSFET. The bootstrap circuit comprises a half-bridge driver chip IR2104S, a bootstrap diode, and a bootstrap capacitor. IR2104S can output a pair of complementary drive levels with a dead zone only by inputting one PWM control signal. The DSP outputs PWM1 and PWM2, two control signals to control two IR2104S chips, achieve the control of the four MOS tubes in the H-bridge circuit, and realize the adjustment of the motor speed and steering.

3.2.3. Digital Signal Sampling Module

A HEDL-5540 1024-line incremental photoelectric encoder is installed at the rear of the RE35 DC motor, as shown in Figure 4.

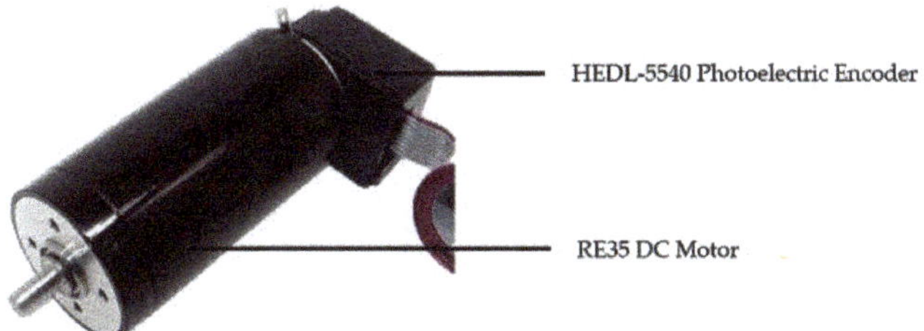

Figure 4. The RE35 DC motor with HEDL-5540 photoelectric encoder.

The digital signal sampling module is used to receive the rotational speed digital signal transmitted by the HEDL-5540 1024-line incremental photoelectric encoder. The

HEDL-5540 encoder is a device that converts the mechanical geometric displacement on the output shaft of the motor into a digital pulse signal through photoelectric conversion. There are three square wave pulse output signals: A, B, and I. Pulse A and B measure the rotation direction and speed, and their phase difference is 90°. Pulse I is used to locate the reference point. The square wave pulses A, B, and I output by the encoder are respectively connected with the QEP1, QEP2, and QEPI pins of the DSP, and the digital signal of the rotational speed detected by the encoder is transmitted to the DSP.

The number of pulses sent by the HEDL-5540 encoder per motor revolution is 1024. Assuming that the total number of pulses measured within a fixed time interval T is m, the calculation formula of the motor speed per minute n is shown in Equation (16):

$$n = \frac{60m}{1024T} \tag{16}$$

In the specific implementation, the DSP timer works in the directional increase/decrease mode, the clock source is set to the QEP circuit module, and the initial value of the timer's count register is set to the intermediate value 0x7FFF. If the phase of the pulse signal A input by the QEP1 pin is ahead of the pulse signal B input by the QEP2 pin, the count register will count up; otherwise, the count register will count down. The motor direction is judged by the sign of the difference between the end value of the count register and the initial value within a fixed time interval, and the speed value is judged by the absolute value. It should be noted that the QEP module of the DSP counts both the upper and lower edges of the pulse, so the clock input frequency generated by the QEP module is 4 times the frequency of the A or B pulse signal.

3.3. Experimental Equipment

To verify the theoretical derivation of the digital micro-drilling resistance method, and compare the difference in detection results between the digital and analog methods under the same condition, a tree ring detection experimental equipment which can complete the detection of the digital method and the analog method simultaneously is designed to carry out comparative experiments.

3.3.1. Mechanical Structure

The experimental equipment adopts the hand-held mechanical structure and dual-motor transmission structure shown in Figures 5 and 6, similar to the mechanical structure proposed by Hu X. in [39].

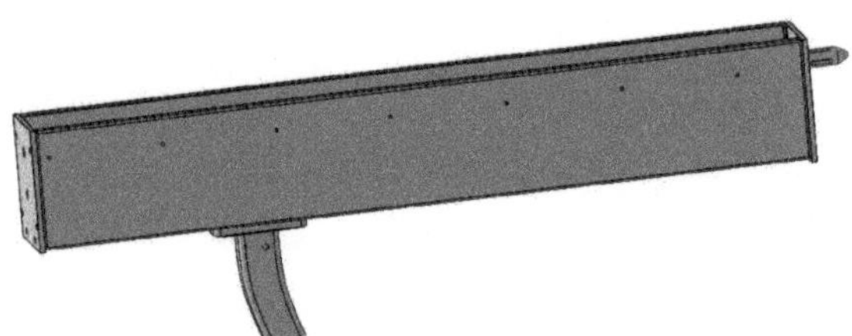

Figure 5. The hand-held mechanical structure.

The DC motor drives the micro-drill to rotate at high speed. The stepping motor drives the forward and backward of the micro-drill. The maximum diameter of the micro-drill is 3 mm. The operator points the equipment at the tree trunk and then uses the button to send instructions and detect tree rings.

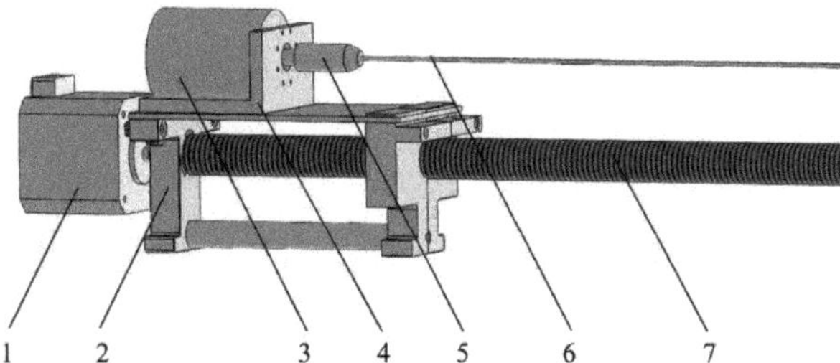

Figure 6. The dual-motor transmission structure. (1. Stepper motor; 2. Sliding base; 3. DC motor; 4. Motor base; 5. Drill clip; 6. Micro-drill; 7. Lead screw).

3.3.2. Hardware Circuit

To compare the difference in detection results between the digital and the analog micro-drilling resistance methods, a special hardware circuit with both detection methods is designed for the experiment. The special hardware circuit can output the detection results of digital and analog methods at the same time, effectively controlling the influence of irrelevant variables on the detection results.

The experimental hardware circuit is based on the hardware implementation of the digital micro-drilling resistance method, and an analog signal sampling module is added. The hardware architecture is shown in Figure 7.

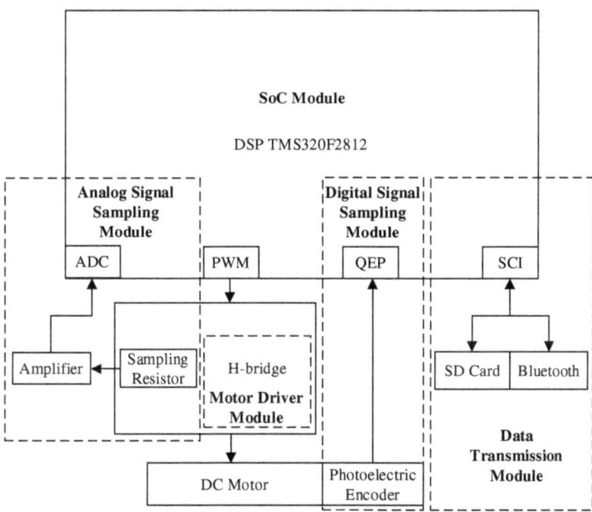

Figure 7. The experimental hardware architecture.

To realize the output of analog detection results, a sampling resistor is added to the armature of the H-bridge motor driver circuit. The INA282 bi-directional current sense amplifier is connected to the sampling resistor. The amplified current signal is transmitted to the ADC pin of the DSP to obtain the detection result. During the detection process, the digital and analog detection circuits operate simultaneously. The analog results sampled by ADC and the digital results calculated by the photoelectric encoder are both stored in the SD card or sent out by SCI. In the detection, the analog and digital detection results are output simultaneously to ensure the consistency of the measurement object and the measurement environment, which reduces the interference of various uncontrollable factors and provides reliable hardware for the comparison experiments carried out in this research.

3.4. Experimental Sample

In this research, tree disk samples were used for the micro-drilling resistance method detection experiment to compare with the actual tree rings. The tree disk samples used in this experiment are sampled from the Jingouling Forest Farm in Wangqing County, Yanbian Prefecture, Jilin Province, China. The Jingouling Forest Farm is located in the upper reaches of the Tumen River on the northwest slope of the Changbai Mountains, located at E $130°5'$ to $130°20'$ and N $43°17'$ to $43°25'$, and the altitude spans from 550 m to 1100 m [58]. The forest vegetation in this area is stratified in the vertical direction. The climate in this region is affected by tropical marine air mass or degenerated marine air mass from May to August every year and is affected by Siberia continental air mass from October to March of the following year, and the seasonal climate cycle changes significantly. The growing season of trees is mainly concentrated in the summer, from July to September, when the temperature is suitable and the rainfall is sufficient. In winter, there is a freezing period of more than 4.5 months. This makes the growth rate of plants in the area cyclically change, so tree rings generally grow one ring per year.

In the experiment, Larch and Fir tree discs from Jingouling Forest Farm were selected as samples. Larch (*Larix gmelinii* (Rupr.) Kuzen) is a deciduous tree of the Pinaceae and Larix genus. Larch is the main forest species in Northeast China, widely used for afforestation and forest regeneration, as well as for various wood and industrial materials. The density of the tree rings changes sharply in the earlywood and latewood, making the growth rings clearly visible, the wood grain is straight, and the structure is thicker, which makes it easy to identify the tree rings. Fir (*Abies nephrolepis* (Trautv.) Maxim.) is a Pinaceae and Abies genus tree with strong adaptability and a preference for cold and wet environments. The density difference between the earlywood and latewood of Fir is smaller than that of Larch, which requires higher detection accuracy.

4. Results

In the experiment, Larch and Fir discs are detected 4 times using the above-mentioned experimental equipment, which can complete the detection of the digital method and the analog method simultaneously. The information of the experimental discs is shown in Table 2, and the detection results of Larch SN. 29-1043-52471 and Fir SN. 30-1013-34894 are selected to show the analysis process.

Table 2. The information of the experimental discs.

Serial Number (SN)	Tree Specie	Diameter/cm
29-1043-52471	Larch	21.5
29-1043-52472	Larch	21.3
29-1257-48683	Larch	17.5
29-1903-44540	Larch	14.5
30-1013-34894	Fir	14.3
30-1512-18696	Fir	15.3
30-1849-16448	Fir	16.8
30-1911-50929	Fir	23.3

4.1. Original Detection Results

Figure 8 shows the waveforms of the original detection results for the digital and analog outputs of the Larch disc (SN. 29-1043-52471). LD refers to the detection result of the Larch disc, shown in red. LA refers to the detection result of the Larch disc, shown in blue. Figure 9 shows the waveforms of the original detection results of the digital and analog outputs of the Fir disc (SN. 30-1013-34894). FD refers to the detection result of the Fir disc, shown in red. FA refers to the detection result of the Fir disc, shown in blue.

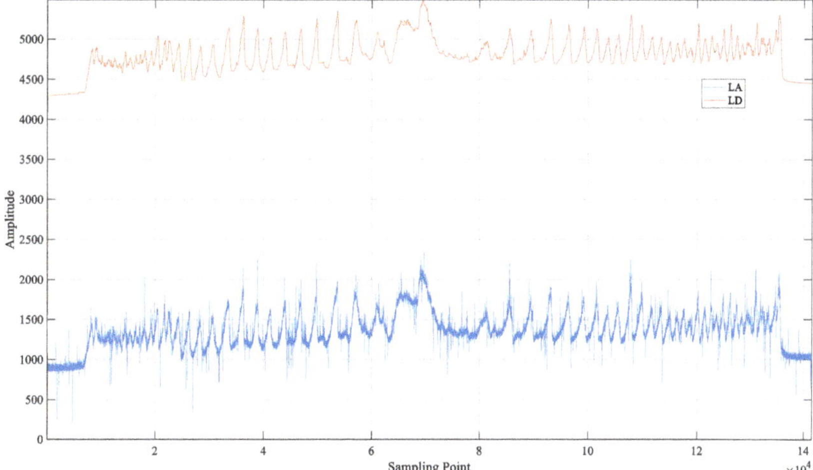

Figure 8. The waveforms of the original detection results for the digital and analog outputs of the Larch disc (SN. 29-1043-52471).

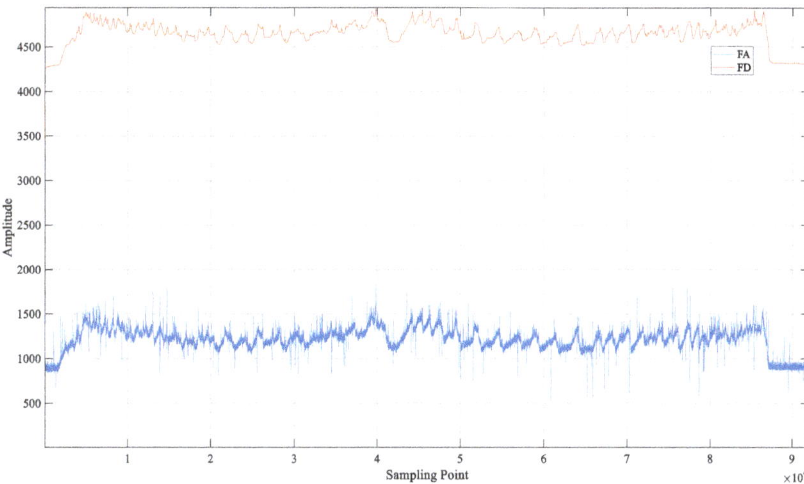

Figure 9. The waveforms of the original detection results of the digital and analog outputs of the Fir disc (SN. 30-1013-34894).

Figure 9 shows the waveforms of the original detection results of the digital and analog outputs of the Fir disc (SN. 30-1013-34894). FD refers to the detection result of the Fir disc, shown in red. FA refers to the detection result of the Fir disc, shown in blue.

4.2. Result of Preprocessing and Correlation Analysis

It can be seen from the original signal waveforms of Figures 8 and 9 that the digital and analog methods obtain the same number of sampling points, indicating the two detection methods are running simultaneously during the experiment. However, the amplitudes of the detection results obtained by the two methods are not the same because of the different sampling methods, which have different physical meanings. According to the analysis of the formula, the detection results of the digital method and the analog method are both proportional to the tree ring resistance. Therefore, this research attempts to convert the amplitudes of the digital and analog detection results to the same benchmark through preprocessing. If the detection results converted to the same benchmark have the same characteristics, the consistency of the two test results and the correctness of the digital method can be proved. The operation of preprocessing is as follows.

The detection signal is first fitted linearly using the least squares method. The fitted target polynomial $p(x)$ is shown in Equation (17):

$$p(x) = p_1 x + p_2 \tag{17}$$

p_1: 1th-degree coefficient; p_2: constant coefficient.

p_1 and p_2 can be calculated by Equation (18):

$$\begin{pmatrix} n & \sum\limits_{i=1}^{n} x_i \\ \sum\limits_{i=1}^{n} x_i & \sum\limits_{i=1}^{n} x_i^2 \end{pmatrix} \begin{pmatrix} p_2 \\ p_1 \end{pmatrix} = \begin{pmatrix} \sum_{i=1}^{n} y_i \\ \sum_{i=1}^{n} x_i y_i \end{pmatrix} \tag{18}$$

The fitted polynomial of LD (p_{LD}) is:

$$p_{LD}(x) = 0.0016x + 4687.3968 \tag{19}$$

The fitted polynomial of LA (p_{LA}) is:

$$p_{LA}(x) = 0.0016x + 1248.1110 \tag{20}$$

Figure 10 shows the digital and analog original signals and the fitted polynomial waveforms of Larch.

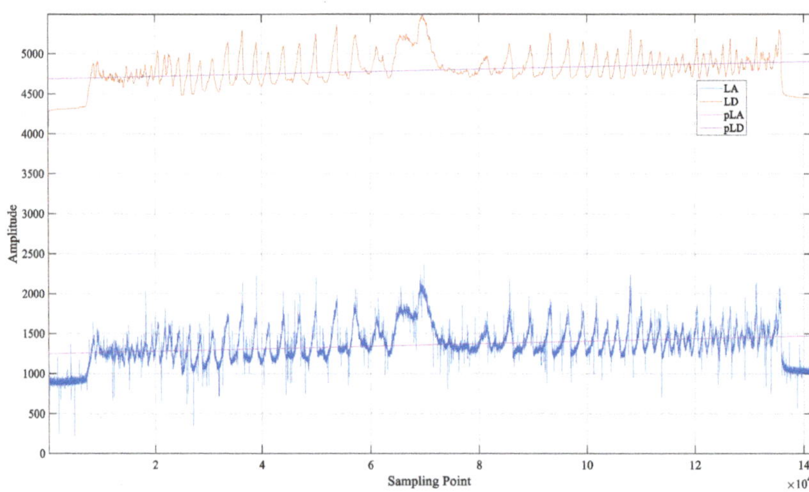

Figure 10. The digital and analog original signals and the fitted polynomial waveforms of Larch (SN. 29-1043-52471).

The fitted polynomial of FD (p_{FD}) is:

$$p_{FD}(x) = -0.0007x + 4677.2159 \tag{21}$$

The fitted polynomial of FA (p_{FA}) is:

$$p_{FA}(x) = -0.0009x + 1255.9182 \tag{22}$$

Figure 11 shows the digital and analog original signals and the fitted polynomial waveforms of Fir.

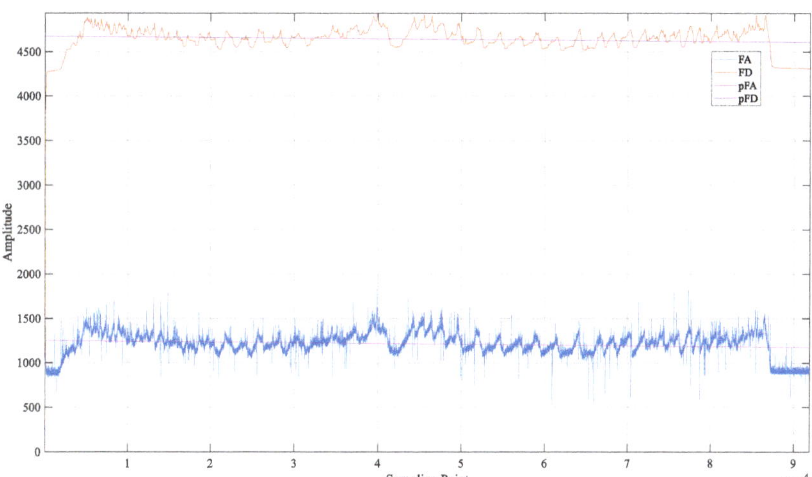

Figure 11. The digital and analog original signals and the fitted polynomial waveforms of Fir (SN. 30-1013-34894).

Subtract the fitted polynomial from the original signal to get the detrended detection signals shown in Figures 12 and 13.

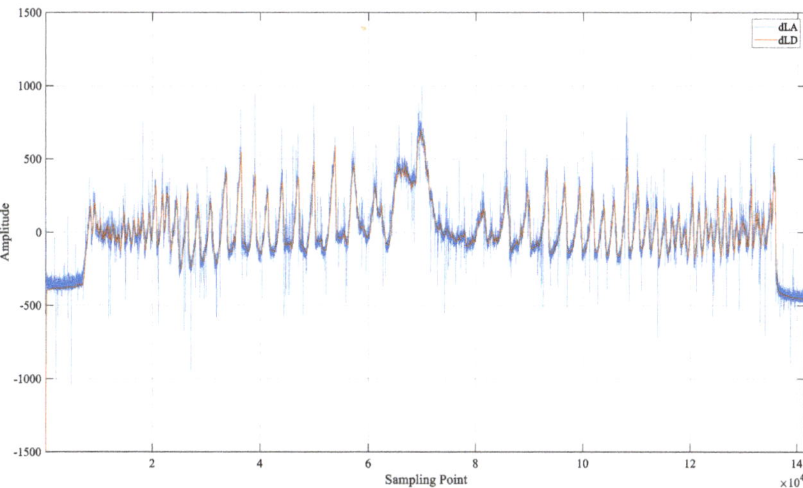

Figure 12. The detrended Larch digital (dLD) detection result waveform and detrended Larch analog (dLA) detection result waveform (SN. 29-1043-52471).

The amplitude and trend of the detrended digital detection results and the detrended analog detection results are very similar. However, it is still difficult to intuitively determine their consistency due to the large amount of interference contained in the detrended analog detection results.

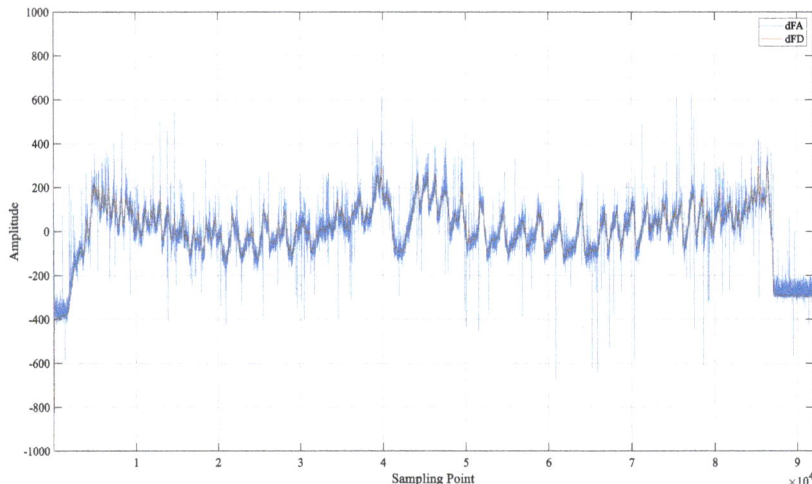

Figure 13. The detrended Fir digital (dFD) detection result waveform and detrended Fir analog (dFA) detection result waveform (SN. 30-1013-34894).

To determine the consistency of the two detection results, a correlation coefficient was introduced for quantitative analysis [59]. The correlation coefficient is defined in Equation (23)

$$\rho_{DA} = \frac{\mathrm{Cov}(D, A)}{\sigma_D \sigma_A} \tag{23}$$

Cov: covariance; σ: standard deviation; D: digital detection result after detrending; A: analog detection result after detrending.

The correlation coefficients ρ_{DA} of each sample are shown in Table 3.

Table 3. The correlation coefficients and average correlation coefficients.

Serial Number (SN)	Correlation Coefficient	Average	
29-1043-52471	0.9576		
29-1043-52472	0.9484	0.9413	
29-1257-48683	0.9287		
29-1903-44540	0.9305		0.9365
30-1013-34894	0.9028		
30-1512-18696	0.9376	0.9317	
30-1849-16448	0.9187		
30-1911-50929	0.9678		

The value range of the correlation coefficient is between -1 to 1. Generally, when the correlation coefficient is greater than 0.9, the two signals can be considered to have a strong positive correlation. The correlation coefficients of the two detection results shown in Table 3 are all greater than 0.9, and the average correlation coefficient of Larch is 0.9413, the average correlation coefficient of Fir is 0.9317, and the overall average is 0.9365. Therefore, the detection results of the digital method and the analog method have a strong correlation, and it can be considered that the detection results output by the two methods are consistent. Based on the above correlation analysis, it can be proved that the detection results of the digital method have the same correctness as the analog method.

4.3. Power Spectrum Analysis and SNR

In the time domain, the digital micro-drilling resistance detection signal is obviously clearer and easier to identify than the analog. The spectrum analysis can be performed on the signal so that it can be seen more intuitively that the digital signal contains less noise interference than the analog signal.

Normalize the power of the digital and the analog detection results according to Equation (24):

$$f(x)' = \frac{f(x)}{\sqrt{XPower}} = \frac{f(x)}{\sqrt{\sum_{i=1}^{n} f(x_i)^2}} \tag{24}$$

The power-normalized digital results are denoted as D', and the power-normalized analog results are denoted as A'. The signal D_G is obtained by filtering the power-normalized digital detection results using a Gaussian filter. Latewood points in the D_G are marked with an asterisk by the identification algorithm. Taking the results of SN.29-1043-52471 as an example, its latewood marking points are shown in Figure 14.

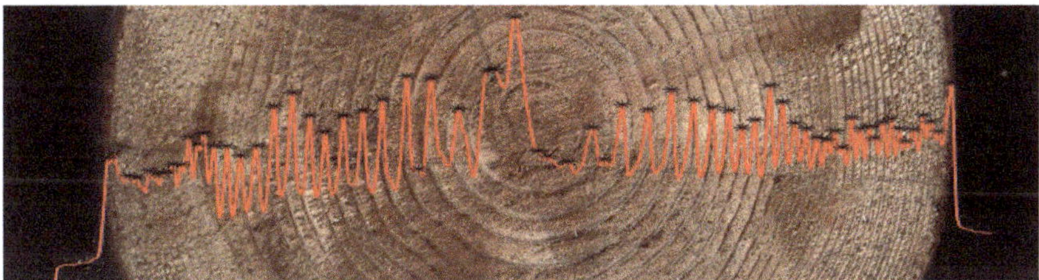

Figure 14. The tree ring latewood identification mark diagram based on the digital detection result after Gaussian filtering (SN. 29-1043-52471).

It can be seen from Figure 14 that the signal is clear and easy to identify, and the latewood points of the tree rings identified automatically by the algorithm have high accuracy and can correspond to the actual tree rings. Therefore, in this study, the D_G is approximately regarded as a noise-free tree ring detection signal, called the desired signal.

Display the power spectrums of the digital method signal D', the analog method signal A', and the useful signal D_G. The result of SN. 29-1043-52471 is shown in Figure 15, and the result of SN. 30-1013-34894 is shown in Figure 16.

It can be seen in Figures 15 and 16 that the power spectrum waveforms characteristics of Larch and Fir are similar. The three waveforms coincide in the low-frequency band, and their energy is concentrated in the low-frequency band, reflecting the change in tree ring resistance. Then the amplitude of the D_G waveform drops rapidly and separates from D' and A'. Since D_G is the desired signal that does not contain noise signals; the separated part represents the noise interference introduced in the detection process. In the separation part, the difference between the D' signal and D_G is significantly smaller than the difference between the A' signal and D_G, indicating that the noise interference level of the digital method is significantly lower than that of the analog method. In addition, there are required noise spikes in A', and the number of noise spikes in D' is also significantly reduced, indicating that the digital method completely avoids noise interference in some frequency bands.

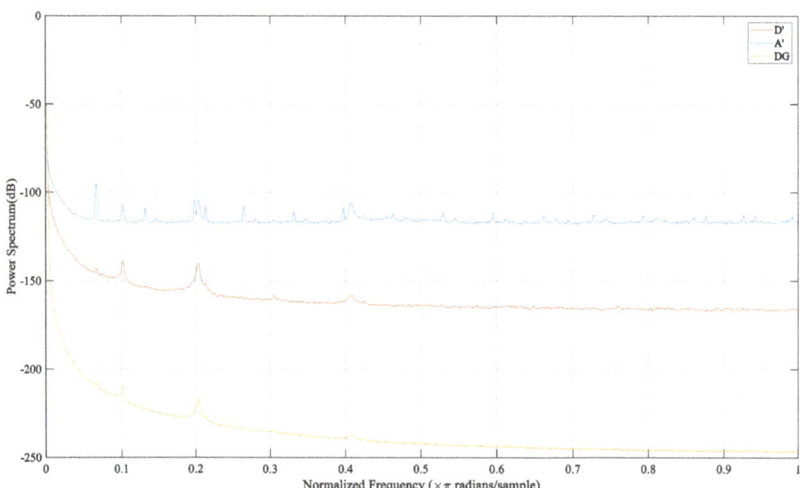

Figure 15. The power spectrums of Larch SN. 30-1013-34894.

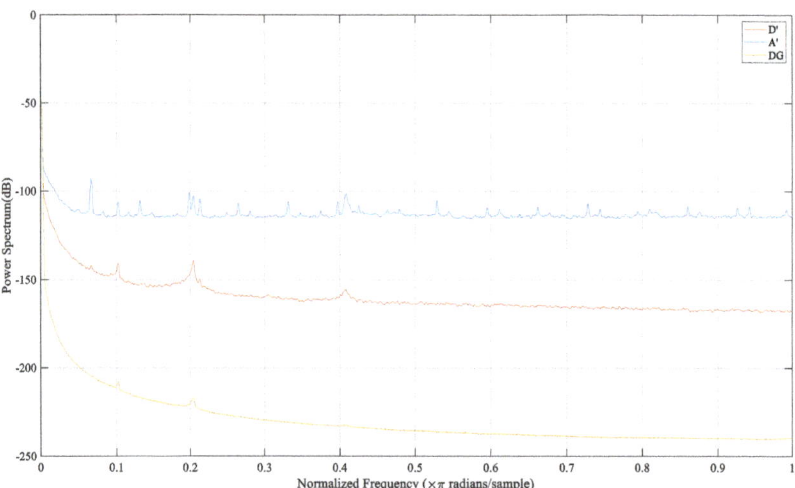

Figure 16. The power spectrums of Fir SN. 29-1043-52471.

To quantify the improvement of the signal quality by the digital method, this research introduces the Signal-to-Noise Ratio (SNR) indicator, which can also be called the Signal-to-Interference and Noise Ratio (SINR) indicator. The SNR refers to the ratio of the desired signal power ($D_G Power$) to the noise and interference signal power ($nPower$), usually in dB. The calculation formula of SNR is shown in Equation (25):

$$SNR = 10lg\frac{D_G Power}{nPower} = 10lg\frac{D_G Power}{nPower} = 10lg\frac{D_G Power}{XPower - D_G Power} \qquad (25)$$

$XPower$ refers to $D'Power$ when calculating SNR for the digital method, $XPower$ refers to $A'Power$ when calculating SNR for the analog method.

The SNR for the digital method and the analog method of each sample are shown in Table 4. All the SNR are in dB. SNR (DM) refers to the SNR for the digital method, and SNR (AM) refers to the SNR for the analog method. SNR Improvement refers to the difference between the SNR of the digital method and the SNR of the analog method.

Table 4. The SNR for the digital method (DM) and the analog method (AM).

Serial Number (SN)	SNR (DM)	Average SNR (DM)	SNR (AM)	Average SNR (AM)	SNR Improvement	Average SNR Improvement
29-1043-52471	39.2524		18.3452		20.9072	
29-1043-52472	39.2172	39.3461	18.6373	18.6280	20.5799	20.7163
29-1257-48683	40.3822		18.7451		21.6371	
29-1903-44540	38.5327		18.7918		19.7409	
		39.0145		19.7555		19.2590
30-1013-34894	40.4539		21.9157		18.5382	
30-1512-18696	38.6922	38.6829	20.3393	20.8812	18.3529	17.8017
30-1849-16448	37.7802		21.1697		16.6105	
30-1911-50929	37.8054		20.1002		17.7052	

The larger the SNR, the smaller the noise interference mixed in the signal and the higher the signal quality; otherwise, the opposite is true. In the eight sample experiments, the average SNR of the digital method is 39.0145 dB, and the average SNR of the analog method is 19.7555 dB. The SNR of the digital method is 19.2590 dB higher than that of the analog method.

Converting the SNR to the percentage P_n of digital method noise interference energy to analog method noise interference energy is shown in Equation (26):

$$P_n = \frac{nPower(DM)}{nPower(AM)} = 10^{\frac{SNR(AM)-SNR(DM)}{10}}\%$$

(26)

The P_n for the digital method and the analog method of each sample are shown in Table 5.

Table 5. The correlation coefficients and average correlation coefficients.

Serial Number (SN)	P_n	Average	
29-1043-52471	0.81%		
29-1043-52472	0.88%	0.86%	
29-1257-48683	0.69%		
29-1903-44540	1.06%		1.27%
30-1013-34894	1.40%		
30-1512-18696	1.46%	1.69%	
30-1849-16448	2.18%		
30-1911-50929	1.70%		

Table 5 shows that the average noise interference energy of the digital method is only 1.27% of that of the analog method. Therefore, the digital method greatly reduces the introduced noise interference and significantly improves signal quality.

5. Discussion

The experimental results can be discussed from two perspectives; time domain and frequency domain. For analysis in the time domain, the first focus is on waveform and amplitude. The waveform of the digital method is clear, which can intuitively identify the tree rings of earlywood and latewood and evaluate the tree age, while the waveform of the analog method fluctuates violently, and the effective tree ring information cannot be identified. This is the signal quality improvement of the digital method, directly reflected from the time domain waveform. There is a significant difference in amplitude between the two signals due to differences in the calculation of the two detection methods. Although their amplitudes are different, they are all used to measure the change of resistance, so if the calculation formula of the digital method is correct, the digital and analog detection signals should have a strong correlation. As the experimental results show, the average correlation coefficient of the two detection methods reaches 0.9365, which means the results obtained by the two have high consistency, and verifies the correctness of the digital method from the perspective of experimental results. Through the flow of the measured signal, the consistency of the detection results can also be proved. The signal flow diagram of the digital method and the analog method is shown in Figure 17.

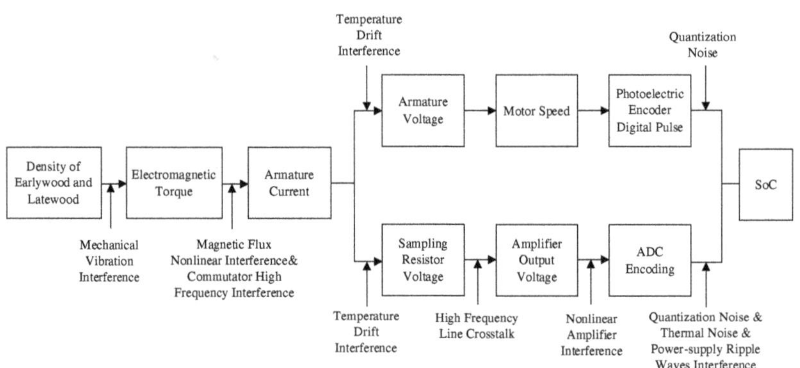

Figure 17. The signal flow diagram of the digital method and the analog method.

It can be seen from Figure 17 that both methods complete the first three steps of signal flow based on the proportional relationship between the armature current and the change in tree ring density. The flow after the armature current starts to differ between the digital method and the analog method, so both detection results should be related to the armature circuit. The analog method is a widely used and correct method of micro-drilling resistance technology, so the detection results of the digital method are strongly related to the analog method, which can verify the correctness of the digital method.

In the analysis of the frequency domain, this study uses SNR to quantify the improvement of signal quality. Compared with the analog method, the digital method has an excellent performance, the average SNR improves by 19.2590 dB, and the average noise interference energy is only 1.27% of the analog method. Such excellent performance can be obtained because the digital method does not take measures to reduce the influence of noise interference but cuts off the way of noise interference in principle. Figure 17 also shows the approach to noise intervention. Comparing the noise intervention of the two methods, the digital method samples the non-electrical quantity speed, which avoids the line crosstalk caused by high-frequency electronic signals such as PWM signals and serial transmission. An amplifier isn't needed in the digital method, so there is no nonlinear amplification interference. At the same time, the digital method uses a photoelectric encoder instead of ADC for sampling. The digital pulse signal output by the photoelectric encoder has a strong anti-interference ability and will not be affected by thermal noise and power-supply ripple waves interference.

There are also some limitations and further research directions on the digital micro-drilling resistance method. The first limitation is the micro-drilling resistance method is used to identify tree rings by density difference. Hence, it's more suitable for the coniferous and ring-porous trees, which have clearly separated tree rings into earlywood and latewood, but not very suitable for the diffuse-porous trees. To compare the performance of the two different micro-drilling resistance methods under the best conditions, the two experimental tree species are coniferous. Experimental on some ring-porous trees or semi-ring-porous trees is a further research direction.

Secondly, the correctness of the digital method was confirmed by the consistency study. To obtain accurate detection more intuitively, the results can also be compared with dendrochronological methods.

The third direction is the changes in reference voltage for digital detection results. In forest operations, batteries are usually used to power equipment, and the output voltage of the batteries will gradually decrease. The output voltage of the battery will be directly connected to the upper and lower bridge arms of the H-bridge motor drive circuit as U_m. The digital method that uses the PWM duty ratio D_{PWM} or pulse width time T_{on} as the output detection result will be affected by the changes in the amplitude of U_m, so U_m can be regarded as a reference voltage. During a single detection process, the power consumption of the battery is very limited, so the change in U_m is very small, and the impact on the average amplitude of the detection results can be ignored. However, for the two detections with a large difference in battery power, the average amplitude of the output results will be greatly affected by U_m. Taking the detection of the same tree disk sample as an example, when the battery power is sufficient, U_m is larger, which makes the average amplitude of the detection results smaller; when the battery power is low, U_m is small, so that the average amplitude of the average detection results increases. For the same detection sample, this will lead to inconsistent results of multiple detection. And for different detection samples, this will lead to misjudgment as the difference is caused by the difference in the average density of the detection samples. Since the experiments carried out in this research all use a constant voltage source for supplying stable power, it is not affected by the above-mentioned voltage reference problem, but this problem should be paid attention to when the digital method is used in the case of battery power supply.

Finally, the small amount of noise still present in the digital method can be a further research direction. Part of the noise is introduced from the first three steps of signal flow due to the digital method starting after the third step. Another reason is the subtle noise interference that may be introduced by the sampling process of digital detection methods. Further research can analyze the above two types of noise interference, and try to reduce or eliminate their influence to obtain a better-quality signal.

6. Conclusions

In this research, a digital micro-drilling resistance method is proposed. The theoretical analysis and comparative experiments show that hardware implementation of the digital micro-drilling resistance method can correctly reflect the tree ring information and significantly improve the signal quality of the micro-drilling resistance technology. This research shows that the digital micro-drilling resistance method has an obvious advantage in signal quality.

Looking forward, the digital micro-drilling resistance method will help improve the identification accuracy of the micro-drilling resistance method, and to develop the application of tree ring micro-destructive detection technology in the high-precision field.

Author Contributions: Conceptualization, X.H. and Y.Z.; methodology, X.H. and D.X.; software, X.H. and D.X.; validation, X.H., D.X. and Q.S.; formal analysis, X.H., D.X. and Q.S.; resources, Y.Z.; data curation, X.H., D.X. and Q.S.; writing—original draft preparation, X.H.; writing—review and editing, X.H. and D.X.; visualization, X.H. and Q.S.; supervision, Y.Z.; project administration, Y.Z.; funding acquisition, Y.Z. All authors have read and agreed to the published version of the manuscript.

Funding: This work is supported by The Fundamental Research Funds for the Central Universities (2021ZY74).

Institutional Review Board Statement: Not applicable.

Informed Consent Statement: Not applicable.

Data Availability Statement: Not applicable.

Conflicts of Interest: The authors declare no conflict of interest. The funders had no role in the design of the study; in the collection, analyses, or interpretation of data; in the writing of the manuscript, or in the decision to publish the results.

References

1. Carlón-Allende, T.; Villanueva-Díaz, J.; Mendoza, M.E.; Pérez-Salicrup, D.R. Climatic Signal in Earlywood and Latewood in Conifer Forests in the Monarch Butterfly Biosphere Reserve, Mexico. *Tree-Ring Res.* **2018**, *74*, 63–75. [CrossRef]
2. Zhang, Y.; Li, J.; Zheng, Z.; Zeng, S. A 479-Year Early Summer Temperature Reconstruction Based on Tree-Ring in the Southeastern Tibetan Plateau, China. *Atmosphere* **2021**, *12*, 1251. [CrossRef]
3. Acosta-Hernández, A.C.; Pompa-García, M.; Camarero, J.J. An Updated Review of Dendrochronological Investigations in Mexico, a Megadiverse Country with a High Potential for Tree-Ring Sciences. *Forests* **2017**, *8*, 160. [CrossRef]
4. Moon, N.H.; Moon, G.H.; Chun, J.H.; Shin, M.Y. Dendroclimatological analysis and tree-ring growth prediction of Quercus mongolica. *For. Sci. Technol.* **2020**, *16*, 32–40. [CrossRef]
5. Li, G.; Harrison, S.P.; Prentice, I.C. Quantifying climatic influences on tree-ring width. *Biogeosci. Discuss.* **2019**. *preprint*. [CrossRef]
6. Gurskaya, M.A. Effect of Summer Monthly Temperatures on Light Tree Ring Formation in Three Larch Species (Larix) in the Northern Forest–Tundra of Siberia. *Russ. J. Ecol.* **2019**, *50*, 343–351. [CrossRef]
7. Sánchez-Calderón, O.D.; Carlón-Allende, T.; Mendoza, M.E.; Villanueva-Díaz, J. Dendroclimatology in Latin America: A Review of the State of the Art. *Atmosphere* **2022**, *13*, 748. [CrossRef]
8. Zou, W.; Jing, W.; Chen, G.; Lu, Y.; Song, H. A Survey of Big Data Analytics for Smart Forestry. *IEEE Access* **2019**, *7*, 46621–46636. [CrossRef]
9. Torresan, C.; Garzón, M.B.; O'Grady, M.; Robson, T.M.; Picchi, G.; Panzacchi, P.; Tomelleri, E.; Smith, M.; Marshall, J.; Wingate, L.; et al. A new generation of sensors and monitoring tools to support climate-smart forestry practices. *Can. J. For. Res.* **2021**, *51*, 1751–1765. [CrossRef]
10. Gallois, E. *Cassiope Tetragona as a Dendroecological Proxy: A Retrospective Analysis of Experimental Warming in the Arctic Tundra*; University of British Columbia: Vancouver, BC, Canada, 2019.
11. García-Hidalgo, M.; García-Pedrero, Á.M.; Caetano-Sánchez, C.; Gómez-España, M.; Lillo-Saavedra, M.; Olano, J.M. ρ-MtreeRing: A Graphical User Interface for X-ray Microdensity Analysis. *Forests* **2021**, *12*, 1405. [CrossRef]
12. García-Hidalgo, M.; García-Pedrero, Á.; Colón, D.; Sangüesa-Barreda, G.; García-Cervigón, A.I.; López-Molina, J.; Hernández-Alonso, H.; Rozas, V.; Olano, J.M.; Víctor, A.-G. CaptuRING: A do-it-yourself tool for wood sample digitization. *Methods Ecol. Evol.* **2022**, *13*, 1185–1191. [CrossRef]
13. Alekseev, A.S.; Sharma, S.K. Long-Term Growth Trends Analysis of Norway Spruce Stands in Relation to Possible Climate Change: Case Study of Leningrad Region, 2020, No3 (375). Available online: https://cyberleninka.ru/article/n/long-term-growth-trends-analysis-of-norway-spruce-stands-in-relation-to-possible-climate-change-case-study-of-leningrad-region (accessed on 16 June 2022).
14. Kozakiewicz, P.; Jankowska, A.; Mamiński, M.; Marciszewska, K.; Ciurzycki, W.; Tulik, M. The Wood of Scots Pine (*Pinus sylvestris* L.) from Post-Agricultural Lands Has Suitable Properties for the Timber Industry. *Forests* **2020**, *11*, 1033. [CrossRef]
15. Oh, J.; Seo, J.-W.; Kim, B.-R. Verifying the Possibility of Investigating Tree Ages Using Resistograph. *J. Korean Wood Sci. Technol.* **2019**, *47*, 90–100. [CrossRef]
16. Karki, J.; Gautam, D.; Thapa, S.; Thapa, A.; Aryal, K.; Sidgel, R. A Century Long Tree-Climate Relations in Manaslu Conservation Area, Central Nepalese Himalaya. *N. Am. Acad. Res.* **2019**, *2*, 49–62.
17. Abiyu, A.; Mokria, M.; Gebrekirstos, A.; Brauning, A. Tree-ring record in Ethiopian church forests reveals successive generation differences in growth rates and disturbance events. *For. Ecol. Manag.* **2018**, *409*, 835–844. [CrossRef]
18. Riechelmann, D.F.C.; Fohlmeister, J.; Kluge, T.; Peter, K.; Richter, D.K.; Deininger, M.; Freidrich, R.; Frank, N.; Scholz, D. Evaluating the potential of tree-ring methodology for cross-dating of three annually laminated stalagmites from Zoolithencave (SE Germany). *Quat. Geochronol.* **2019**, *52*, 37–50. [CrossRef]
19. Kagawa, A.; Fujiwara, T. Smart increment borer: A portable device for automated sampling of tree-ring cores. *J. Wood Sci.* **2018**, *64*, 52–58. [CrossRef]
20. Marin, G.; Abrudan, I.V.; Strimbu, B. Increment Cores of the National Forest Inventory from Romania. *Math. Comput. For. Nat. Resour. Sci.* **2019**, *11*, 294–302.
21. Gärtner, H.; Cherubini, P.; Fonti, P.; von Arx, G.; Schneider, L.; Nievergelt, D.; Verstege, A.; Bast, A.; Schweingruber, F.H.; Buntgen, U. A technical perspective in modern tree-ring research-how to overcome dendroecological and wood anatomical challenges. *J. Vis. Exp.* **2015**, *97*, e52337. [CrossRef]
22. Patrut, A.; Woodborne, S.; Patrut, R.T.; Rakosy, L.; Lowy, D.A.; Hall, G.; von Reden, K.F. The demise of the largest and oldest African baobabs. *Nat. Plants* **2018**, *4*, 423–426. [CrossRef]
23. Soge, A.O.; Popoola, O.I.; Adetoyinbo, A.A. Detection of wood decay and cavities in living trees: A review. *Can. J. For. Res.* **2021**, *51*, 937–947. [CrossRef]
24. Fabiánová, A.; Šilhán, K. The Growth Responses of *Picea abies* (L.) Karst. to Increment Borer Wounding. *Tree-Ring Res.* **2021**, *77*, 74–85. [CrossRef]
25. Allison, R.B.; Wang, X.; Senalik, C.A. Methods for Nondestructive Testing of Urban Trees. *Forests* **2020**, *11*, 1341. [CrossRef]
26. Houjiang, Z. Inspection of growth quality for urban trees. *Sci. Silvae Sin.* **2005**, *41*, 198.
27. Tomazello, M.; Brazolin, S.; Chagas, M.P.; Oliveira, J.T.S.; Ballarin, A.W.; Benjamin, C.A. Application of X-ray Technique in Nondestructive Evaluation of Eucalypt Wood. *Maderas Cienc. Tecnol.* **2008**, *10*, 139–149. [CrossRef]
28. Jacquin, P.; Longuetaud, F.; Leban, J.M.; Mothe, F. X-ray Microdensitometry of Wood: A Review of Existing Principles and Devices. *Dendrochronologia* **2017**, *42*, 42–50. [CrossRef]

29. Schönfelder, O.; Zeidler, A.; Borůvka, V.; Bílek, L.; Vítámvás, J. Effect of Shelterwood and Clear-Cutting Regeneration Method on Wood Density of Scots Pine. *Forests* **2020**, *11*, 868. [CrossRef]

30. Rinn, F.; Schweingruber, F.H.; Schär, E. Resistograph and X-ray Density Charts of Wood. Comparative Evaluation of Drill Resistance Profiles and X-ray Density Charts of Different Wood Species. *Holzforschung* **1996**, *50*, 303–311. [CrossRef]

31. İÇEL, BİLGİN and GÜLER, GÜRCAN Nondestructive determination of spruce lumber wood density using drilling resistance (Resistograph) method. *Turk. J. Agric. For.* **2016**, *40*, 10. [CrossRef]

32. Rinn, F. Typical trends in resistance drilling profiles of trees. *Arborist News* **2014**, *47*, 42–47.

33. Sharapov, E.; Brischke, C.; Militz, H.; Smirnova, E. Prediction of modulus of elasticity in static bending and density of wood at different moisture contents and feed rates by drilling resistance measurements. *Eur. J. Wood Prod.* **2019**, *77*, 833–842. [CrossRef]

34. Walker, T.D.; Isik, F.; McKeand, S.E. Genetic Variation in Acoustic Time of Flight and Drill Resistance of Juvenile Wood in a Large Loblolly Pine Breeding Population. *For. Sci.* **2019**, *65*, 469–482. [CrossRef]

35. Tomczak, K.; Tomczak, A.; Jelonek, T. Measuring Radial Variation in Basic Density of Pendulate Oak: Comparing Increment Core Samples with the IML Power Drill. *Forests* **2022**, *13*, 589. [CrossRef]

36. Downes, G.M.; Harrington, J.J.; Drew, D.M.; Lausberg, M.; Muyambo, P.; Watt, D.; Lee, D.J. A Comparison of Radial Wood Property Variation on Pinus radiata between an IML PD-400 'Resi' Instrument and Increment Cores Analysed by SilviScan. *Forests* **2022**, *13*, 751. [CrossRef]

37. Rinn, F. Basics of typical resistance-drilling profiles. *West. Arborist* **2012**, *17*, 30–36.

38. Cao, Y.; Wang, D.; Wang, Z.; Tian, L.; Zheng, C.; Tian, Y.; Liu, Y. Research on Tree Pith Location in Radial Direction Based on Terrestrial Laser Scanning. *Forests* **2021**, *12*, 671. [CrossRef]

39. Hu, X.; Zheng, Y.; Liang, H.; Zhao, Y. Design and Test of a Microdestructive Tree-Ring Measurement System. *Sensors* **2020**, *20*, 3253. [CrossRef]

40. Chen, X. Design of Needle Measurement System for Tree Annual Ring. Master's Thesis, Beijing Forestry University, Beijing, China, 2019. [CrossRef]

41. Hong, P.; Jun, L.; Xiangdong, L.; Xuzhan, G.; Jianfeng, Y.; Shouzheng, T. Tree Age Estimation Based on Resistograph Stationary Kalman Filter. *Sci. Silvae Sin.* **2021**, *57*, 14–23. [CrossRef]

42. Jianfeng, Y.; Yandong, Z.; Jun, L.; Yili, Z.; Ruidong, G.; Shouzheng, T. Annual-ring Measurement Method Based on Adaptive Filtering Algorithm. *Trans. Chin. Soc. Agric. Mach.* **2020**, *51*, 216–222.

43. Traversari, S.; Giovannelli, A.; Emiliani, G. Wood Formation under Changing Environment: Omics Approaches to Elucidate the Mechanisms Driving the Early-to-Latewood Transition in Conifers. *Forests* **2022**, *13*, 608. [CrossRef]

44. Berdanier, A.B.; Miniat, C.F.; Clark, J.S. Predictive models for radial sap flux variation in coniferous, diffuse-porous and ring-porous temperate trees. *Tree Physiol.* **2016**, *36*, 932–941. [CrossRef] [PubMed]

45. Fabijańska, A.; Danek, M. DeepDendro—A tree rings detector based on a deep convolutional neural network. *Comput. Electron. Agric.* **2018**, *150*, 353–363. [CrossRef]

46. Rinn, F. Intact-decay transitions in profiles of density-calibratable resistance drilling devices using long thin needles. *Arboric. J.* **2016**, *38*, 204–217. [CrossRef]

47. IML > Products > Wood testing Systems > IML-RESI Systems > IML-RESI MD300. Available online: https://www.iml-service.com/product/iml-resi-md300/ (accessed on 5 June 2022).

48. Guller, B.; Guller, A.; Kazaz, G. Is Resistograph an appropriate tool for the annual ring measurement of Pinus brutia. In Proceedings of the 42nd International Conference NDE Safety, Sec, Czech Republic, 30 October–1 November 2012; pp. 89–94.

49. Oh, J.-A.; Seo, J.-W.; Kim, B.-R. Determinate the Number of Growth Rings Using Resistograph with Tree-Ring Chronology to Investigate Ages of Big Old Trees. *J. Korean Wood Sci. Technol.* **2019**, *47*, 700–708. [CrossRef]

50. Radecki, A.; Rybicki, T. An Accurate State Visualization of Multiplexed and PWM Fed Peripherals in the Virtual Simulators of Embedded Systems. *Appl. Sci.* **2022**, *12*, 3137. [CrossRef]

51. Choi, W.; Kang, F. H-bridge based multilevel inverter using PWM switching function. In Proceedings of the INTELEC 2009—31st International Telecommunications Energy Conference, Incheon, Korea, 18–22 October 2009; pp. 1–5. [CrossRef]

52. Hernández-Alvarado, R.; García-Valdovinos, L.G.; Salgado-Jiménez, T.; Gómez-Espinosa, A.; Fonseca-Navarro, F. Neural Network-Based Self-Tuning PID Control for Underwater Vehicles. *Sensors* **2016**, *16*, 1429. [CrossRef]

53. Zhu, R.; Wu, H. Dc motor speed control system based on incremental pid algorithm. *Instrum. Tech. Sens.* **2017**, *7*, 121–126.

54. Wang, J.; Li, M.; Jiang, W.; Huang, Y.; Lin, R. A Design of FPGA-Based Neural Network PID Controller for Motion Control System. *Sensors* **2022**, *22*, 889. [CrossRef]

55. Zhang, Q.; Pei, W. DSP Processer-in-the-Loop Tests Based on Automatic Code Generation. *Inventions* **2022**, *7*, 12. [CrossRef]

56. Xu, J.; You, B.; Ma, L. Research and development of DSP based servo motion controller. In Proceedings of the 2008, 7th World Congress on Intelligent Control and Automation, Chongqing, China, 25–27 June 2008; pp. 7720–7725. [CrossRef]

57. Chen, H.-C. An H-bridge driver using gate bias for DC motor control. In Proceedings of the 2013 IEEE International Symposium on Consumer Electronics (ISCE), Hsinschu, Taiwan, 3–6 June 2013; pp. 265–266. [CrossRef]

58. Zhou, Z.; Fu, L.; Zhou, C.; Sharma, R.P.; Zhang, H. Simultaneous Compatible System of Models of Height, Crown Length, and Height to Crown Base for Natural Secondary Forests of Northeast China. *Forests* **2022**, *13*, 148. [CrossRef]

59. Fujii, S.; Ishii, Y.; Katsuyama, S.; Akai, K.; Ueda, T.; Inada, M.; Fukushima, A.; Nakajima, K. Research and development of anti-maceration laparoscopic surgical cotton swabs. *Minim. Invasive Ther. Allied Technol.* **2022**, *31*, 587–594. [CrossRef] [PubMed]

 *forests*

Article

Development of a Robust Machine Learning Model to Monitor the Operational Performance of Fixed-Post Multi-Blade Vertical Sawing Machines

Stelian Alexandru Borz [1,*], Gabriel Osei Forkuo [1], Octavian Oprea-Sorescu [1] and Andrea Rosario Proto [2]

[1]   Department of Forest Engineering, Forest Management Planning and Terrestrial Measurements, Faculty of Silviculture and Forest Engineering, Transilvania University of Brasov, Şirul Beethoven 1, 500123 Brasov, Romania; gabriel.forkuo@student.unitbv.ro (G.O.F.); octavian.oprea@student.unitbv.ro (O.O.-S.)

[2]   Department of AGRARIA, Mediterranean University of Reggio Calabria, Feo di Vito snc, 89122 Reggio Calabria, Italy; andrea.proto@unirc.it

*   Correspondence: stelian.borz@unitbv.ro; Tel.: +40-742-042-455

Abstract: Monitoring the operational performance of the sawmilling industry has become important for many applications including strategic and tactical planning. Small-scale sawmilling facilities do not hold automatic production management capabilities mainly due to using obsolete technology which is an effect of low financial capacity and focus their strategy on increasing value recovery and saving resources and energy. Based on triaxial acceleration data collected over five days at a sampling rate of 1 Hz, a robust machine learning model was developed with the purpose of using it to infer the operational events based on lower sampling rates adopted as a strategy to collect long-term data. Among its performance metrics, the model was characterized in its training phase by a very high overall classification accuracy (CA = 98.7%), F1 score (98.4%) and a very low error rate (LOG LOSS = 5.6%). For a three-class problem, it worked very well in classifying the main events related to the operation of the machine, with active work being characterized by an F1 score of 99.6% and an error of 3.6%. By accounting for the same metrics, the model was proven to be invariant to the sampling rates of up to 0.05 Hz (20 s) and produced even better results in the testing phase (CA = 98.9%, F1 = 98.6%, LOG LOSS = 5.5%, for a testing sample extracted at 0.05 Hz), while there were no differences in the share of class data irrespective of the sampling rate. The developed model not only preserves a high classification performance in the training and testing phases but it also seems to be invariant to lower sampling rates, making it useful for prediction over data collected at low sampling rates. In turn, this would enable the use of cheap data collectors to be operated for extended periods of time in various locations and will save human resources and money associated with data collection. Further tests would be required only for validation and they could be supported by collecting and feeding new data to the model to infer the long-term performance of similar sawmilling machines.

Keywords: forestry 4.0; automation; artificial intelligence; wood technology; sawmilling; productivity; prediction; long-term

Citation: Borz, S.A.; Forkuo, G.O.; Oprea-Sorescu, O.; Proto, A.R. Development of a Robust Machine Learning Model to Monitor the Operational Performance of Fixed-Post Multi-Blade Vertical Sawing Machines. *Forests* **2022**, *13*, 1115. https://doi.org/10.3390/f13071115

Academic Editors: Milan Gaff and Gianni Picchi

Received: 26 May 2022
Accepted: 14 July 2022
Published: 15 July 2022

Publisher's Note: MDPI stays neutral with regard to jurisdictional claims in published maps and institutional affiliations.

1. Introduction

Sawmilling facilities represent one of the key components of the wood supply chain, because they enable the first important transformation of roundwood into finite products, acting as a hub between the provision of raw wood materials and the markets [1]. With the growing demand for wood products and globalization in a relatively stable market, important changes occurred in the technology used to process the wood, favoring the establishment of large stakeholders who followed a trend of automating their business to a large extent. However, to be both resilient and efficient, such facilities depend largely on a steady supply and resource availability, therefore, they could be less flexible to significant fluctuations in provision.

On the opposite side, there are the small to medium-sized sawmilling facilities characterized by low processing capacities [2] and, typically, by the absence of automation [3,4]. With some exceptions, these use rather obsolete technology, are operated by a small number of workers and do not hold production planning and management capabilities [5–8]. Still, they support local economies by providing added value and by diversification in opportunities for employment, while also complementing the sawmilling capabilities of a region to valorize wood assortments which are less demanded by large processing facilities [9,10].

Due to the lack of production monitoring systems that comes with using obsolete technology, as well as due to limited financial ability to procure updated technology, such small to medium businesses may become fragile in a dynamic, changing and competitive market. The main reasons for still operating are those related to cutting down the investments in updated technology and struggling in value recovery and energy saving, which generally characterize the sawmilling operations [11,12] and which have lately turned out to be important parameters for optimization, particularly when facing new challenges in energy and resource security. In addition, these challenges might not be area- or technology-specific since some have found efficiency issues in rather well-established wood industries [13]. Consequently, the above-mentioned were among the reasons for which the usefulness of cheap solutions was researched in previous studies with the aim of providing data and tools to support operational planning and management, although they were externally generated and implemented. The rationale behind developing them was that once data could be extracted, or part of the functions of a system could be automated, they would positively be contributing to the overall efficiency, either by developing models in a traditional way with the aim of predicting efficiency or as an effect of function automation, as proved by other studies [14]. For instance, concepts of using different kinds of sensors to automate the extraction of useful data were described for rather more complicated operations and equipment [15,16]. In relation to sawmilling operations, the beginning of operational monitoring was most likely characterized by the use of manual solutions based on the concepts of traditional time-and-motion studies [17,18], which aimed at characterizing the productive performance in small-to-medium sawmilling operations of various configurations [3,5–8,19]. However, these became less practical to implement due to the resources spent to collect and process large amounts of data [20], the reluctance of operators when facing observation, and the capability limits of the observers [21]; in addition, manual data collection procedures hold a limited ability to capture the pattern in operational performance over a long term, restricting the applicability of the developed statistics in a range of variations from which they were built [18]. As potential solutions to the above limitations, cheap sensor systems [4] and methods of computer vision [22] associated with machine learning techniques were tested. Although they were found to be very useful and accurate, by their application, they were intended to externally monitor the sawmilling performance, serving more the science, although practical applications could have been supported, assuming that business holders were willing to internally implement such systems to monitor their operations. While this may not change, and despite the fact that it is not formally acknowledged, the interest in monitoring the long-term performance of several facilities and operations has increased lately, mainly because such information is required in strategic and tactical planning of the wood processing sector. In addition, getting long-term data at a low cost would support optimization or at least help in identifying and characterizing in the time frame those factors which cause variation in sawmilling efficiency. However, this would require several systems or data loggers and significant resources to be spent by researchers conducting observations in several locations when opting for a very fine sampling rate.

A typical example is that of using accelerometer data loggers, which were found to be very sensitive to motion and vibration, making them very versatile in getting useful information in many disciplines. There are already many examples of studies using accelerometer data, which were implemented in forestry and other sectors with the aim of solving specific problems, mainly those related to operational activity recognition [4,23–30], proving that acceleration data may be successfully used in many tasks. As a fact, electronics have increasingly been used in forestry [31] to find efficient solutions to the current challenges. For some applications which are known to yield high vibrations, which are typical of many sawmilling configurations, patterns in acceleration were found to be useful in inferring specific events, as well as to feed machine learning algorithms able to predict the operational behavior in sawmilling operations [24,25,32,33]. Once an accurate machine learning model is trained, more data could be brought on a regular basis to feed the models and to build an overview of performance over very long time periods and at low costs. This would require a robust model which nowadays can be built by using freely available tools [34,35] as well as some lower degree programming in commonly used office tools such as Microsoft Excel® (Microsoft, Redmond, WA, USA). In relation to data collectors, however, this would require enough memory on a given device and a sufficient life span of the power source to be able to run the observation in the long term. Currently, most of the affordable offline data loggers come with rather a limited memory availability and time span of the batteries. With the above-mentioned in mind, solutions need to be researched so as to build a robust machine learning model able to accurately classify the most important events in the time domain while being invariant to the data collection location. In addition, it is important to check what sampling rate would be sufficient to enhance the performance in memory and battery use while preserving the timeshare of events, which is typically dependent on the monitored equipment and underlying process [36].

Previous studies integrating signal data collection and machine learning have focused on rather more flexible band saws [4,22], which tend to replace the older, fixed sawmilling equipment due to the possibility of enhancing value recovery. However, they could require more energy to be spent per processed unit, given their typical operational pattern which requires returning the cutting frames before starting new cuts and readjusting the logs during processing [3,4,22]. On the other hand, fixed-post cutting frame equipment holds the advantage of feeding the logs into the sawing blades, therefore log processing is carried out in one turn, although the sawing speed may be lower. Such sawmilling machines are made of a steel frame that supports the vertical cutting device, enabling the adjustment of the sawing thickness by the distance at which the blades are fixed on a vertically moving frame, therefore they require a rather exact sawing pattern that is established a priori.

The goal of this study was to explore the possibility of building a robust machine learning model able to accurately work in classifying triaxially collected acceleration data to predict three main operational events which are characteristic of fixed-post, multi-blade vertical sawing machines while providing the opportunity for collecting long-term data. The first objective was that of inferring the best machine learning model and its architecture by a trial-and-error hyperparameter tuning of two popular machine learning model classes, namely Neural Network (NN) and Random Forest (RF). The second objective was to see how much variation would be in the classification performance in relation to the amount of data used to train the model, and if there are significant differences in classification performance between the training and testing phases brought by the amount of data used, with the aim of validating the general model. The third objective was to check if there are high variations in classification performance due to the variation in sampling rate with the aim of extending data collection capabilities to longer periods of time.

2. Materials and Methods

2.1. Machine Description, Observed Functions and Data Collection

A fixed-post electrically-powered, multi-blade vertical sawing machine (Figure 1) was selected for observation based on two reasons. First of all, in Romania, this kind of equipment still accounts for an important share in use in both private and state companies; secondly, this kind of equipment is lacking the functions of operational monitoring and production management.

Figure 1. Sawmilling machine used in study and the placement of the triaxial acceleration data logger; (**a**) Main components of the machine: 1—multi-blade steel frame, 2—degrees of movement restricted to the vertical plane, 3—steel blades, 4—exhausting and guiding rollers, 5—sawn wood, 6—exhaust direction; (**b**) Data logger placement: 1—triaxial acceleration data logger placed on the steel frame at the log-feeding part of the machine, 2—log feeding direction.

For this type of machine, during operation, the logs are continuously fed into a vertical go-forth displacing blade frame that converts them into pre-processed lumber. Width of the resulting lumber can be adjusted by the way and distance at which the blades are fixed on the frame. For a given time window, the machine can be identified in four possible states related to its operation, namely off, when the engine is off and the blades are not moving (hereafter, OFF), turning on—the engine is turned on and the blades start to move until full speed and displacement (hereafter, TON), on—the engine is turned on and the blades are moving at full speed and displacement, which is the operational state in which the machine is working and the logs are sawn (hereafter, WORK), and turning off—the engine is turned off and the speed and displacement of the blades start to decrease until full stop (hereafter, TOFF). For simplicity, TON and TOFF events were merged together in a machine state switching event (hereafter SWITCH).

Machine monitoring data were collected over five operational days by the means of a VB300 tri-axial acceleration data logger (Extech® Instruments, FLIR Commercial Systems Inc., Nashua, NH, USA). The data logger was set to collect time-labeled acceleration data at 1 Hz, and it was placed on the machine's frame (Figure 1) in a location that was chosen by considering several criteria such as that of collecting a good signal characterizing the underlying process (closeness to the active blades), possibility to reproduce the experiment in each day of observation as well as on long term, avoiding the variance in the collected signal by using the same location of the data logger and avoiding the obstruction of operations. In parallel, an HD 1080 Pro Black Box digital camera (Shenzen AISHINE Electronics Co. Ltd., Shenzen, China) was set up and used to continuously collect time-labeled video data over the observed period. It was placed in a location that enabled convenient monitoring of the machine.

2.2. Data Processing

The data collected by the accelerometer and video camera were downloaded to a personal computer at the end of each day of observation. A Microsoft Excel® spreadsheet (Microsoft, Redmond, USA) was used to merge, store the original tri-axial acceleration data and label the machine's operational state for each observation. Following the removal of that data covering the setup and placement, as well as taking down the data logger from the machine, the labeled dataset contained a number of 78,707 observations, accounting for approximately 22 h of observation (Table 1). It included the identification number, responses on the x, y and z axis, vector magnitude (Euclidian Norm, which is the squared root of the sum of squared axial responses), measurement unit of acceleration (g) and a time and date label for each observation.

Table 1. Main statistics of the triaxial acceleration data in the labeled dataset.

Day of Observation	Number of Observations (Samples Taken at 1 Hz)	Share in the Dataset (%)	Time Covered (h)
Day 1	15,646	19.88	4.35
Day 2	15,817	20.09	4.39
Day 3	15,780	20.05	4.38
Day 4	15,701	19.95	4.36
Day 5	15,763	20.03	4.38
Total	78,707	100.00	21.86

As shown in Table 1, the size of the daily collected datasets was relatively even in terms of number of observations collected and share in the labeled dataset. The data also show a rather low machine utilization rate (approximately 50% of the shift time), which is typical for such facilities and level of technology used. Data collected in the five days of observation were merged into a single file by keeping the order of data collection. Data labeling comprised a visual analysis of video files in the sequence used to collect them at the sawmilling facility as well as of the patterns in data (magnitude of Euclidian Norm plotted in the time domain), followed by data coding to account for the three operational states (OFF, SWITCH, WORK).

In data processing and machine learning tasks, the data in the form of Euclidian Norm (hereafter EN, g) was used as a feature and the classes OFF, SWITCH and WORK were used as target variables. Hereafter, this dataset was called the initial dataset (ID).

To answer the first objective of the study, a first data processing workflow (hereafter WF1, Figure 2) was that of using the initial dataset (ID) for checking which machine learning algorithm and what kind of architecture set for it could produce the best classification performance. For that reason, the ID's data were fed into two popular machine learning algorithms (neural network and random forest, respectively) which were tuned by a trial-and-error approach, as described in Section 2.3, and the results were evaluated based on the performance indicators described in Section 2.4.

The best-performing machine learning architecture was then used to achieve the second objective of the study by implementing a second data processing workflow (hereafter WF2, Figure 3) which consisted of an iterative splitting of the initial dataset into a training (hereafter TRAIN) and a testing (hereafter TEST) subset. Data partitioning was based on a step of 10% of the data and was applied over the same sequence of data contained in the initial dataset. The procedure started by allocating the first 20% of the initial data to the TRAIN and the rest (80%) to the TEST subset, then it added and subtracted 10% of the data to and from the TRAIN and TEST datasets, respectively, resulting in a proportion of 30 to 70%, and so forth until reaching a proportion from 80 to 20% of the data in the TRAIN and TEST datasets, respectively. By doing so, in total, 7 new pairs of subsets were created and each time the best machine learning architecture was trained and tested on the respective subsets (Figure 3). Evaluation of the classification performance was carried out by the metrics described in Section 2.4 for both the training and testing phases.

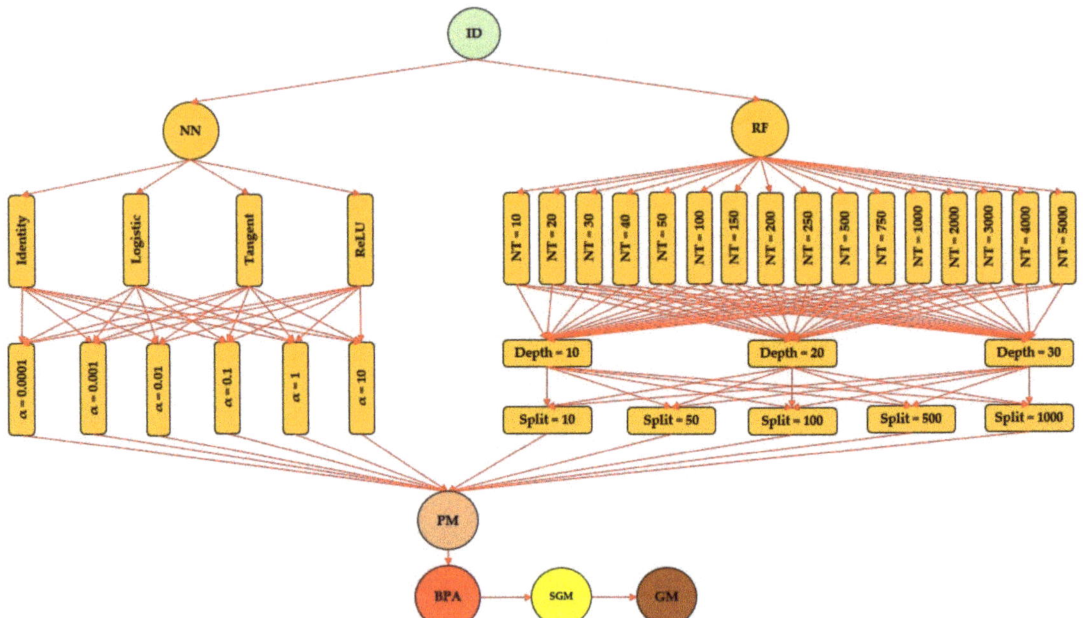

Figure 2. Description of Workflow 1 (WF1) used to infer the best model architecture. Legend: ID—input dataset; NN—neural network; RF—random forest; Identity, Logistic, Tangent and ReLU—activation functions used for NN; α—regularization parameter used for NN, NT—number of trees used for RF; Depth—parameter controlling the depth of the RF, Split—parameter controlling the number of observations when splitting the data; PM—performance metrics; BPA—best-performing architecture; SGM—save general model; GM—general model characterizing the best-performing architecture, trained over all data from ID. Note: input datasets are represented in green, architecture of the machine learning algorithms including hyperparameter tunning options are represented in orange, performance metrics are represented in light brown, the purpose of the workflow is represented in red, actions taken are represented in yellow and the produced models are represented in dark brown. Architecture of the machine learning options used is described in Section 2.3 and performance metrics used to choose the best architecture are described in Section 2.4.

Finally, to check the last objective of the study, a third data processing workflow (hereafter WF3, Figure 4) was implemented. The initial dataset was systematically resampled at 0.500, 0.333, 0.250, 0.200, 0.167, 0.143, 0.125, 0.111, 0.100, 0.067 and 0.050 Hz (from 2 to 10 s at a step of 1 s, 15 and 20 s, respectively). Then, the best-performing machine learning model obtained from WF1 (Figure 1) was used for testing the data from the newly created datasets (11 datasets); the evaluation of the classification performance considered the performance indicators described in Section 2.4.

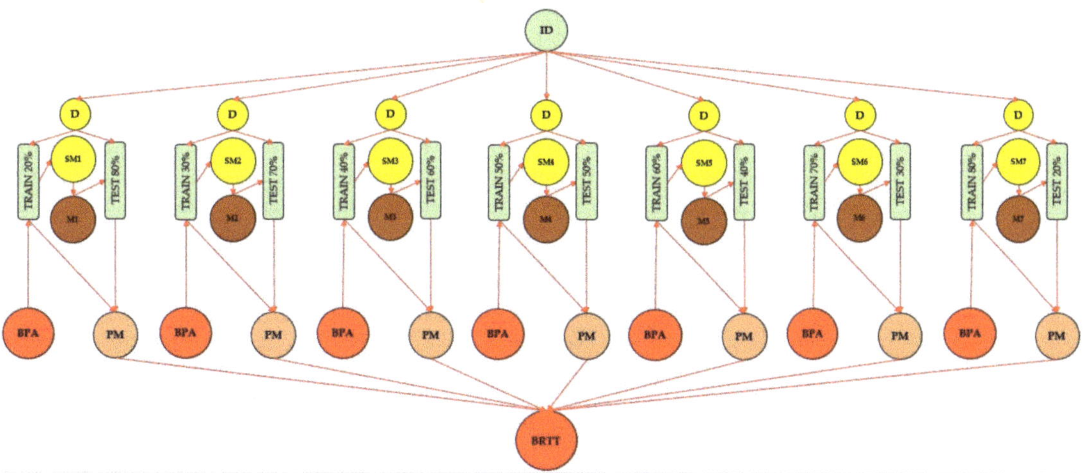

Figure 3. Description of Workflow 2 (WF2) used to infer the best ratio of data partitioning in training and testing subsets. Legend: ID—input dataset; D—division of input dataset; TRAIN—training sample (the shares following the TRAIN word stand for the amount of data used in the training samples); TEST—testing sample (the shares following the TEST word stand for the amount of data used in the training samples); BPA—best-performing architecture; SM1 to SM7—saving models 1 to 7; M1 to M7—saved models 1 to 7, trained on their respective training datasets; PM—performance metrics; BRTT—best ratio of training to testing datasets. Note: input, training and testing datasets are represented in green, architecture of the algorithm is represented in red; performance metrics are represented in light brown, the purpose of the workflow is represented in red at the end of the workflow, actions taken are represented in yellow and the produced models are represented in dark brown. Architecture of the machine learning options used is described in Section 2.3 and performance metrics used to choose the best architecture are described in Section 2.4.

2.3. Machine Learning Algorithms

Two machine learning algorithms were considered, namely the Artificial Neural Networks (hereafter, NN) and Random Forests (hereafter, RF). The choice was based on the popularity of these two machine learning techniques [4,22–24,26,30,37] as well as on the capabilities and functionalities of the software used [34] to tune, train and test the models (Section 2.4). By the software used, NN models are implemented in the form of multilayer perceptrons with backpropagation [34,38]. They require tunning of several parameters, many of which were developed so as to increase the computational performance. RF is a machine learning algorithm proposed by Ho [39] and further developed by Breiman [40]. It has the advance of working well on high dimensional data and fast training. Both machine learning algorithms may be used for classification tasks.

Architecture of the NN machine learning algorithms is commonly described by the depth and width, where the depth stands for the number of hidden layers and the width stands for the number of neurons stored in the hidden layers. Recent findings on testing the performance of NNs over acceleration signal data [41] have indicated that developing the architecture towards a maximal one (i.e., increasing the number of neurons and of hidden layers) may contribute to increments in classification performance. In addition, neural nets were found to increase their representational capacity by increasing the number of neurons in them [42]. The maximum depth and width of the NN was used, as enabled by the used software [34], namely the number of hidden layers was set at 3 and the number of neurons was set at 100 per hidden layer. Providing a better chance to learn was also considered by setting the number of iterations to the maximal one enabled by the software [34], that is 1,000,000 iterations. Adam solver (the stochastic gradient descent optimizer) was set and used for all scenarios due to its enhanced performance [43]. Learning process of the NNs is typically controlled by the type of activation (transfer) function used, and by the value set for the regularization parameter. The activation function controls whether or not a neuron will produce an output. The software used for training and testing purposes enables the use of both linear and nonlinear activation functions [34]. In the first category, the software provides the Identity (Linear) activation function (hereafter Identity) whose output is not confined in a given range [44]. In the second category, the software enables the use of Logistic (hereafter Logistic), Hyperbolic Tangent (hereafter Tangent) and Rectified Linear Unit (ReLU, hereafter ReLU) activation functions. Logistic and Tangent activation functions hold output ranges between 0 and 1 and -1 and 1, respectively [44]. ReLU has become the most used activation function due to its high performance [45,46]. For values less than 0 it returns 0 and for values higher than 0, it returns the actual value [44]. All of the above-described functions were considered in the first workflow used to infer the best architecture of the NN model. The second component of the learning process is the regularization parameter, a hyperparameter which controls the shape of decision functions [44]. For the NN machine learning model, and for all the activation functions, the parameter of the regularization term (α) was tuned to take values of 0.0001, 0.001, 0.01, 0.1, 1 and 10 (Figure 2).

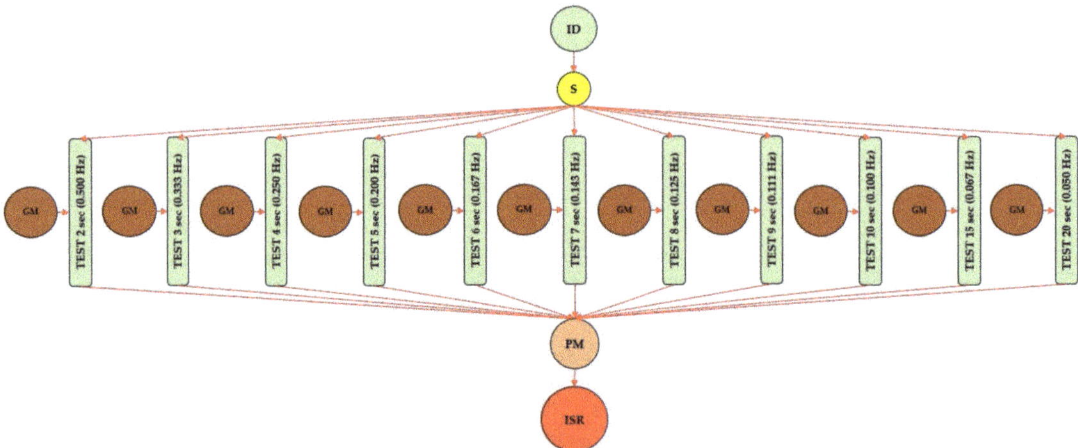

Figure 4. Description of Workflow 3 (WF3) used to evaluate the invariance of the best model architecture to the sampling rate. Legend: ID—input dataset; S—systematic sampling of input dataset; GM—general model developed in WF1 (Figure 2); TEST—systematically sampled testing sample (sampling rate is given both in seconds and Hz); PM—performance metrics; ISR—invariance in performance to sampling rate. Note: input and testing datasets are represented in green, performance metrics are represented in light brown, the purpose of the workflow is represented in red, actions taken are represented in yellow and the input models are represented in dark brown. Architecture of the machine learning options used is described in Section 2.3 and performance metrics used to choose the best architecture are described in Section 2.4.

Typically, the architecture of the RF algorithm may be controlled at two levels, namely the tree and the forest level. There is a set of hyperparameters that can affect the performance of the model [47]. For instance, the depth of the RF algorithm is characterized by longest path between the root and leaf nodes, and higher depths may contribute to performance enhancement in the training phase but may also overfit the model. If not controlled, the number of splits that can happen in a model may reach to nodes which are completely pure, resulting in tree growth and model overfitting. Number of trees is an important parameter in RF, as more trees would help producing a more generalized result [47]. However, as the number of trees increases, similar to the depth and size of the NNs, the time complexity of the model will increase [47]. In this study, the number of attributes considered at each split was kept at the default value provided by the software, which is the square root of the number of attributes present in the data [48]. The models were trained by controlling two parameters of tree growth, namely the depth, which was set successively at 10, 20 and 30 nodes, and the subset splitting restriction, which was set successively at 10, 50, 100, 500 and 1000 observations (Figure 2). Acknowledging that significant changes in performance could be produced when using smaller numbers of trees, this parameter was varied from 10 to 50 with a step of 10, from 50 to 250 with a step of 50, from 250 to 1000 with a step of 250 and from 1000 to 5000 with a step of 1000 (Figure 2).

In total, 24 (4 activation functions $\times$ 6 values set for α) models were trained in the case of NN algorithm, and 240 (6 options for the number of trees $\times$ 3 values for the tree depth $\times$ 5 values for the split control) models were trained for RF which, together, took a computational time of close to 97 h. In both cases a cross-validation by five folds was used to evaluate the training performance. Evaluation of the best model architecture from each class as well as choosing the best model architecture were mainly based on the overall values of the log loss error function which was the first criterion to differentiate among the 264 models. The minimum values were those indicating the model to choose and, in case of ties, the values of F1 score were used for differentiation (maximum values). When there was a tie also for F1 score, the selection algorithm was repeated at the class level in the order WORK–OFF–SWITCH. Although the log loss error and F1 metric were used for selection, several other performance metrics such as the classification accuracy, precision, recall, and sensitivity were estimated as well (Section 2.4). Once the best architecture was inferred it was used over the training datasets from WF2 (Figure 3). For this purpose, the tuned parameters of the best-performing architecture were kept the same during the tests. Each time, and for each ratio of data in the training and testing datasets, a new model was saved with the purpose of testing it. Additionally, a general model was saved characterizing the inferred best architecture (Figure 2), which was then used to test the invariance of classification performance to systematic data sampling (Figure 4). For this purpose, the general model was tested over the systematically sampled datasets.

2.4. Computer Architecture and Software Used—Performance Evaluation

The tasks of training and testing the machine learning models were performed on a computer architecture that included the following features: system type—Alienware 17 R3, processor—Intel® Core™ i7-6700HQ CPU, 2.60 GHz, 2592 MHz, 4 cores, 8 Logical Processors, installed physical memory (RAM)—16 GB, operating system—Microsoft Windows 10 Home. Microsoft Excel ® (Microsoft, Redmond, WA, USA) was used to store and preprocess the data, including the tasks of dividing data into the necessary subsets, performing simple computations, and of resampling the data. Part of the artwork used in this study was built with the same software. The software used to train and test the machine learning models, as well as to build a part of the artwork, was the Orange Visual Programming Software, version 3.31.1 [34], which holds the necessary functionalities for building and running NN (multi-layer perceptron models with backpropagation) and RF models based on the creation of widget-based workflows. Data, Neural Network, Random Forest, Test and Score, Save Model, Load Model and Predictions widgets were used for training and testing purposes. Based on the multidimensional input data, Scatter Plot widget including its "color regions" and "jittering" graphical features were used to depict the relations between the parameter tuning options and key-selected classification performance metrics for both, NN and RF architectures.

Orange Visual Programming Software enables the computation of several classification performance metrics. The full list of metrics computed for the training and testing phases includes the training and testing time, area under the ROC (receiver operating characteristic, hereafter AUC), classification accuracy (hereafter CA), F1 score, which is the harmonic mean of the classification's precision and recall (hereafter F1), precision (hereafter PREC), recall (hereafter REC), log loss (cross-entropy) error (hereafter LOG LOSS) and specificity (hereafter SPEC). All of these metrics were computed for all the training and testing tasks and the most important ones were reported where appropriate. For exemplification, however, performance metrics such as the LOG LOSS, F1 score and CA, were compared in more detail in the results section of the paper, including their differences as an effect of the parameters used for tuning (WF1), partition of the data in training and testing subsamples (WF2) and data resampling in the testing subsets (WF3), respectively. For reference, the performance of classifiers is discussed, for instance, in [49]. References on definitions and explanations of the classification performance metrics are given in [50]. Classification accuracy (CA) is one of the important metrics used for evaluating classification models. It is defined as the ratio of correct predictions to the total number of predictions. Recall (REC), or the hit rate [49,50], is defined as the ratio of correctly classified true positives (true positives) to the total number of positives (true positives + false negatives). Precision (PREC) is the ratio of correctly classified positives (true positives) to the total predicted positives (true positives + false positives) [49,50]. F1 score is a metric that balances precision and recall, being better adapted to class imbalance [50]. Orange Visual Programming software computes the LOG LOSS according to the equation given in [51]. Class level and overall ratio values of the performance metrics which characterize the training and testing datasets are typically multiplied by 100 to obtain a percent-based overview of the classification performance [50], an approach that was used in the Results and Discussion section.

In the testing phase, a given model operates over the test data by providing the probability of each instance being classified in a true class. Given the one-dimensionality of the input data, such probabilities were assigned to EN (Euclidian Norm) of the data in the testing datasets to map each instance in a given class (WORK, SWITCH, OFF).

3. Results

3.1. Best-Performing Model Architecture

The variation of the main classification performance metrics as a function of the models' architecture is shown in Figures 5–9. Figure 5, for instance, plots the LOG LOSS values against the activation functions and regularization terms used in the NN architecture.

There were no major differences in LOG LOSS except those returned by the Logistic activation function when using regularization terms set at 1 and 10 (less complex decision functions). In general, the values of LOG LOSS were in the range of between 5.6 and 6.3% for the 24 trained NN models. Figure 6 shows the variation in F1 score as a function of the models' architecture, indicating a similar trend. In general, the values of the F1 score varied between 97.9 and 98.4%, indicating a high classification performance of the NN models. A similar data organization is shown in Figures 7–9 for the RF model, where each dot stands for a model of a given architecture by jittering the data so as to be visible in the plots. Lower depths (Depth) of the RF model coupled with higher amounts of data preserved at node splitting (Split) were among the highest contributors to lower LOG LOSS errors (Figure 7), which was also generally true for the highest values of the F1 score (Figure 8). In terms of model specificity (Figure 9), there were found two classes, in which highly specific models were generally shaped by lower amounts of data preserved at splits. In general, the LOG LOSS values decreased as a function of the number of trees used to train the models. LOG LOSS and F1 score values varied between 5.8 and 6.9% and 98.3 and 98.4%, respectively. Based on the fixed parameters and criteria described in the Materials and Methods section, the best-performing architecture was identified for a NN machine learning model when using the ReLU activation function and the regularization parameter set at $\alpha = 0.01$.

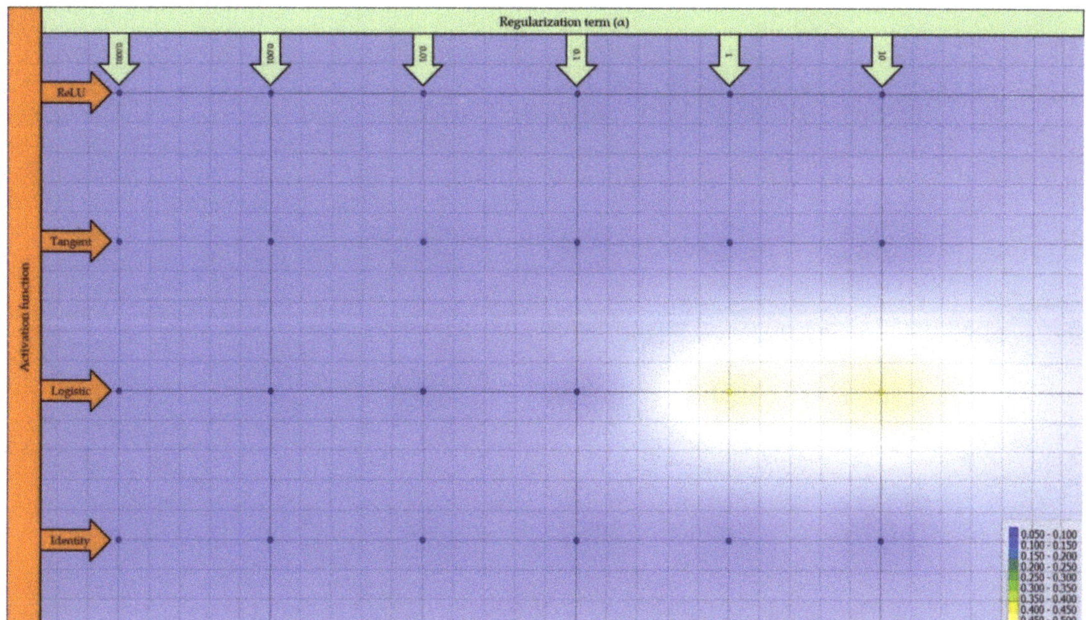

Figure 5. Variation in LOG LOSS as a function of NN architecture in WF1. Note: legend at the bottom right part of the figure describes the values of the LOG LOSS error.

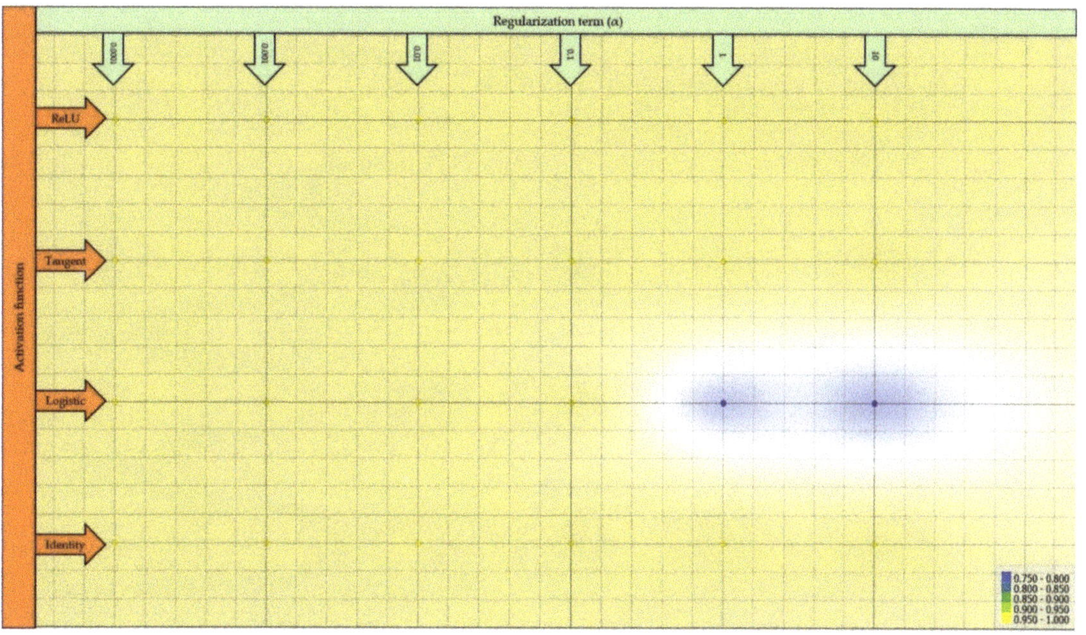

Figure 6. Variation in F1 score as a function of NN architecture in WF1. Note: the legend at the bottom right part of the figure describes the values of the F1 score.

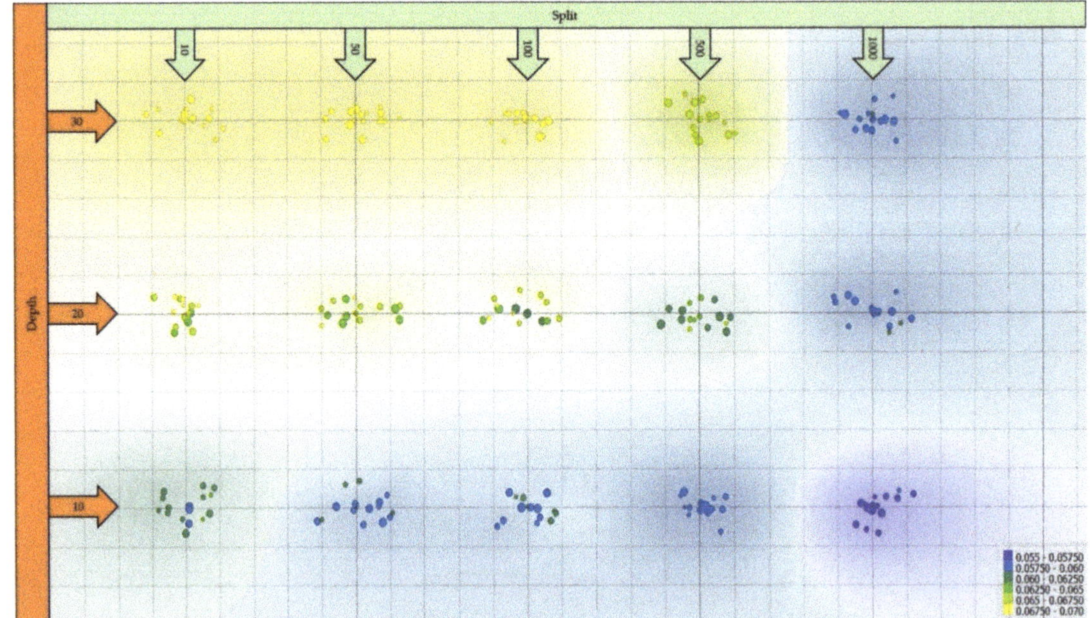

Figure 7. Variation in LOG LOSS as a function of RF architecture in WF1. Note: legend at the bottom right part of the figure describes the values of the LOG LOSS error; size of the dots stands for the number of trees used (i.e., more trees correspond to bigger dots).

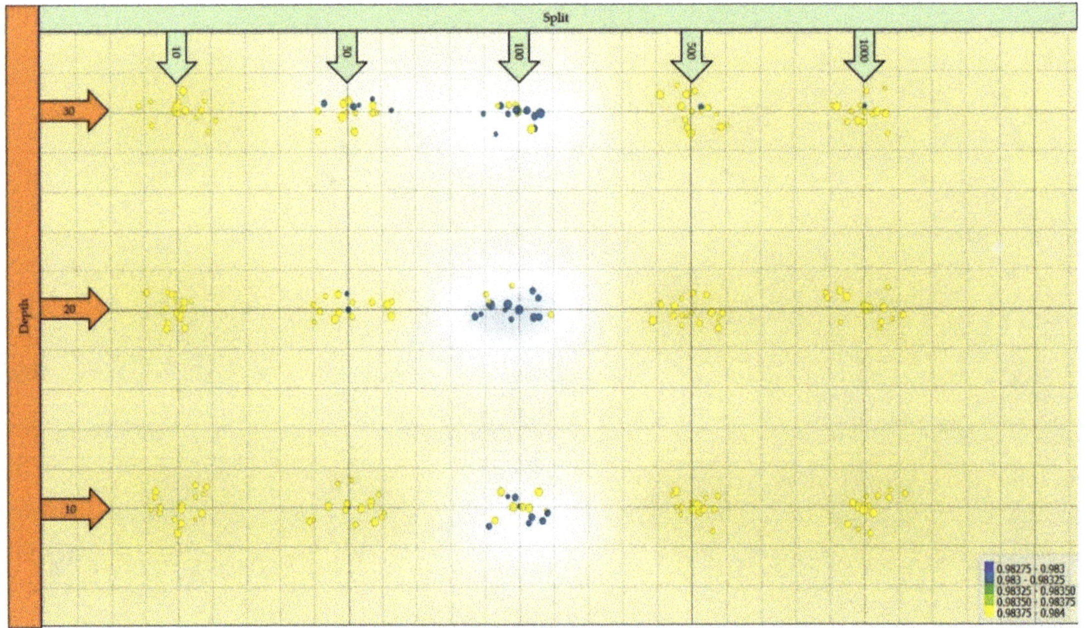

Figure 8. Variation in F1 score as a function of RF architecture in WF1. Note: the legend at the bottom right part of the figure describes the values of the F1 score; size of the dots stands for the number of trees used (i.e., more trees correspond to bigger dots).

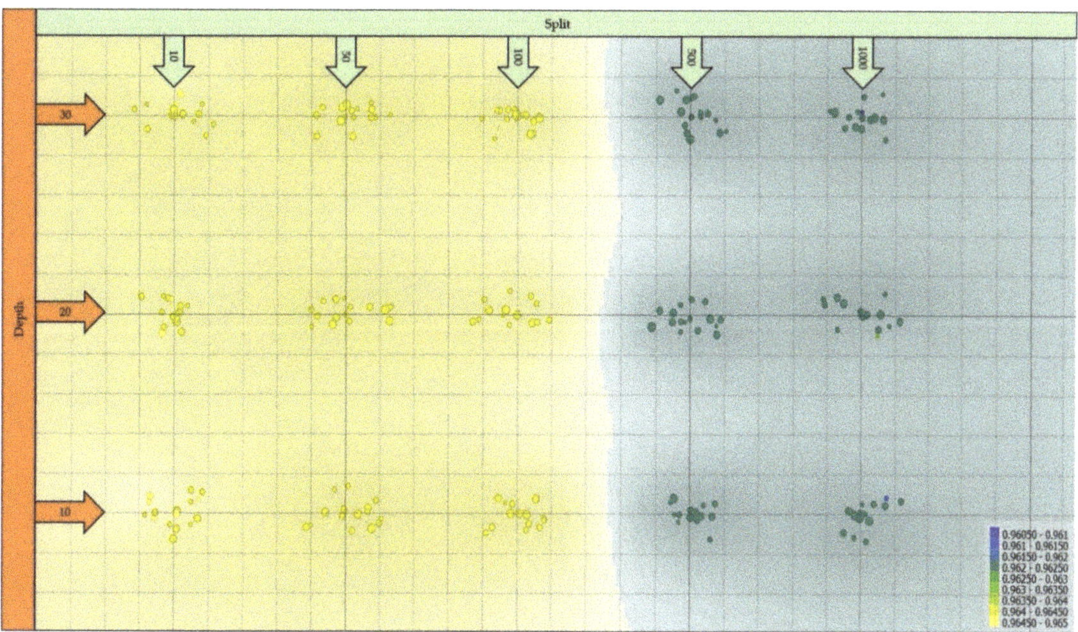

Figure 9. Variation in specificity (SPEC) as a function of RF architecture in WF1. Note: the legend at the bottom right part of the figure describes the values of the specificity; size of the dots stands for the number of trees used (i.e., more trees correspond to bigger dots).

3.2. Effect of Data Share in the Training and Testing Subsets on Classification Performance

The results shown in Figures 10–12 and in Table 2 are consistent with the rule of thumb of using a high share of data in the training (TRAIN, TR) as opposed to the testing (TEST, TE) dataset. For instance, Figure 10 shows the variation in LOG LOSS error in the training and testing datasets as well as the absolute differences in values as a function of the share of data used in the subsets divided according to WF2 (Figure 3). A share of 20%–80% has produced the most contrasting results in terms of LOG LOSS error, which was close to 11% in the dataset used for training and close to 6% in the dataset used for testing. However, a model trained on a small data partition (20%) was able to generalize very well on the rest of the data, as proved by the absolute differences between the training and testing values of LOG LOSS error, which was the highest among the tested options (5.4%).

As the share of data used for training increased, the classification error decreased, reaching a value of 6% for a share of 80%–20% in the training and testing datasets in the training phase, respectively (Figure 10). The range of LOG LOSS values also decreased as the amount of data used in the training sample increased. Irrespective of the share of data used for training and testing, the results show that the errors of the testing phase were much lower compared to those of the training phase (Figure 10), ranging from 2 (80%TR-20%TE) to 5.5% (20%TR-80%TE), a fact that proves that the learned models had a high generalization ability. This can be seen also in the variation of F1 (Figure 11) and CA (Figure 12) values which followed a similar trend of improvement as more data were fed into the training sample, with the most important differences occurring up to a data share of 50 to 50%. However, the absolute differences were smaller accounting for 0.7 to 2.3% in the case of F1 score (Figure 11) and for 0.5 to 1.6% in the case of classification accuracy (CA, Figure 12).

Table 2. Summary of the classification performance metrics and of their differences as a function of the data shared in the training (TRAIN) and testing (TEST) datasets.

Dataset and Share of Data	Classification Performance Metrics						
	AUC	CA	F1	PREC	REC	LOGLOSS	SPEC
TRAIN 20%	0.979	0.973	0.966	0.967	0.973	0.109	0.969
TEST 80%	0.991	0.989	0.989	0.988	0.989	0.054	0.970
Difference	−0.012	−0.016	−0.023	−0.021	−0.016	0.055	−0.001
TRAIN 30%	0.981	0.979	0.973	0.974	0.979	0.087	0.964
TEST 70%	0.991	0.988	0.987	0.987	0.988	0.051	0.962
Difference	−0.010	−0.009	−0.014	−0.013	−0.009	0.036	0.002
TRAIN 40%	0.983	0.980	0.974	0.975	0.980	0.083	0.965
TEST 60%	0.991	0.991	0.991	0.990	0.991	0.041	0.964
Difference	−0.008	−0.011	−0.017	−0.015	−0.011	0.042	0.001
TRAIN 50%	0.985	0.981	0.977	0.977	0.981	0.078	0.967
TEST 50%	0.991	0.993	0.992	0.991	0.993	0.034	0.963
Difference	−0.006	−0.012	−0.015	−0.014	−0.012	0.044	0.004
TRAIN 60%	0.985	0.982	0.978	0.978	0.982	0.074	0.967
TEST 40%	0.990	0.994	0.993	0.992	0.994	0.030	0.956
Difference	−0.005	−0.012	−0.015	−0.014	−0.012	0.044	0.011
TRAIN 70%	0.986	0.985	0.981	0.981	0.985	0.065	0.966
TEST 30%	0.990	0.992	0.991	0.991	0.992	0.035	0.959
Difference	−0.004	−0.007	−0.010	−0.010	−0.007	0.030	0.007
TRAIN 80%	0.988	0.986	0.982	0.983	0.986	0.060	0.967
TEST 20%	0.983	0.991	0.989	0.991	0.991	0.040	0.932
Difference	0.005	−0.005	−0.007	−0.008	−0.005	0.020	0.035

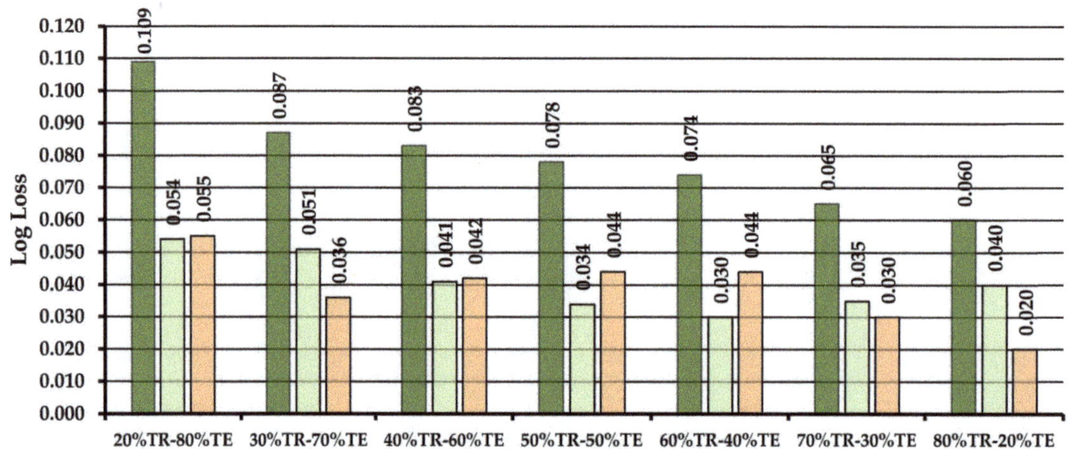

Figure 10. Variation of LOG LOSS error and of its absolute difference in the training (TRAIN) and testing (TEST) datasets as a function of data shared in the training (TR) and testing (TE) datasets.

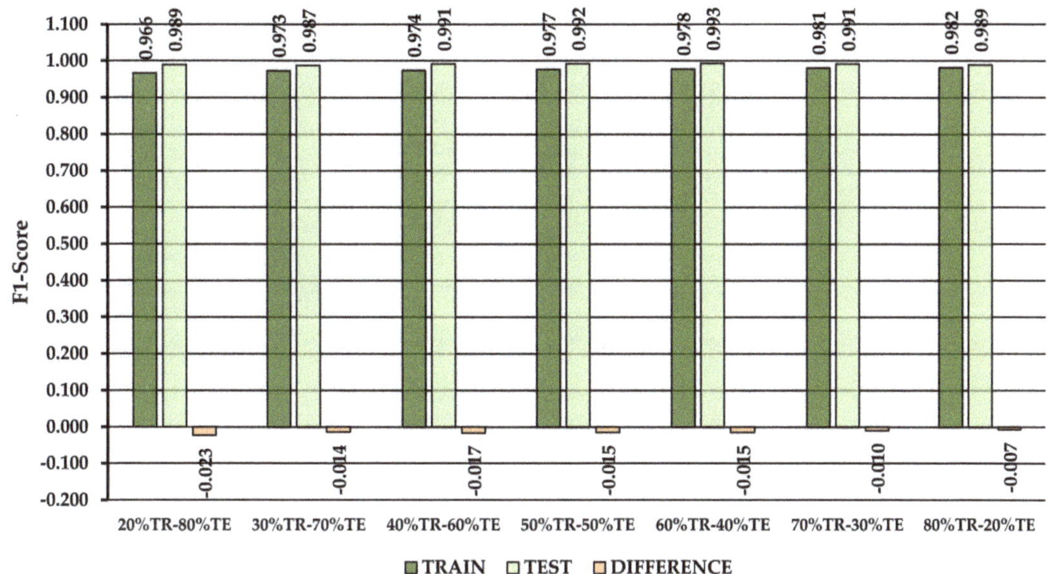

Figure 11. Variation of F1 score and of its absolute difference in the training (TRAIN) and testing (TEST) datasets as a function of data shared in the training (TR) and testing (TE) datasets.

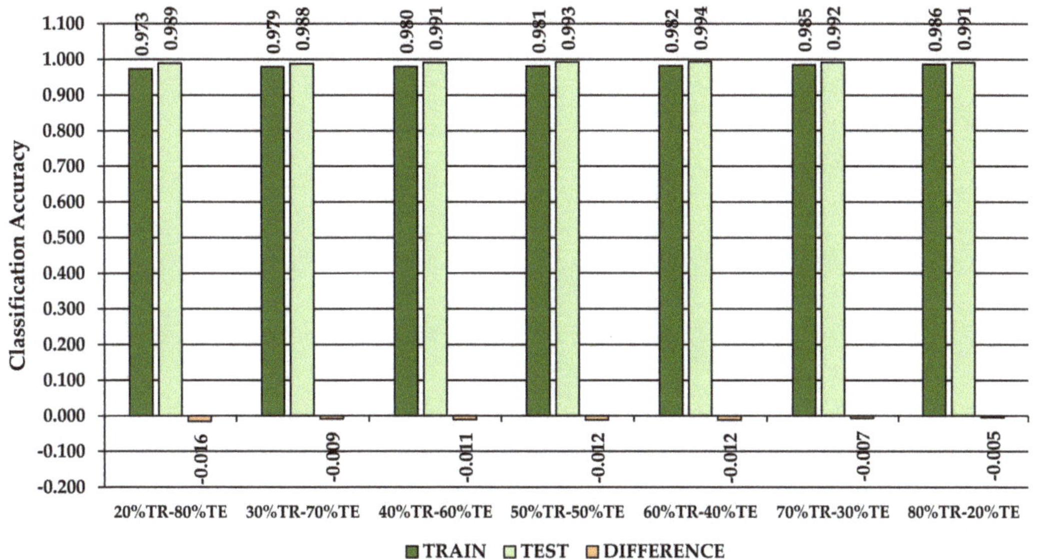

Figure 12. Variation of classification accuracy (CA) and of its absolute difference in the training (TRAIN) and testing (TEST) datasets as a function of data shared in the training (TR) and testing (TE) datasets.

Table 2 summarizes the values of the main classification performance metrics computed for the two datasets and data sharing strategy, including the differences found between the values of the metrics.

Precision (PREC) and recall (REC), which are used to compute the F1 score, followed a similar trend in values and in differences as the F1 and CA did. In all, CA, F1, PREC, REC and LOG LOSS, the most important differences were between using a share of 20 to 80%. For the rest of the data partitioning strategies, the differences were much less, typically less than 1% in the case of TRAIN datasets and less than 0.5% in the case of TEST datasets. Altogether, these trends indicate important improvements, sustain the practice rules of data partitioning, show the variance of classification performance as a function of the strategy used for data partitioning and are useful in providing hints for data partitioning attempts.

3.3. Effect of Sampling Rate on Classification Performance

By resampling, the original dataset (ID) used to train the general model (WF1, Figure 2) was progressively reduced in size from 78,707 (100%) to 3953 instances (5%). The reduction in size compared to ID is illustrated in Figure 13, which shows the 11 newly created testing datasets (WF3, Figure 4) plotted in the time domain. Figure 14, on the other hand, shows the size of the systematically sampled datasets relative to the original dataset (ID).

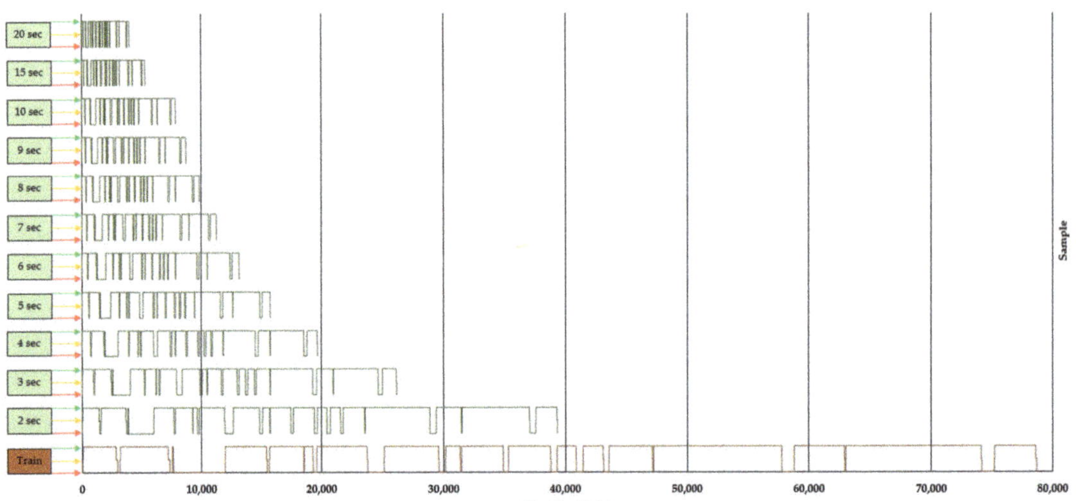

Figure 13. A representation of the original (brown) and resampled (green) datasets in the time domain. Legend: in brown is given the original dataset (ID, Train) and in green (2 to 20 s) are represented the systematically resampled datasets; red arrows placed near the bottom of each dataset indicate the occurrence of the OFF event, while yellow (middle) and green (up) arrows indicate the occurrence of the SWITCH and WORK events, respectively.

For example, by systematically resampling ID at 15 and 20 s, respectively, the amount of data in the testing sets was reduced to less than 7%. However, the used sampling procedure has preserved the share of data on true classes which differed between the original (ID) and sampled datasets by less than 0.1% in all classes. Therefore, a relative data share of ca. 14%–1%–85% was preserved in all datasets for the OFF, SWITCH and WORK classes, respectively.

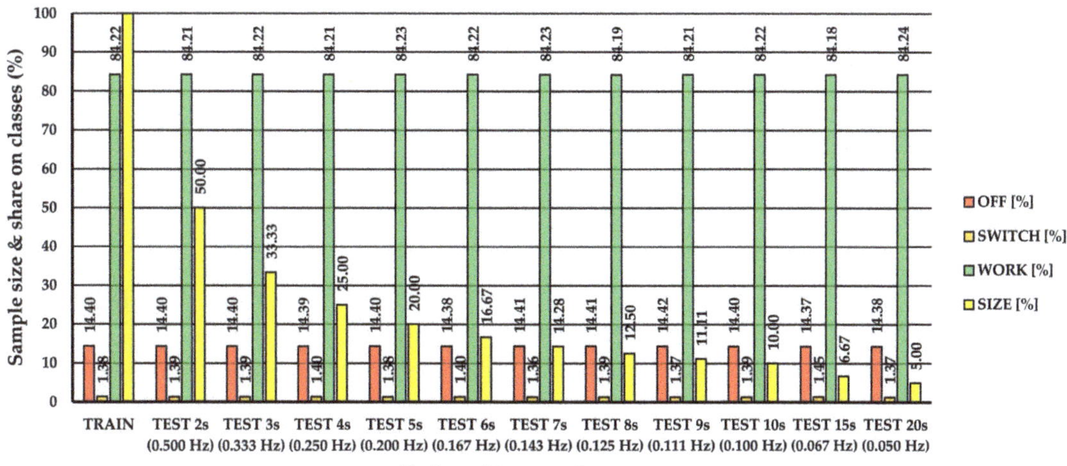

Figure 14. Share of the data in the original (ID, TRAIN) and resampled (TEST) datasets and their relative size.

The results summarized in Table 3 prove the invariance of classification performance metrics of the testing phase to the amount of data used in the testing datasets. Differences brought by the testing sample size in LOG LOSS error are illustrated in Figure 15. Although they were both positive and negative, they were only minor, accounting for a maximum absolute value of 0.2%. Figures 16 and 17 indicate similar trends of differences in F1 score and classification accuracy, which did not exceed an absolute value of 0.2%. As observed (Table 3, Figures 15–17), there were no evident increasing or decreasing trends in the differences between LOG LOSS, F1 and CA as a function of the sampling rate used, although higher differences were found for F1 and CA in the TEST 20 s data subset as compared to the initial dataset (ID, TRAIN).

Table 3. Summary of the classification performance metrics in the systematically sampled testing (TEST) datasets.

Dataset	Share	Number of Observations	Classification Performance Metrics						
			AUC	CA	F1	PREC	REC	LOGLOSS	SPEC
TRAIN	100.00	78,707	0.987	0.987	0.984	0.984	0.987	0.056	0.964
TEST 2 s (0.500 Hz)	50.00	39,353	0.988	0.987	0.984	0.985	0.987	0.054	0.964
TEST 3 s (0.333 Hz)	33.33	26,235	0.987	0.987	0.984	0.985	0.987	0.055	0.964
TEST 4 s (0.250 Hz)	25.00	19,676	0.988	0.988	0.985	0.986	0.988	0.055	0.965
TEST 5 s (0.200 Hz)	20.00	15,741	0.988	0.987	0.984	0.985	0.987	0.055	0.965
TEST 6 s (0.167 Hz)	16.67	13,117	0.987	0.988	0.985	0.986	0.988	0.054	0.964
TEST 7 s (0.143 Hz)	14.28	11,243	0.988	0.988	0.985	0.986	0.988	0.056	0.964
TEST 8 s (0.125 Hz)	12.50	9838	0.988	0.988	0.985	0.986	0.988	0.055	0.965
TEST 9 s (0.111 Hz)	11.11	8745	0.986	0.987	0.984	0.984	0.987	0.057	0.961
TEST 10 s (0.100 Hz)	10.00	7870	0.987	0.988	0.985	0.986	0.988	0.057	0.966
TEST 15 s (0.067 Hz)	6.67	5247	0.986	0.987	0.984	0.985	0.987	0.058	0.963
TEST 20 s (0.050 Hz)	5.00	3935	0.986	0.989	0.986	0.988	0.989	0.055	0.969

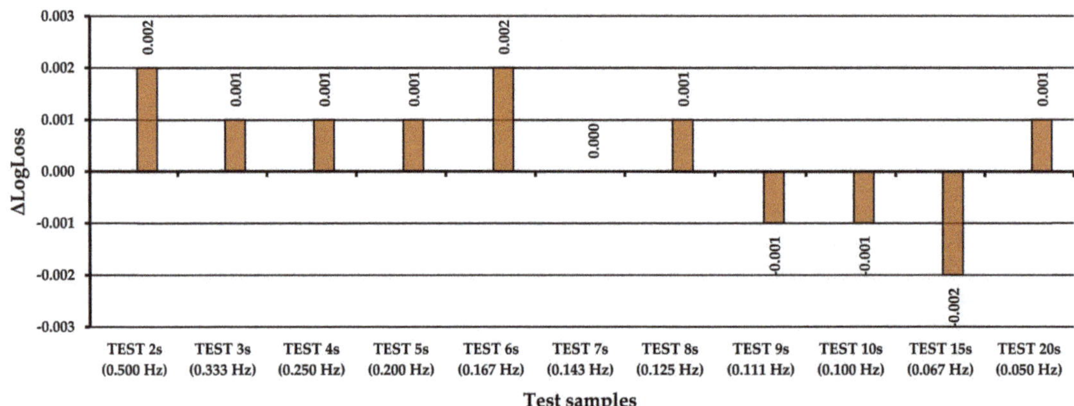

Figure 15. Differences in LOG LOSS error of the testing datasets as opposed to the initial dataset (ID, TRAIN).

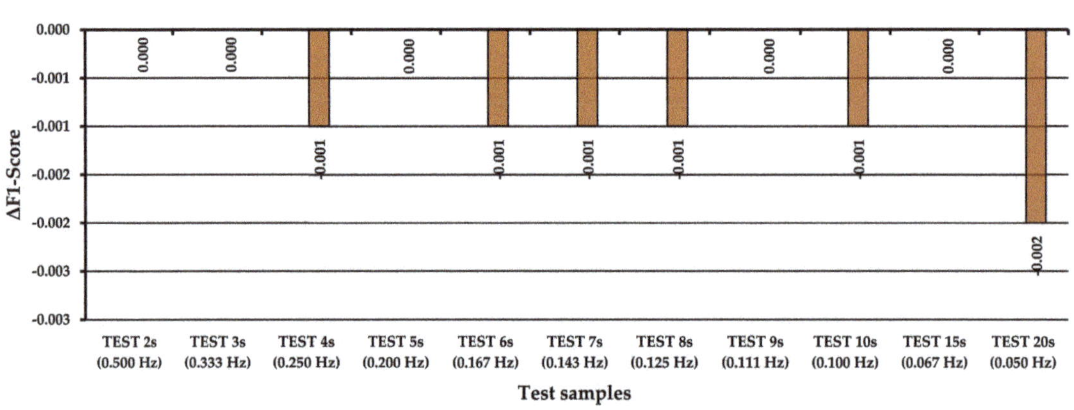

Figure 16. Differences in F1 score of the testing datasets as opposed to the initial dataset (ID, TRAIN).

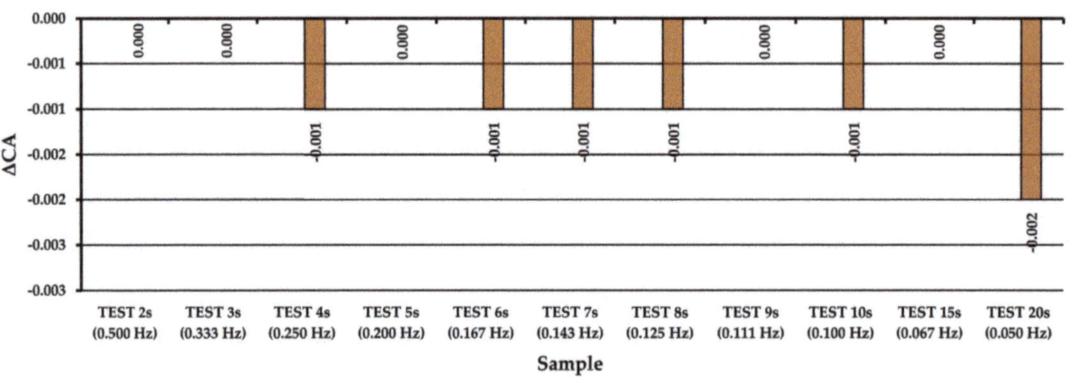

Figure 17. Differences in classification accuracy (CA) of the testing datasets as opposed to the initial dataset (ID, TRAIN).

These results illustrate well the invariance of the general model in terms of classification performance when performing on newly generated datasets, although they were artificially created. In particular, the differences in terms of LOG LOSS, F1 and CA were minor, accounting for up to 0.2% for that case in which the data sample used for testing was the smallest one. Taking as a reference the general model built by the WF1, it correctly classified close to 99% of the data. The F1 score which balances the precision (PREC) and recall (REC) also indicated a high performance, accounting for 98.4%, while LOGLOSS was 5.6%. With minor differences (up to 0.2%) this classification performance was preserved in all of the testing models.

4. Discussion

Several studies have tested the capability of NN models in correctly predicting class-based outcomes with various applications in forestry, some of which have focused on using acceleration signals to make predictions by machine learning [4,24,52]. Most of them agree that general classification accuracies of up to 100% may be achieved depending on several factors such as the complexity of classification, signal quality and accuracy of data labeling. In contrast, some have opted for using RF machine learning algorithms for classification purposes [23,26], finding also highly accurate classifications when collecting data multimodally. Unless a given device holds the capabilities and can be used to collect data multimodally by integrating several sensors, the procurement of separate devices would incur more costs, limiting the economic efficiency of data collection. Nevertheless, by the use of a multimodal approach and RF algorithms, the classification outcomes were found to be similar to those provided by NN, with values of between 97.7 and 99.6% [23,26]. Therefore, it is obvious that when several machine learning algorithms enable classification over a given signal typology, several options need to be checked to evaluate their performance.

With some minor exceptions in regard to the activation function used for NN and the amount of data preserved at a node and the depth of the trees, the classification performance measured by the LOG LOSS error, F1 score and classification accuracy did not vary widely and returned no evident contrasts as an effect of the machine learning architectures used. Therefore, the best model architecture was selected using the first criterion, namely the error during training which was found to be the lowest for an NN architecture when using the ReLU activation function and α set at 0.01. It turns out that the best-performing models reported in other studies checking the effect of classification performance on acceleration data had similar architectures, placing the use of the ReLU activation function and of the regularization terms of up to 0.1 among the best options in terms of classification performance [24,52]. However, the performance of NN depends also on several other factors [53,54], including signal quality and other issues specific to classification tasks such as intra-class variability and inter-class similarity. Altogether, the classification performance of the selected architecture was very high, with a classification accuracy of 98.7%, an F1 score of 98.4%, a precision of 98.4% a recall of 98.7% and an error of 5.6%. Transition parts in the acceleration signal (SWITCH) were poorly classified compared to OFF (CA = 99.4%, F1 = 98.0%, PREC = 99.4%, REC = 99.4%, LOG LOSS = 2.2%) and WORK (CA = 99.3%, F1 = 99.6%, PREC = 99.2%, REC = 99.9%, LOG LOSS = 3.6%) events, although their classification error was still low (LOG LOSS = 5.2%). In addition, there was an evident class imbalance with a relative data share of ca. 14%–1%–85% for OFF, SWITCH and WORK events, respectively. This may raise the question of how the developed model would perform in cases in which the data collection will be deployed for longer periods of time. In this regard, it is likely that the share of SWITCH events will decrease in the data samples, mainly at the expense of increasing the share of OFF events since the data loggers would also need to operate during the night, at weekends and during legal holidays. As such, the two classes characterized by the highest performance will dominate the data, potentially making the model more effective in classification. In relation to the SWITCH events which were poorly classified, previous studies have already described how the inter-class similarity, which was typical to this event, may affect the classification performance [4,24,52].

The general rule of using most of the data in the training phase held true. However, the differences in performance were not particularly contrasting in relation to the share of data used in the training and testing subsets, although the LOG LOSS error in the training dataset decreased as it contained more data. Along with improved values of the F1 score and of the classification accuracy as more data were added to the training data subset, this indicates that a model trained over all data would hold better predictive capabilities. What is to be emphasized is that the generalization ability (testing phase) measured by the value of the LOG LOSS error, F1 score and classification accuracy was improved. Accordingly, the LOG LOSS was lower in the testing phase from 2 to 5.5%, while the F1 score and classification accuracy were higher in the testing phase from 0.7 to 2.3 and 0.5 to 1.6%, respectively. All of these results are coming in high contrast to those of the previous studies which have found either similar [4] or higher values of classification errors and lower values in terms of classification accuracy in the testing phase [24,52].

The battery life of a data logger such as that used herein is assumed to be ca. 1000 h while its internal memory (4 MB) can hold 168,042 readings made in the normal data collection mode [55]. More or less this means that, in theory, one can cover close to two days of observation at a sampling rate of 1 s (1 Hz) before downloading the data. However, the data collection time frame can be effectively managed by assuming that a higher sampling rate would still accurately reflect the operational pattern in the collected data. By their functional construction, machines such as that described herein may be characterized by relatively long events of working intercalated by short events of switching and long events of being off. With the capabilities of the data logger in mind, a sampling rate of 2 s would double data collection capabilities, while sampling rates of 10 and 20 s would extend the data collection period by ca. 5 to 20 times, meaning that sampling at 20 s would cover a timeframe of more than one month. Therefore, increasing the value of the sampling rate will only prolong further the memory and battery availability. In this regard, by systematically sampling the initial dataset at rates of 2 to 20 s, the share of classes was preserved. Moreover, the general model performed better and similarly in the testing phase, irrespective of the testing dataset used, which is an indication that further data sampled at different rates may be fed into the model which would be able to output a high classification performance. Overall, the generalization errors were improved in the testing phases by up to 0.2% and only in three cases did the testing phase yield higher values of the LOG LOSS. Similar patterns were found in the F1 score and classification accuracies which were generally either the same or higher in values at the testing phase.

As in any other studies on the topic, there are some limitations to be addressed. A first limitation is that of collecting the acceleration data by considering only coniferous logs. As known, a given acceleration signal contains three important components: movement, gravity and noise [56]. Therefore, in the case of processing hardwoods, it is likely to obtain a more differentiated (higher) response in the magnitude of acceleration during the WORK events as the interaction between the logs and blades will produce more vibration. If this does occur, the performance of the model could be an issue that may need additional checking. However, the NN tools of the software used perform by default a normalization procedure over the data before feeding it to the model [57], a fact that serves in weighting the importance of high magnitude data and still preserving the relationships between the original data [58,59]. The same may apply to the variance in log dimensions, particularly to their diameters by reducing or increasing the contact area between the blades and logs which, in theory, would decrease or increase the amount of vibration. On the opposite side, a validation of the model would be required by feeding it with long-term collected, unseen data. In this regard, and based on the performance of the selected machine learning architecture, it is likely to obtain a high classification performance in such a validation phase. Once proved to have a high performance on new datasets, the rest of the steps required to automatically extract and systematize the data, as well as those required for prediction, could be easily managed by the software components described in the Material and Methods section.

Last but not least, the applicability of the described methods may be extended to other sawmilling machines assuming a similar operational pattern in the time domain. This is because, by their construction, they produce vibration during working events. However, the quality of the milled lumber has become of great concern lately [60] and this will possibly become a driver of technological change. Until such changes occur, the proposed model could solve the problems of long-term operational monitoring while after that it could serve, by adaptation, in monitoring operations when such capabilities are not embodied in the sawmilling equipment.

5. Conclusions

Monitoring the operational performance of sawmilling facilities is important for both science and practice. Accordingly, the tools used to obtain useful information need to be adapted to extend data collection and inference capabilities. A robust machine learning model was developed with the purpose of using it to infer the operational events based on lower sampling rates, so as to be able to extend data collection capabilities by low-cost acceleration sensors.

The results indicate a high performance of the model which was less sensitive to the amount of data used to train it, although some variation was found. In this regard, neural networks performed better than random forest algorithms in terms of classification performance. Indeed, they needed more training time, but at this point, this cannot be seen as a limitation since the model is readily available for feeding with new data. The developed model not only preserves a high classification performance in the training and testing phases but it also seems to be invariant to lower sampling rates, making it useful for prediction over long-term collected data. These model properties indicate a high degree of stability to the data potentially fed to the model, as well as a capability enhancement in the sense of lowering the sampling rate of the data to be fed into it.

Altogether, the proposed approach is promising in enabling the use of cheap data collectors to be operated for extended periods in various locations and has the capability of saving human resources and money associated with data collection. Further tests would be required to validate the model, which could be straightforward given the relatively high differentiation which was found at the class level, enabling a visual judgment of predicted classes.

Author Contributions: Conceptualization, S.A.B. and A.R.P.; Data curation, S.A.B., G.O.F. and O.O.-S.; Formal analysis, S.A.B., G.O.F. and O.O.-S.; Investigation, G.O.F. and O.O.-S.; Methodology, S.A.B. and A.R.P.; Project administration, S.A.B. and A.R.P.; Resources, S.A.B., G.O.F. and A.R.P.; Software, S.A.B.; Supervision, S.A.B. and A.R.P.; Visualization, S.A.B.; Writing—original draft, S.A.B., G.O.F. and A.R.P.; Writing—review and editing, S.A.B. and A.R.P. All authors have read and agreed to the published version of the manuscript.

Funding: Some activities of this work were funded by the inter-institutional agreement between the Transilvania University of Braşov (Romania) and the Mediterranean University of Reggio Calabria (Italy).

Data Availability Statement: Data supporting this study may be provided upon reasonable request to the first author of the study.

Acknowledgments: The authors would like to thank the Department of Forest Engineering, Forest Management Planning and Terrestrial Measurements, Faculty of Silviculture and Forest Engineering, Transilvania University of Brasov, for providing the equipment needed to carry on this work. The authors would like to thank the company which supported this research and which wished to remain anonymous.

Conflicts of Interest: The authors declare no conflict of interest.

References

1. Shahi, S.K.; Dia, M.; Yan, P.; Choudhury, S. Developing and training artificial neural networks using bootstrap data envelopment analysis for best performance modeling of sawmills in Ontario. *J. Model. Manag.* **2022**, *17*, 788–811. [CrossRef]
2. Schreiner, F.; Thorenz, B.; Schmidt, J.; Friedrich, M.; Döpper, F. Development of a measuring and control unit for an intelligent adjustment of process parameters in sawing. *MM Sci. J.* **2021**, *5*, 5203–5210. [CrossRef]
3. Borz, S.A.; Oghnoum, M.; Marcu, M.V.; Lorincz, A.; Proto, A.R. Performance of small-scale sawmilling operations: A case study on time consumption, productivity and main ergonomics for a manually-driven bandsaw. *Forests* **2021**, *12*, 810. [CrossRef]
4. Cheța, M.; Marcu, M.V.; Iordache, E.; Borz, S.A. Testing the capability of low-cost tools and artificial intelligence techniques to automatically detect operations done by a small-sized manually driven bandsaw. *Forests* **2020**, *11*, 739. [CrossRef]
5. Cedamon, E.D.; Harrison, S.; Herbohn, J. Comparative analysis of on-site free-hand chainsaw milling and fixed site mini-bandsaw milling of smallholder timber. *Small Scale For.* **2013**, *12*, 389–401. [CrossRef]
6. De Lasaux, M.J.; Spinelli, R.; Hartsough, B.R.; Magagnotti, N. Using a small-log mobile sawmill system to contain fuel reduction treatment cost on small parcels. *Small Scale For.* **2009**, *8*, 367–379. [CrossRef]
7. Gigoraş, D.; Borz, S.A. Factors affecting the effective time consumption, wood recovery rate and feeding speed when manufacturing lumber using a FBO-02 CUT mobile bandsaw. *Wood Res.* **2015**, *60*, 329–338.
8. Venn, T.J.; McGavin, R.L.; Leggate, W.W. Costs of portable sawmilling timbers from the acacia woodlands of Western Queensland, Australia. *Small Scale For. Econ. Manag. Policy* **2004**, *3*, 161–175. [CrossRef]
9. Telford, C.G. *Small Sawmill Operator's Manual. Agriculture Handbook No 27—January 1952*; United States Department of Agriculture: Washington, DC, USA, 1952.
10. FAO. *Small and Medium Sawmills in Developing Countries*; Forestry Paper No. 28; FAO: Rome, Italy, 1981.
11. Grönlund, A. *Sågverksteknik del 2—Processen*; Sveriges Skogsindustriförbund: Markaryd, Sweden, 1992; p. 270.
12. Lundahl, C.G. Optimized Processes in Sawmills. Licentiate Thesis, Luleå University of Technology, Skellefteå, Sweden, 2007.
13. Hyytiäinen, A.; Viitanen, J.; Mutanen, A. Production efficiency of independent Finnish sawmills in the 2000's. *Balt. For.* **2011**, *17*, 280–287.
14. Spinelli, R.; Visser, R.; Magagnotti, N.; Lombardini, C.; Ottaviani, G. The effect of partial automation on the productivity and cost of a mobile tower yarder. *Ann. For. Res.* **2020**, *63*, 3–14. [CrossRef]
15. Borz, S.A. Turning a winch skidder into a self-data collection machine using external sensors: A methodological concept. *Bull. Transilv. Univ. Braşov* **2016**, *9*, 1–6.
16. Cheța, M.; Borz, S.A. Automating data extraction from GPS files and sound pressure level sensors with application in cable yarding time and motion studies. *Bull. Transilv. Univ. Braşov* **2017**, *10*, 1–10.
17. Björheden, R.; Apel, K.; Shiba, M.; Thompson, M. *IUFRO Forest Work Study Nomenclature*; Swedish University of Agricultural Science, Department of Operational Efficiency: Grapenberg, Sweden, 1995.
18. Acuna, M.; Bigot, M.; Guerra, S.; Hartsough, B.; Kanzian, C.; Kärhä, K.; Lindroos, O.; Magagnotti, N.; Roux, S.; Spinelli, R.; et al. *Good Practice Guidelines for Biomass Production Studies*; Magagnotti, N., Spinelli, R., Eds.; CNR IVALSA: Sesto Fiorentino, Italy, 2012. Available online: http://www.forestenergy.org/pages/cost-action-fp0902/good-practice-guidelines/ (accessed on 15 April 2018).
19. Ištvanić, J.; Lučić, R.B.; Jug, M.; Karan, R. Analysis of factors affecting log band saw capacity. *Croat. J. For. Eng.* **2009**, *30*, 27–35.
20. Muşat, E.C.; Apăfăian, A.I.; Ignea, G.; Ciobanu, V.D.; Iordache, E.; Derczeni, R.A.; Spârchez, G.; Vasilescu, M.M.; Borz, S.A. Time expenditure in computer aided time studies implemented for highly mechanized forest equipment. *Ann. For. Res.* **2016**, *59*, 129–144. [CrossRef]
21. Spinelli, R.; Laina-Relano, R.; Magagnotti, N.; Tolosana, E. Determining observer effect and method effects on the accuracy of elemental time studies in forest operations. *Balt. For.* **2013**, *19*, 301–306.
22. Borz, S.A.; Păun, M. Integrating offline object tracking, signal processing, and artificial intelligence to classify relevant events in sawmilling operations. *Forests* **2020**, *11*, 1333. [CrossRef]
23. Becker, R.M.; Keefe, R.F. A novel smartphone-based activity recognition modelling method for tracked equipment in forest operations. *PLoS ONE* **2022**, *17*, e0266568. [CrossRef]
24. Borz, S.A. Development of a modality-invariant multi-layer perceptron to predict operational events in motor-manual willow felling operations. *Forests* **2021**, *12*, 406. [CrossRef]
25. Borz, S.A.; Talagai, N.; Cheța, M.; Chiriloiu, D.; Gavilanes Montoya, A.V.; Castillo Vizuete, D.D.; Marcu, M.V. Physical strain, exposure to noise and postural assessment in motor-manual felling of willow short rotation coppice: Results of a preliminary study. *Croat. J. Eng.* **2019**, *40*, 377–388. [CrossRef]
26. Keefe, R.F.; Zimbelman, E.G.; Wempe, A.M. Use of smartphone sensors to quantify the productive cycle elements of hand fallers on industrial cable logging operations. *Int. J. For. Eng.* **2019**, *30*, 132–143. [CrossRef]
27. Marogel-Popa, T.; Cheța, M.; Marcu, M.V.; Duță, C.I.; Ioraş, F.; Borz, S.A. Manual cultivation operations in poplar stands: A characterization of job difficulty and risks of health impairment. *Int. J. Environ. Res. Public Health* **2019**, *16*, 1911. [CrossRef] [PubMed]
28. McDonald, T.P.; Fulton, J.P.; Darr, M.J.; Gallagher, T.V. Evaluation of a system to spatially monitor hand planting of pine seedlings. *Comput. Electron. Agric.* **2008**, *64*, 173–182. [CrossRef]

29. Strandgard, M.; Mitchell, R. Automated time study of forwarders using GPS and a vibration sensor. *Croat. J. Eng.* **2015**, *36*, 175–184.
30. Muşat, E.C.; Borz, S.A. Learning from acceleration data to differentiate the posture, dynamic and static work of the back: An experimental setup. *Healthcare* **2022**, *10*, 916. [CrossRef]
31. Picchio, R.; Proto, A.R.; Civitarese, V.; Di Marzio, N.; Latterini, F. Recent contributions of some fields of the electronics in development of forest operations technologies. *Electronics* **2019**, *8*, 1465. [CrossRef]
32. Wescoat, E.; Krugh, M.; Henderson, A.; Goodnough, J.; Mears, L. Vibration analysis using unsupervised learning. *Procedia Manuf.* **2019**, *34*, 876–888. [CrossRef]
33. Borz, S.A.; Talagai, N.; Cheţa, M.; Gavilanes Montoya, A.V.; Castillo Vizuete, D.D. Automating data collection in motor-manual time and motion studies implemented in a willow short rotation coppice. *Bioresources* **2018**, *13*, 3236–3249. [CrossRef]
34. Demsar, J.; Curk, T.; Erjavec, A.; Gorup, C.; Hocevar, T.; Milutinovic, M.; Mozina, M.; Polajnar, M.; Toplak, M.; Staric, A.; et al. Orange: Data Mining Toolbox in Python. *J. Mach. Learn. Res.* **2013**, *14*, 2349–2353.
35. R Interface to Keras. Available online: https://keras.rstudio.com/ (accessed on 19 May 2022).
36. Smith, S.W. *Digital Signal Processing—A Practical Guide for Engineers and Scientists*, 1st ed.; Newnes—Elsevier Science: Burlington, MA, USA, 2003.
37. Proto, A.R.; Sperandio, G.; Costa, C.; Maesano, M.; Antonucci, F.; Macri, G.; Scarascia Mugnozza, G.; Zimbalatti, G. A three-step neural network artificial intelligence modeling approach for time, productivity and costs prediction: A case study in Italian forestry. *Croat. J. For. Eng.* **2020**, *41*, 15. [CrossRef]
38. Neural Network Models (Supervised). Multi-Layer Perceptron. Available online: https://scikit-learn.org/stable/modules/neural_networks_supervised.html (accessed on 19 May 2022).
39. Ho, T.K. Random decision forests. In Proceedings of the 3rd International Conference on Document Analysis and Recognition, Montreal, QC, Canada, 14–16 August 1995; pp. 278–282.
40. Breiman, L. Random forests. *Mach. Learn.* **2001**, *45*, 5–32. [CrossRef]
41. Cheţa, M.; Marcu, M.V.; Borz, S.A. Effect of training parameters on the ability of artificial neural networks to learn: A simulation on accelerometer data for task recognition in motor-manual felling and processing. *BUT Ser. II For. Wood Ind. Agric. Food Eng.* **2020**, *13*, 19–36. [CrossRef]
42. Goodfellow, J.; Bengio, Y.; Courville, A. *Deep Learning*; MIT Press: Cambridge, MA, USA, 2016; Available online: https://www.deeplearningbook.org/ (accessed on 19 May 2022).
43. Kingma, D.P.; Ba, J.L. 2015: ADAM: A method for stochastic optimization. In Proceedings of the 3rd International Conference on Learning Representations (ICLR 2015), San Diego, CA, USA, 7–9 May 2015.
44. Activation Functions in Neural Networks. Available online: https://towardsdatascience.com/activation-functions-neural-networks-1cbd9f8d91d6 (accessed on 19 May 2022).
45. Maas, A.L.; Hannun, A.Y.; Ng, A.Y. Rectifier nonlinearities improve neural network acoustic models. In Proceedings of the 30th International Conference on Machine Learning (ICML 2013), Atlanta, GA, USA, 16–21 June 2013.
46. Nair, V.; Hinton, G.E. Rectified linear units improve restricted Boltzmann machines. In Proceedings of the 27th International Conference on Machine Learning (ICML 2010), Haifa, Israel, 21–24 June 2010.
47. A Beginner's Guide to Random Forest Hyperparameter Tunning. Available online: https://www.analyticsvidhya.com/blog/2020/03/beginners-guide-random-forest-hyperparameter-tuning/?utm_source=blog&utm_medium=decision-tree-vs-random-forest-algorithm (accessed on 19 May 2022).
48. Orange Visual Programming. Random Forest. Available online: https://orange3.readthedocs.io/projects/orange-visual-programming/en/latest/widgets/model/randomforest.html (accessed on 19 May 2022).
49. Fawcett, T. An introduction to ROC analysis. *Pattern Recogn. Lett.* **2006**, *27*, 861–874. [CrossRef]
50. Kamilaris, A.; Prenafeta-Boldu, F.X. Deep learning in agriculture: A survey. *Comput. Electron. Agric.* **2018**, *147*, 70–90. [CrossRef]
51. Sklearn Metrics. Log Loss. Available online: https://scikit-learn.org/stable/modules/generated/sklearn.metrics.log_loss.html (accessed on 19 May 2022).
52. Castro Pérez, S.N.; Borz, S.A. Improving the event-based classification accuracy in pit-drilling operations: An application by neural networks and median filtering of the acceleration input data. *Sensors* **2021**, *21*, 6288. [CrossRef]
53. Chen, K.; Zhang, D.; Yao, L.; Guo, B.; Yu, Z.; Liu, Y. Deep learning for sensor-based human activity recognition: Overview, challenges and opportunities. *J. ACM* **2018**, *37*, 111. [CrossRef]
54. Bulling, A.; Blanke, U.; Schiele, B. A tutorial of human activity recognition using body-worn inertial sensors. *ACM Comput. Surv.* **2014**, *46*, 33. [CrossRef]
55. Extech®User Manual. 3-Axis G-Force Datalogger. *Model VB300*. Available online: http://www.extech.com/products/resources/VB300_UM-en.pdf (accessed on 19 May 2022).
56. van Hees, V.T.; Gorzelniak, L.; León, E.C.D.; Eder, M.; Pias, M.; Taherian, S.; Ekelund, U.; Renström, F.; Franks, P.W.; Horsch, A.; et al. Separating movement and gravity components in an acceleration signal and implications for the assessment of human daily physical activity. *PLoS ONE* **2013**, *8*, e61691. [CrossRef]
57. Orange Visual Programming. Neural Network. Available online: https://orange3.readthedocs.io/projects/orange-visual-programming/en/latest/widgets/model/neuralnetwork.html (accessed on 19 May 2022).
58. Aggarwal, C.C. *Data Mining. The Textbook*; Springer: New York, NY, USA, 2015; 734p.

59. Han, J.; Kamber, M.; Pei, J. *Data Mining: Concepts and Techniques*, 3rd ed.; Morgan Kaufmann Publishers: Burlington, MA, USA, 2006.
60. Kamperidou, V.; Aidinidis, E.; Barboutis, I. Impact of structural defects on the surface quality of hardwood species sliced veneers. *Appl. Sci.* **2020**, *10*, 6265. [CrossRef]

Article

Potential of Measure App in Estimating Log Biometrics: A Comparison with Conventional Log Measurement

Stelian Alexandru Borz *, Jenny Magaly Morocho Toaza, Gabriel Osei Forkuo and Marina Viorela Marcu

Department of Forest Engineering, Forest Management Planning and Terrestrial Measurements, Faculty of Silviculture and Forest Engineering, Transilvania University of Brasov, Şirul Beethoven 1, 500123 Brasov, Romania; jenny.morocho@student.unitbv.ro (J.M.M.T.); gabriel.forkuo@student.unitbv.ro (G.O.F.); viorela.marcu@unitbv.ro (M.V.M.)

* Correspondence: stelian.borz@unitbv.ro; Tel.: +40-742-042-455

Citation: Borz, S.A.; Morocho Toaza, J.M.; Forkuo, G.O.; Marcu, M.V. Potential of Measure App in Estimating Log Biometrics: A Comparison with Conventional Log Measurement. *Forests* 2022, 13, 1028. https://doi.org/10.3390/f13071028

Academic Editor: Henning Buddenbaum

Received: 29 April 2022
Accepted: 28 June 2022
Published: 30 June 2022

Publisher's Note: MDPI stays neutral with regard to jurisdictional claims in published maps and institutional affiliations.

Abstract: Wood measurement is an important process in the wood supply chain, which requires advanced solutions to cope with the current challenges. Several general-utility measurement options have become available by the developments in LiDAR or similar-capability sensors and Augmented Reality. This study tests the accuracy of the Measure App developed by Apple, running by integration into Augmented Reality and LiDAR technologies, in estimating the main biometrics of the logs. In a first experiment (E1), an iPhone 12 Pro Max running the Measure App was used to measure the diameter at one end and the length of 267 spruce logs by a free-eye measurement approach, then reference data was obtained by taking conventional measurements on the same logs. In a second experiment (E2), an iPhone 13 Pro Max equipped with the same features was used to measure the diameter at one end and the length of 200 spruce logs by a marking-guided approach, and the reference data was obtained similar to E1. The data were compared by a Bland and Altman analysis which was complemented by the estimation of the mean absolute error (MAE), root mean squared error (RMSE) and normalized root mean square error (NRMSE). In E1, nearly 86% of phone-based log diameter measurements were within ±1 cm compared to the reference data, of which 37% represented a perfect match. Of the phone-based log length measurements, 94% were within ±5 cm compared to the reference data, of which approximately 22% represented a perfect match. MAE, RMSE, and NRMSE of the log diameter and length were of 0.68, 0.96, and 0.02 cm, and of 1.81, 2.55, and 0.10 cm, respectively. Results from E2 were better, with 95% of the phone-based log diameter agreeing within ±1 cm, of which 44% represented a perfect match. As well, 99% of the phone-based length measurements were within ±5 cm, of which approximately 27% were a perfect match. MAE, RMSE, and NRMSE of the log diameter and length were of 0.65, 0.92, and 0.03 cm, and of 1.46, 1.93, and 0.04 cm, respectively. The results indicated a high potential of replacing the conventional measurements for non-piled logs of ca. 3 m in length, but the applicability of phone-based measurement could be readily extended to log-end diameter measurement of the piled wood. Further studies could check if the accuracy of measurements would be enhanced by larger samples and if the approach has good replicability. Finding a balance between capability and measurement accuracy by extending the study to longer log lengths, different species and operating conditions would be important to characterize the technical limitations of the tested method.

Keywords: wood; diameter; length; close-range sensing; LiDAR; Augmented Reality; comparison; accuracy; effectiveness; potential

1. Introduction

Measurement and grading are important activities in the wood value chain because they provide essential quantitative and qualitative information for transactions. The wood is commonly delivered to the industry as roundwood [1–5], for which a volume estimation is required to document the delivered quantity and to form the basis for payment. Ideally, a single measurement done in the forest could act as a transaction interface between the suppliers and customers [6–8] and would ease the effort and resources spent in these activities. However, in complex wood supply chains (e.g., [9]) characterized by a low integration of technology, as well as for reasons such as removing public suspicion, preventing illegal logging, enabling traceability, and building trust, supplementary checks of the wood may be required, particularly by third parties such as the public authorities. In Romania, for instance, a typical example is that of taking custody over the wood by a carrier, which requires detailed measurement, grading, and reporting in a wood tracking system at the landing, before making the delivery [10] and, in case of suspicion, additional checks may be in question within the wood value chain. There are many other examples in practice and science in which wood measurement is required. In felling areas or at the landings, a pre-grading of the wood which includes measurement, is required for optimal bucking by motor-manual means [1]. In forestry-related scientific applications of time and motion study wood measurement is a prerequisite for estimating the amount of production and characterizing the productivity in relation to operational factors [11–16].

There are many methods that can be used to estimate the volume of individual logs. Depending on the procedures used to measure the required parameters, and in particular, on the type of contact between the measuring instrument and the log under measurement, they can be fairly categorized into two groups: direct and remotely-sensed measurements. Direct measurement methods include the estimation of wood volume by hydrostatics [17], gravimetry [17], and water displacement [17–19], as well as conventionally by a tape and a caliper. Measurement by hydrostatics, gravimetry, and water displacement is typically constrained to scientific applications and it is limited by the size of the logs and infrastructure needed [17]. Estimating the volume of a log by conventional methods is commonly used in practice and requires biometric information about the length and diameter(s) of the log. Depending on the concept used to estimate the volume, log diameter may be measured at both ends or at the middle, then the log is typically assimilated to a cylinder when making the mathematical computations to estimate its volume [17]. Measuring wood biometrics such as the diameter, height, or volume by remote sensing includes the use of photogrammetric [20,21], time of flight (ToF) [21,22], and Light Detection and Ranging (LiDAR) [23] based methods, many of which still share some limitations, namely the need to post-process the data and the high data acquisition costs incurred by the instruments and software used. Although phone-based compact solutions integrating remote sensing technologies were developed specifically to provide real- or near-real time estimates on some biometrics of standing trees [24] and logs [22], in our knowledge, solutions dedicated to replace measuring tapes and calipers for length and diameter measurement of the logs were not developed, meaning that in many geographical areas this activity still needs to rely on manual, direct-contact measurements, which are typically done in challenging environments such as in the felling areas, or in conditions which constrain the access to logs, such as those characterizing the wood piled at the landing.

The developments in mobile phone technology by the integration of common-use measuring applications based solely on the performance of phone cameras or integrating also the capabilities of LiDAR technology and Augmented Reality (AR) environments has opened new doors for potential applications in forestry. Apple's Measure app [25], for instance, could be a potential alternative to measuring log biometrics (diameter and length), which could provide the benefits of excluding rather uncomfortable to carry equipment such as the forest tapes and calipers while including their capabilities into a single device. Currently, the Measure app uses AR technology and it was first released for free without the support of LiDAR sensors, which became an integrant part of the iPhone devices starting with the 12th version, namely the iPhone 12 Pro and iPhone 12 Pro Max; in turn, the integration of LiDAR sensors resulted in improved accuracy and quicker measurement capabilities as claimed by the producing company. The application enables making several measurements, and copying and pasting the results into external applications for further use [25], making it suitable for saving the results of measurement. Still, its suitability in providing accurate results in log measurement was not tested. In this regard, only one study was found on the topic of using the Measure app for estimating the breast height diameter of the trees [26], but the device used has not been equipped with a LiDAR sensor.

The goal of this study was to compare the measurements taken on the diameter and length of coniferous logs by the use of the Measure app supported by a LiDAR sensor and AR (hereafter phone measurement, PM) with those taken by conventional means (hereafter conventional measurements, CM), namely a tape and a caliper, in two experiments. A first experiment was set up to emulate the real-world measurement conditions by taking the PM by a free-eye approach, meaning that no guiding marks were placed on the logs to support picking up the starting and ending points of a given PM. In a second experiment, PM were guided by marks placed on the logs at the points at which the CM on diameter and length were taken. Two objectives were pursued by this study. The first was to check the agreement between PM and CM in both experiments having as a reference the CM data and the second was to check the accuracy of PM in both experiments, having as a reference the CM data.

2. Materials and Methods

2.1. Study Location and Data Collection

Data used in this study were collected in the sawmilling facility of the HS Timber Productions Reci S.R.L., which is located near the village of Reci, Covasna, Romania at the coordinates of 45°51′01″ N–25°56′52″ E. The company processes coniferous logs, mostly of Norway spruce (*Picea abies*, L. (Karst)), which is one of the dominant coniferous species in Romania [27]. Typically, the logs are delivered at the factory gate in lengths of 3 to 4 m. In the period used to collect the data, the sky was mostly clear and the weather was relatively cold.

Two experiments (hereafter E1 and E2) were set up to collect the data for this study. In E1, the main approach of phone-based measurements was that of emulating the real-world measurement conditions in which no guiding marks are available on the logs, therefore the measurement needs to rely on the experience of the operator in setting the starting and ending points to measure the diameters and the lengths of the logs. For this experiment, the field data collection was done on the 14, 15, and 17 February 2022, by considering a total number of 267 logs.

To facilitate log measurement by using both, a measurement tape and a forest caliper (CM) and the phone-based Measure app (PM), the logs measured in each of the three days were placed on transversal logs and spaced at ca. 60 cm apart (Figure 1). Then, an identification number (hereafter ID) was painted on each log taken into the study (Figure 1) and the logs were marked at half-meter intervals starting from the painted end with the aim of preventing accidental measurement errors and supporting data collection. Where/When was the case, the additional length which was less than half a meter was painted on the opposite end. Conventional measurement (CM) was done to the nearest centimeter by a field researcher using a forestry caliper and a measurement tape. For comparison purposes, one diameter was measured for each log at the end painted with the identification number (hereafter Dman, cm), by approaching the log with the caliper placed perpendicularly on the log axis, in a vertical plane, with arms oriented downwards. The length (hereafter Lman, cm) was measured by a forestry tape on the upper part of each log. Data on log ID, Dman, and Lman was noted in a field book. To estimate the volume of each log by conventional measurement and formulae, the diameter at the middle (cm) and at the second end (cm) of each log were measured by the same procedures and noted accordingly in the field book. As the conventional measurement progressed, the Measure app installed on an iPhone 12 Pro Max smart phone device developed by Apple was used to measure the diameter at the end painted with the log ID (hereafter Dmeas, cm) and the length of the log (hereafter Lmeas, cm). The measurements were taken by a free-eye approach, meaning that no marking signs were placed on the logs to delimitate the starting and ending points of the measurement. The functionalities of the application used are given in Figure 2 by some examples. Dmeas was taken in a direction that was as close as possible to the parallel to the ground, by a line that followed the diameter of the log, and Lmeas was taken between the middle points of the log's ends located at the upper part of each log.

Figure 1. An example of log placement to facilitate measurements. Note: numbers painted in red stand for the identification numbers of the logs.

(a)   (b)   (c)

Figure 2. Examples of diameter and length measurements by the Measure app of iPhone in the first experiment (E1): (**a**) an example of using the Measure app and AR for diameter measurement; (**b**) an example of using the Measure app and AR for length measurement; (**c**) perspective in AR over a group of logs and a measurement on log length.

Diameters (Dmeas) were measured from a distance of up to 0.5 m and the lengths (Lmeas) were measured by walking along the log at a slow walking speed. Initial and final measurement points of the Dmeas and Lmeas were taken from a close perspective to the log. For Dmeas, these measurement points were taken as close as possible to the log ends on its diameter (over the bark) while for Lmeas they were taken as close as possible to the log ends. Although the AR environment enables adjustments of the measurements, for simplicity and for keeping the procedure as close as possible to that eventually used in practice, such adjustments were not done over the measurements. Once the measurements over Dmeas and Lmeas were done, their results were noted in the field book.

In E2, the main approach of the experiment was to guide the researcher in placing the initial and final PM points for both diameter and length measurements. The data of E2 was collected in the same location between the 11 and 15 April 2022 and accounted for a number of 200 logs. The procedures used to conventionally measure the log diameters and lengths were complemented by marking with dots (ca. 1 cm in diameter, Figure 3) the points at which the caliper arms were tangent to the log end when measuring the diameters, as well as the end points located at the top of the log, which were used as starting and ending points to measure its length by the tape.

(a)   (b)

Figure 3. Examples of diameter measurements by the Measure app of iPhone in the second experiment (E2): (**a**) an example of log with marks placed; (**b**) an example of measurement over the log's end diameter: 1—starting and ending points of measurement, 2—dot marked as the starting point of measurement for log length.

In this experiment, measurements on the mid and opposite diameters of the log were disregarded. However, the rest of the experimental design used for conventional measurement was kept the same and it included the activities of placing marks at a 0.5 m interval and painting the excess length on the opposite end when it was less than 0.5 m. Also, the platform used for PM was an iPhone 13 Pro Max. In both experiments and for both methods, the measurements were taken at the nearest centimeter.

2.2. Data Processing and Statistical Analysis

In experiment 1 (E1) the data collected by the two measurement methods were manually transferred into a Microsoft Excel® spreadsheet equipped with a Real Statistics add-in, where further processing steps were taken to estimate the volume of each log by the Huber's (hereafter VH, m^3) and Smalian's (hereafter VS, m^3) formulae. The statistical steps used to compare the data consisted of running a normality check by the means of a Shapiro-Wilk test, developing the main descriptive statistics for the variables taken into study (Dman, Lman, Dmeas, Lmeas, VH, and VS) followed by a graphical comparison of the volume estimates, and a comparison of the two measurement methods by the means of Bland and Altman's method [28] applied to the diameters (Dmeas vs. Dman) and lengths (Lmeas vs.

Lman) and having as a reference the values collected by the conventional method (CM). Where relevant, a confidence level of 95% ($\alpha < 0.05$) was assumed.

The method developed by Bland and Altman is typically used to compare two measurements of the same variable in terms of agreement assuming that each measurement is affected by errors [29]. As such, the method may be used when one attempts to test or introduce a new measurement method, procedure, or instrument, being applicable when the acceptable limits of agreement can be defined a priori [28,29]. At its core, the method is based on a plot that compares the means of each pair of measurements against the differences between them in a space characterized by the mean of differences (bias) and a 95% prediction interval called the limits of agreement (upper and lower limits of agreement) [28]. Measurement agreement between the methods is typically achieved when the values of differences are clustered around the bias within two standard deviations of their mean [28,29]. The method assumes that the values of differences between the compared pairs are normally distributed, although failing a normality test is seen to be not as serious as in other statistical contexts [28], and requires checking and applying various methods to deal with heteroskedastic data [29].

The procedures used to run the analysis consisted in calculating the differences between paired measurements of diameters (hereafter ∆D) and lengths (hereafter ∆L), computing the bias as the mean of differences, setting the limits of agreement within two standard deviations of differences, checking for normality in differences and plotting the data. In addition to the development and visual analysis of the Bland and Altman plots, testing for homoskedasticity was done by plotting the squared residuals of the CM (Dman, Lman) data against the predicted values of PM data (Dmeas, Lmeas), followed by a Breusch-Pagan test for homoskedasticity [30,31]. In addition to Bland and Altman plots, graphs showing the absolute frequencies of differences were developed to characterize their frequency and magnitude. Also, the data collected by CM and PM were pairwise compared in graphs showing the equality lines [28] and regression through the origin (RTO) which fitted the PM as response and CM as explanatory variables.

Finally, error metrics such as the mean absolute error (MAE), root mean square error (RMSE) and the normalized root mean squared error (NRMSE) were estimated having as a reference the datasets collected by CM, with the aim of quantifying the differences between the two methods. MAE is defined as the ratio of the sum of absolute differences between the reference and measured data to the number of observations in a given sample, RMSE takes the square root of the ratio of squared differences between the reference and measured data to the number of observations in a given sample and the NRMSE is the ratio of RMSE to the data range in a given sample. These error metrics are commonly used to compare among paired values of the same variable as they stand for the average difference rather than average error when no set of estimates is known to be the most reliable [32]. As such, in this study, they were used to point out the differences between the two measurement methods. Excluding the volume estimation and comparison, processing and statistical analysis of the data from E2 followed the same procedural steps.

3. Results

3.1. Experiment 1 (E1): Free-Eye Measurement

3.1.1. Descriptive Statistics of E1

Table A1 shows the results of the normality check which was carried out by the means of the Shapiro–Wilk test. As shown, none of the variables taken into study followed a normal distribution. The main descriptive statistics of the log volume estimates by Huber's (HV) and Smalian's (VS) formulae are given in Figure 4b in the form of a boxplot. On average, the values estimated by the two formulae were close (VH = 0.132 and VS = 0.135 m^3), but the data range was wider in the case of VS (0.655 m^3) as compared to VH (0.501 m^3). This came largely from the maximum values which were higher in the case of VS (0.688 m^3) as compared to VH (0.531 m^3).

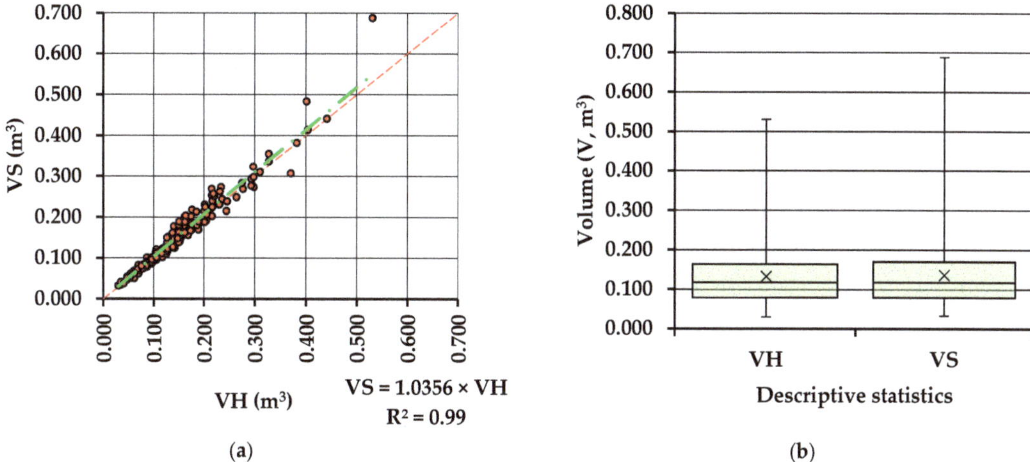

(a)   (b)

Figure 4. Descriptive statistics of the estimated log volume (E1): (**a**) A comparison between VS and VH, where the red dashed line stands for perfect agreement (equality line) and the green dot-dash line stands for the dependence relation between VS and VH fitted by RTO; (**b**) boxplots showing the summary of data distribution and the main descriptive statistics of the volume estimates, including the minimum, mean, median and maximum value.

As shown in Figure 4a, there was a relative agreement between the two volume estimates based on the same measurements, at least for the data range from ca. 0.030 to ca. 0.130 m^3. Beyond this threshold, the disagreement started to increase relative proportionally to the magnitude of volume.

Figure 5 shows the main descriptive statistics of the diameter and length measurements done by the two methods. The mean values of Dman and Dmeas diameters were of 22.86 and 22.89 cm, respectively, and the median values were of 22 cm. In the same order, the minimum values were of 10 and 9 cm, respectively, while the maximum ones were of 61 and 58 cm, respectively. On average, the values of log lengths were close, with a mean value of 306.27 in the case of Lman and a mean value of 305.92 cm in the case of Lmeas. Minimum and maximum values were also close, with values of 291, 290, 314, and 315 cm for Lman and Lmeas, respectively. For both variables, data ranges were close as values among the methods.

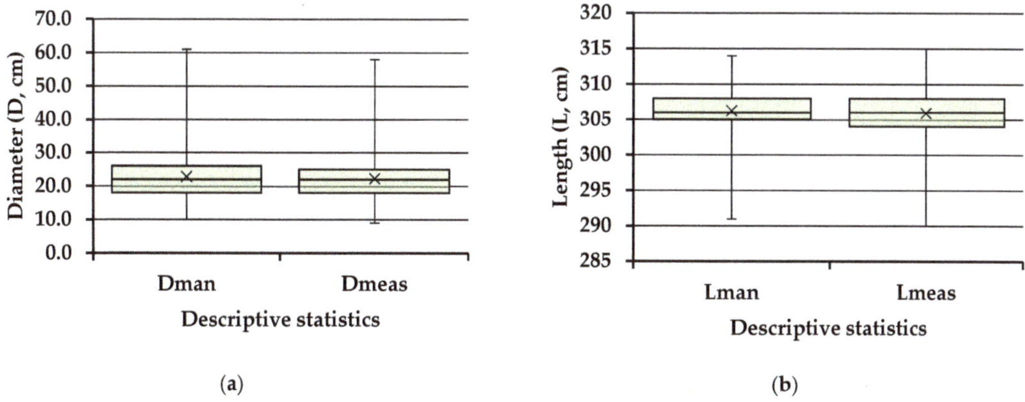

(a)   (b)

Figure 5. Descriptive statistics of the variables measured in E1: (**a**) Boxplots showing the main descriptive statistics for diameters; (**b**) Boxplots showing the main descriptive statistics for lengths.

3.1.2. Agreement between the Measurement Methods in E1

In terms of diameters, and taking as a reference the CM data, ca. 37% (99) of the observations were found in perfect agreement, 43% (116) were underestimated by 1 cm and ca. 6% (15) were overestimated by 1 cm (Figure 6). Approximately 86% of the observations were found in a difference range of ±1 cm, while the maximum absolute difference between the two methods was of 4 cm, in the form of an overestimation produced by the PM. The bias of the measurements was of 0.6 cm, meaning that the PM measured, on average, 0.6 cm less than the CM, and ca. 98% of the measurements were found between the limits of agreement. As proved by a Shapiro-Wilk test, data on differences did not follow a normal distribution.

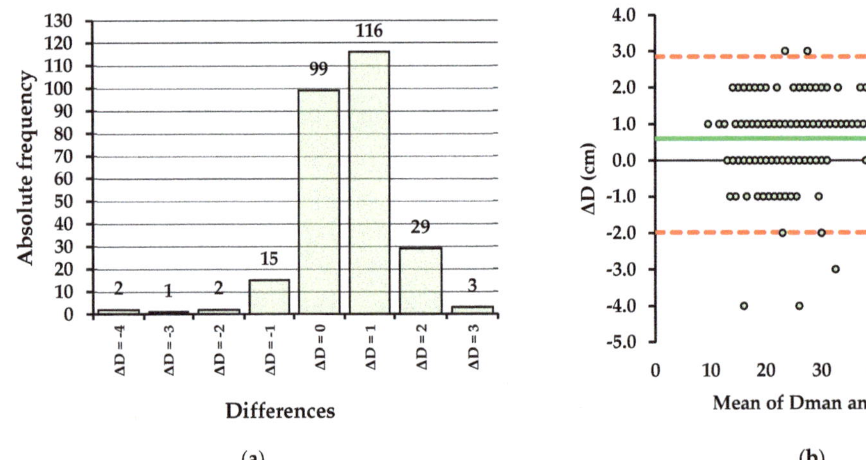

(a)

(b)

Figure 6. Agreement between the methods in terms of diameter measurement (E1): (**a**) Absolute frequency of differences (ΔD) between Dman and Dmeas; (**b**) Bland-Altman plot showing the differences plotted against the mean of paired measurements, the bias (green line), and the lower and upper limits of agreement (red dashed lines).

Figure 7, on the other hand, shows two important findings of data comparison. First of them is that the two datasets were strongly linearly-related as proven by the coefficient of determination (R^2 = 99.9), while the slope of regression through the origin equation was close to that of 1:1 represented by the red dashed line which stands for a perfect agreement (line of equality) between the two methods. Checking the data for heteroskedasticity by the Breusch-Pagan test indicated that the data was homoskedastic ($p > 0.05$), therefore it can be said with a confidence of 95% that the differences between measurements were not affected by other factors than the measurement methods themselves. This can be seen in Figure A1a which plots the squared residuals against the predicted Dmeas; as shown there was no increasing, decreasing, or other kind of trend in data as a function of predicted Dmeas.

Results of agreement between the two methods in terms of length are given following a similar data representation in Figures 8 and 9. More than 22% (60 observations) of the data indicated a perfect agreement between the two methods, close to 56% of the data (149 observations) were found to disagree by up to ±1 cm, and more than 95% (254 observations) were found to disagree by up to ±5 cm. The bias was of 0.3 cm, meaning that, on average, PM measured less by 0.3 cm. Approximately 94% (251 observations) were found in between the limits of agreement. Although there was a strong dependence relation between the length measured by the two methods (R^2 close to 1) and the slope of the regression line was very close to that of the perfect agreement, the data was quite spread indicating a higher degree of disagreement as compared to diameter measurement.

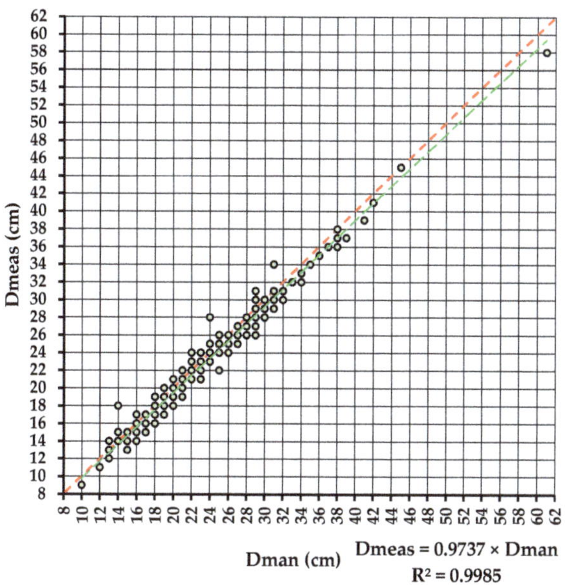

Figure 7. Relation between the diameters measured by the two methods (E1). Legend: red dashed line stands for perfect agreement and the green dashed line stands for the dependence relation between Dmeas and Dman fitted by RTO.

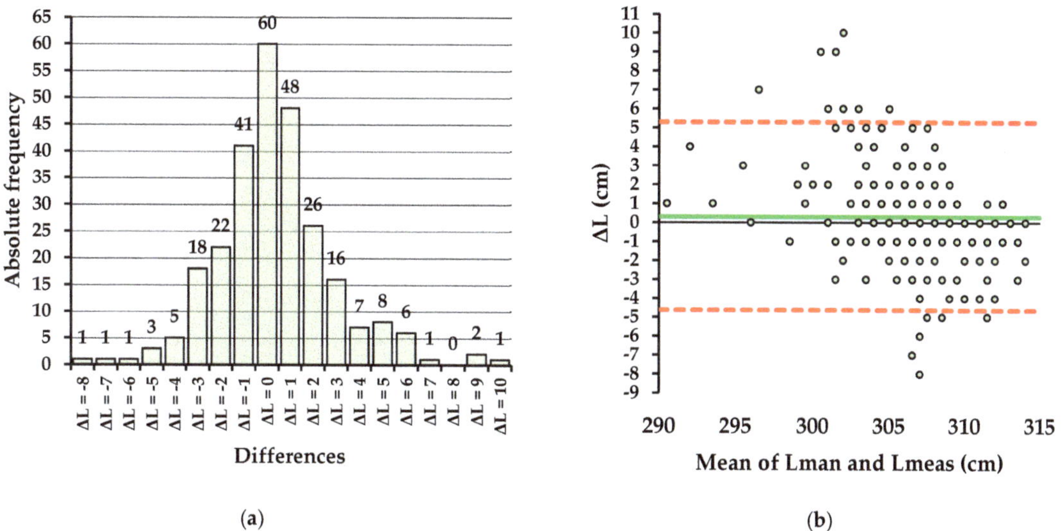

Figure 8. Agreement between the methods in terms of length measurement (E1): (**a**) Absolute frequency of differences (ΔL) between Lman and Lmeas; (**b**) Bland-Altman plot showing the differences plotted against the mean of paired measurements, the bias (green line), and the lower and upper limits of agreement (red dashed lines).

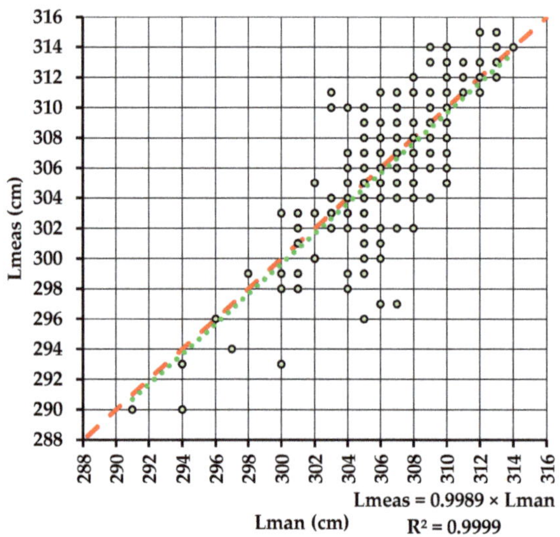

Figure 9. Relation between the lengths measured by the two methods (E1). Legend: red dashed line stands for perfect agreement and green dashed line stands for the dependence relation between Lmeas and Lman fitted by RTO.

Similar to the diameter measurement, the data on differences between the measurements of length did not follow a normal distribution; by the results of the Breusch-Pagan test the data was found to be homoskedastic ($p > 0.05$, Figure A1b).

3.1.3. Measurement Errors in E1

Table 1 shows the results of the error metrics for the first experiment (E1). Mean absolute error (MAE), Root Mean Squared Error (RMSE) and the Normalized Root Mean Squared Error (NRMSE) had values of less than 1 cm in the case of diameter measurements.

Table 1. Results on errors for the first experiment (E1).

Error Metric	Diameter (cm)	Length (cm)
MAE	0.68	1.81
RMSE	0.96	2.55
NRMSE	0.02	0.10

For length measurement, MAE was less than 2 cm, RMSE was close to 2.5 cm and NRMSE was less than 1 cm. On average, these results indicate very low differences, therefore supporting a high agreement between the two methods, when approaching the problem at the sample size.

3.2. Experiment 2 (E2): Guided Measurement

3.2.1. Descriptive Statistics of E2

Table A2 shows the results of the normality check over the diameter and length variables for the second experiment (E2). Similar to the first experiment, none of the variables met the normality assumption. Figure 10, on the other hand, shows the descriptive statistics of the diameter and length variables as specific to the second experiment.

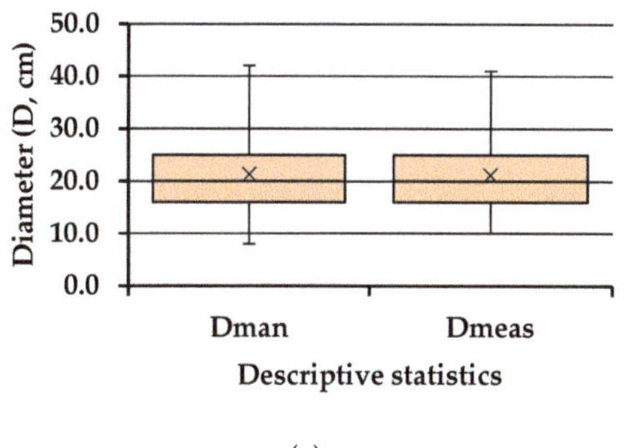

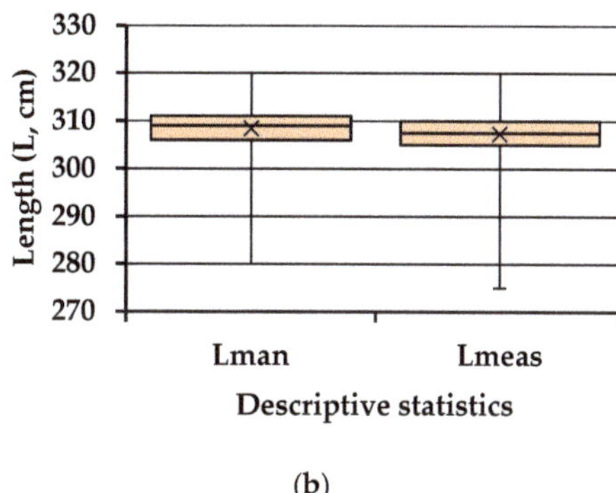

(a)

(b)

Figure 10. Descriptive statistics of the measured variables (E2): (**a**) Boxplots showing the main descriptive statistics for diameters; (**b**) Boxplots showing the main descriptive statistics for lengths.

The mean diameters from CM and PM were close in value (21.3 and 21.2 cm, respectively); they were lower by approximately 1 cm compared to those from the first experiment, and varied in a lower range compared to E1. Lengths were also close in mean values between CM and PM (308.3 and 307.3 cm, respectively) and higher by approximately 2 cm compared to their counterparts from the first experiment. They varied in a wider range as opposed to E1.

3.2.2. Agreement between the Measurement Methods in E2

Figure 11 shows the results on absolute differences and agreement between diameter measurements as specific to the second experiment (E2). In terms of absolute differences, approximately 44% of the data was in perfect agreement, and close to 95% of the data was in a difference range of ±1 cm. These results indicate a greater agreement as opposed to that from E1. As proved by a Shapiro-Wilk test, data on differences did not follow a normal distribution (Table A2).

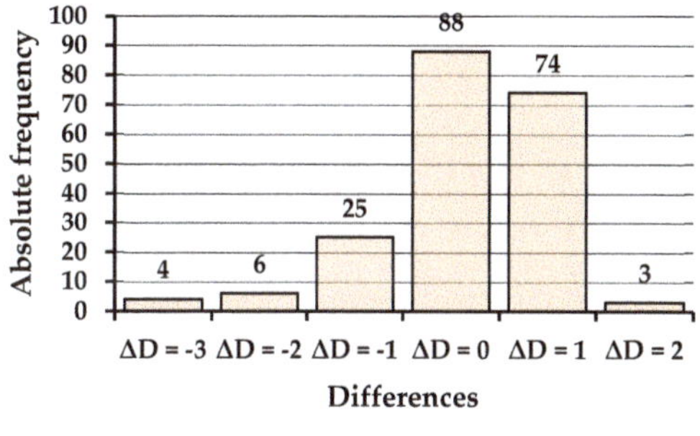

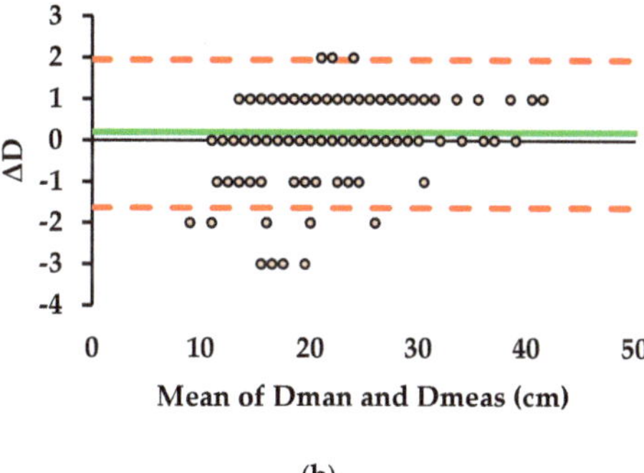

(a)

(b)

Figure 11. Agreement of diameter measurement methods in E2: (**a**) Absolute frequency of differences (ΔD) between Dman and Dmeas; (**b**) Bland-Altman plot showing the differences plotted against the mean of paired measurements, the bias (green line), and the lower and upper limits of agreement (red dashed lines).

The bias was of 0.2 (Figure 11b), meaning that, on average, PM measured less by 0.2 cm, which was better than in E1 (one-third of the bias in E1), and close to 94% of the observations were found between the limits of agreement, which was similar to E1. The two datasets (Dman, Dmeas, Figure 12) were strongly linearly-related as proven by the coefficient of determination ($R^2 = 99.8$), while the slope of regression through the origin equation was close to 1 which indicates a high agreement between the two methods. Similar to E1, data was found to be homoskedastic (Figure A2a).

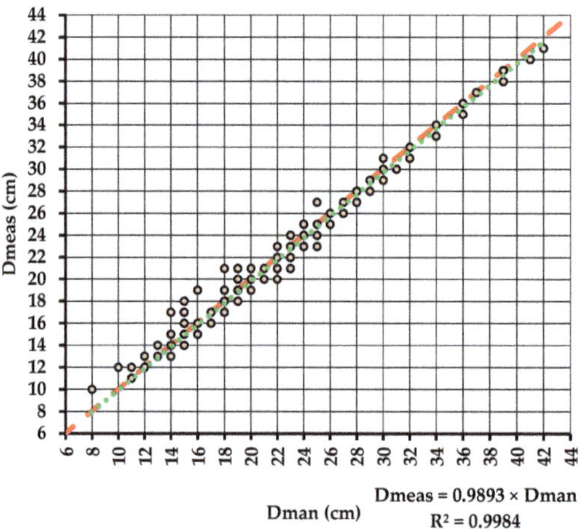

Figure 12. Relation between the diameters measured by the two methods (E2). Legend: red dashed line stands for perfect agreement and green dashed line stands for the dependence relation between Dmeas and Dman fitted by RTO.

Similar to diameters, the results of length measurement were better in terms of agreement (Figures 13 and 14) compared to those from E1. Close to 26% (53 observations) were in full agreement, which was higher compared to E1, 55% of the data (110 observations) were found to disagree by up to ±1 cm, which was close to E1, and 99% (198 observations) were found to disagree by up to ±5 cm. The bias was of 0.3 cm, meaning that, on average, PM measured less by 0.3 cm, which was the same as in E1, and 97% of the data (194 observations) were found within the limits of agreement. Similar to E1, there was a strong dependence relation between the length measured by the two methods (R^2 close to 1), and the slope of the regression line was very close to that of the perfect agreement; however, the data was quite spread indicating a higher degree of disagreement as compared to diameter measurement. The data on differences between the measurements of length did not follow a normal distribution; by the results of the Breusch-Pagan test the data was found to be homoskedastic ($p > 0.05$, Figure A2b).

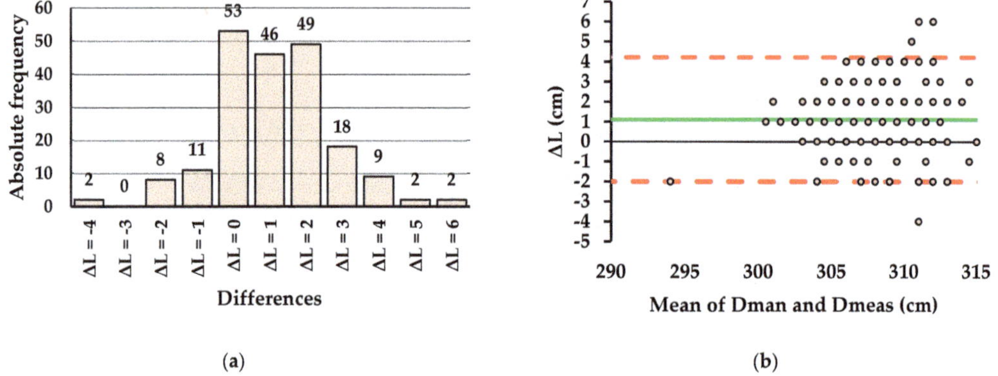

(a)   (b)

Figure 13. Agreement between the length measurement methods in E2: (**a**) Absolute frequency of differences (ΔL) between Lman and Lmeas; (**b**) Bland-Altman plot showing the differences plotted against the mean of paired measurements, the bias (green line), and the lower and upper limits of agreement (red dashed lines).

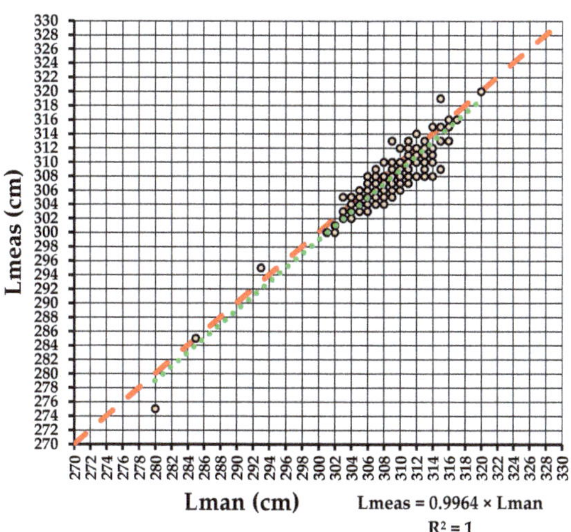

Figure 14. Relation between the lengths measured by the two methods (E2). Legend: red dashed line stands for perfect agreement and green dashed line stands for the dependence relation between Lmeas and Lman fitted by RTO.

3.2.3. Measurement Errors in E2

The results of measurement errors estimated for the second experiment (E2) are shown in Table 2. For both, diameter and length measurements they were lower although close to that of E1, with a diameter MAE, RMSE, and NRMSE of 0.65, 0.92, and 0.03 cm, respectively, and a length MAE, RMSE, and NRSME of 1.46, 1.93 and 0.04 cm, respectively.

Table 2. Results on errors for the second experiment (E2).

Error Metric	Diameter (cm)	Length (cm)
MAE	0.65	1.46
RMSE	0.92	1.93
NRMSE	0.03	0.04

The highest differences in errors of the E2 were those of length measurements. However, they accounted for less by 0.35 (MAE), 0.62 (RMSE), and 0.06 (NRMSE) cm, respectively, as compared to E1. Differences in the measurement errors of diameters were in range of 0.01 to 0.04 cm. Altogether, these results indicate that, at a sample level, the disagreement between the two ways of measuring the logs by the Measure app was rather small.

4. Discussion

The last decade has been characterized by a significant diversification in the development and testing of the non-conventional measurement tools, particularly in the tree biometrics measurement. LiDAR technology has been increasingly used in the form of highly-accurate expensive equipment to detect and measure biometrical characteristics of the individual trees, although several challenges related to the protocols to be used and to the diversity in forest conditions still remain [23,33], in addition to its high costs. Most likely, lower costs, equipment compactness, large-scale availability, and polyfunctionality, have led scholars to testing small-sized alternatives in tree [21,26,34,35] and log measurement [36]. Only some limited research has documented the accuracy of equipment capable of instantly provide the measurement results (e.g., [26]), and there is a lack of studies on the applicability and accuracy of close-range compact solutions to log measurement [36]. As a fact, most of the tested platforms still require more or less complicated workflows to process the data before producing the biometric estimates. The above-mentioned motivated this study to check if the general-purpose Measure application integrated with AR and LiDAR technologies could be a feasible solution to log biometrics measurement.

A first performance parameter that needs to be discussed in the measurement bias, characterizing the agreement between the conventional (CM) and phone-based (PM) measurements. For diameter measurement in E1, and E2 the bias was 0.6 and 0.2 cm, respectively, meaning that, on average, PM measured less (underestimated) than CM, irrespective of the experimental setup. This difference could be largely attributed to the fact that it was almost impossible to place the starting and ending points of the measurement right at the opposed ends of the log, characterizing a given diameter line. However, this may be changed by adjusting the positions of the starting and ending measurement points right after a measurement is done, therefore would require more time for measuring. While this approach was not taken in this study, it could hold the ability of improving the measurement agreement, by providing estimates close to those measured conventionally. In E1, the bias (0.6) was three times higher compared to E2 (0.2). Excluding random errors which could have been characterizing the CM and PM, as well as the effect of rounding the results of CM to the nearest centimeter, in E1 the higher bias value can largely come from the fact that there was not a perfect agreement among the points at which the log was touched by the caliper and those used to measure the diameter by the phone. As such, for scientific reference, the bias from E2 could be closer to the real agreement. For length measurements, there were no apparent differences between E1 and E2 in terms of bias. In both cases it was 0.3, which can be explained by fewer dimensional deviations as the end-edges of the log at its top were easier to appreciate visually, irrespectively of setting or not guiding marks. Still, there was an underestimation which can be attributed to at least two factors: the impossibility to place the measurement points exactly at the edges and the length of the log along the taper which could have been systematically higher in the mechanical measurement by tape. In relation to both, diameter and length estimates, ovality, curvature, and buttress of the logs may be additional factors explaining the differences found by this study.

There were significant differences in terms of diameter measurement agreement between E1 and E2. Guided measurement (E2) has led to 95% of the observations falling in a ±1 cm agreement range, while free-eye measurement (E1) accounted for 86% of the observations in this range. On the other hand, close to 98% of the data has fallen in an agreement range of ±2 cm, irrespective of the experiment, therefore the effect of the experimental setup was lower for this range of agreement. The frequency of observations falling in a ±5 cm agreement range for length measurement was close between E1 (95%) and E2 (99%). Since the data on both, diameter and length measurement differences has been proved to be homoskedastic, higher differences between measurements could be due to random errors in CM or to an improper setup of the starting and ending measurement points in PM. In addition, the condition of the logs under measurement could have been influenced by the PM accuracy since some logs were either wet or partially covered by snow (Figure 1).

Error metrics used in this study have indicated that there were no high differences between the experiments in terms of diameter measurement. Mean absolute error (MAE) was found to be of 0.68 and 0.65 cm for E1 and E2, respectively. This means that, on average, there was an absolute difference of close to 0.7 cm between CM and PM. Root mean squared errors (RMSE), on the other hand, were higher, accounting for close to 1 cm. However, RMSE error metrics are known to be driven in their magnitude by outlying data such as that characterizing high differences, as well as by the number of observations in a given sample [32]. Therefore, the values of MAE could be more closer estimates of the real differences. While for diameter measurement they were less than 0.7, for length measurement they were approximately two times higher, accounting for 1.81 and 1.46 cm in E1 and E2, respectively.

Compared to the results reported by other studies using mobile general-use platforms, the differences in terms of bias or error metrics found by this study were, in general, less. The study of [26] has reported biases of 0.3 to 0.36 cm for DBH measurements taken by a phone at a distance of 1.5 m. Their results agree with those of length measurements from this study, but were higher compared to those from the guided experiment (E2) for diameter measurement. RMSE values of 1.12 and 1.83–1.91 cm were found by [21] when estimating the DBH based on close-range photogrammetry and Google Tango technology embodied in mobile platforms, respectively. Also, by tree reconstruction from photographs taken with a mobile phone, the study of [35] indicated a general RMSE of 1.9 cm in DBH measurement, while the study of [34] has found RMSE and bias values in the range of 3.13–4.51 and −0.58 to 1.03, respectively, when extracting the DBH from point clouds collected with three applications installed on an iPad device. Having in mind the limitations of the RMSE as an error metric [32], a comparison with the above-mentioned studies indicates that the PM errors for diameter and length measurement were less and close, respectively, to the values indicated by other studies for DBH measurement.

Finally, a general evaluation of the applicability of the Measure app in log measurement needs to be considered. Given the results of this study, it seems that the PM option would be suitable for producing general estimates on log volume assuming that logs are piled and there is no access to them so as to properly implement CM. This is typical of landings where log piles are formed to save space and to facilitate loading and transportation of wood, and where a quantitative estimation of the wood could be in question before loading [1]. For such log grouping conditions, PM could prove indeed valuable for quickly measuring the diameters at the log ends. Assuming a bias of 0.6 cm, the measurements taken by PM could be corrected to better reflect reality, or other calibrations between the measurements could be in question depending also on the personal skills of the operators. However, for piled wood, there are some limits in capability which need to be addressed. First of all, the LiDAR range of the used platform is of 5 m, while the used app may take measurements only at ca. 2.5 m, as proved by some indoor experiments carried out by the authors. This means that the diameter of piled logs which are not accessible in this range cannot be measured by the tested solution. Secondly, for tall piles, it would be impossible to accurately locate both ends of a given log. Producing the final assortments at the felling area requires information about diameters and lengths of the logs [1], therefore PM could be used to pre-grade the delimbed stems before bucking. Of course, this would require diameter measurement taken over the felled stems, a measurement option which is different from that explored in this study. However, the accuracy for such applications seems to be acceptable based on the results of [26]. Accuracy in length, on the other hand, needs extra caution since failing to provide a length required for a given assortment may lead to downgrading a given log.

Log measuring is commonly done in hazardous work environments and it may burden the workers with the need to carry, use and store rather uncomfortable equipment. From these points of view, the use of the Measure app can add value by integrating the commonly used log measurement tools in a single lightweight device in which the personal functionalities can be extended by integrating those required by the measuring job. It also gives the possibility of saving the measurements and, more importantly, its use does not require direct contact with the logs, therefore it can contribute to safety enhancements. Compared to other apps and platforms such as those able to collect point clouds [22,24,36–38], the tested solution still shares some limitations in terms of data transfer, wood traceability, and transparency in the wood supply.

In relation to the experiments of this study, there are several other research directions that may be approached in the future. First of all, similar experiments should be set up to extend the sample size and to better infer the disagreements between the methods. Close to 500 logs were used in this study; however, it is likely that larger sample sizes will improve the rate of agreement between the methods by moving more pairwise measurements in higher agreement ranges. Accordingly, it is less likely that the bias between the two methods would change significantly, due to the reasons discussed above. Since replicability is of first importance in extending the PM, further studies could focus on the agreement of measurements as done by different operators. The logs measured in this study were limited to approximately 3 m in length and they were all processed from coniferous trees. To what extent the measuring capability may be extended to longer lengths, and how the measurement accuracy would respond to different species and operating conditions, need to be checked in the future to find a balance between the accuracy and capabilities of the measuring device.

5. Conclusions

Moving to digital solutions in log measurement is of the first importance for an efficient wood supply. Through two experiments, this study evaluated the agreement between the conventional and phone-based measurement having at its core the general-purpose Measure app developed by Apple, Augmented Reality, and LiDAR sensing capabilities of the iPhone 12 and 13 Pro Max platforms. The results indicate a good agreement between measurements, making this digital solution useful for several log-measurement applications, mainly by providing accurate results, improving ergonomics, and safety of measuring operations. Further studies could check if the accuracy of measurements would be enhanced by larger log samples and if there would be a good replicability of the method as of different operators. Also, finding a balance between capability and measurement accuracy by extending the study to longer log lengths, different species and operating conditions would be important to characterize the technical limitations of the tested method.

Author Contributions: Conceptualization, S.A.B.; Data curation, S.A.B. and J.M.M.T.; Formal analysis, S.A.B. and J.M.M.T.; Funding acquisition, S.A.B.; Investigation, S.A.B., J.M.M.T., G.O.F. and M.V.M.; Methodology, S.A.B.; Project administration, S.A.B.; Resources, S.A.B., J.M.M.T., G.O.F. and M.V.M.; Software, S.A.B.; Supervision, S.A.B. and M.V.M.; Validation, S.A.B.; Visualization, S.A.B.; Writing—original draft, S.A.B., J.M.M.T., G.O.F. and M.V.M.; Writing—review and editing, S.A.B. All authors have read and agreed to the published version of the manuscript.

Funding: This work was supported by a grant of the Romanian Ministry of Education and Research, CNCS—UEFISCDI, project number PN-III-P4-ID-PCE-2020-0401, within PNCDI III. An objective of the Hypercube 4.0 project is to test and develop alternative solutions to conventional wood measurement, in particular those based on LiDAR. Part of the activities of this work were funded by the grant "Proiectul meu de diplomă 2022" awarded to Jenny Magaly Morocho Toaza by the Transilvania University of Brasov. The APC was funded by the Transilvania University of Brasov.

Institutional Review Board Statement: Not applicable.

Informed Consent Statement: Not applicable.

Data Availability Statement: Data supporting this study may be provided upon a reasonable request to the first author of the study.

Acknowledgments: The authors would like to thank to the Hypercube 4.0 team members for helping with the data collection activities as specific to this study. The authors acknowledge the support provided by the Department of Forest Engineering, Forest Management Planning and Terrestrial Measurements, Faculty of Silviculture and Forest Engineering, Transilvania University of Braşov, which was essential in collecting the data and writing this paper. Also, the authors would like to thank to the management and staff of HS Timber Productions Reci S.R.L., a member of HS Timber Group, for providing the logistics needed to carry on this study.

Conflicts of Interest: The authors declare no conflict of interest.

Appendix A

Table A1. Results of normality check for the first experiment (E1).

Statistic	VH	VS	Dman	Dmeas	Lman	Lmeas
W-stat	0.861984	0.831085	0.931375	0.938037	0.942384	0.966199
p-value	<0.001	<0.001	<0.001	<0.001	<0.001	<0.001
α	0.05	0.05	0.05	0.05	0.05	0.05
Diagnose	no	no	no	no	no	no

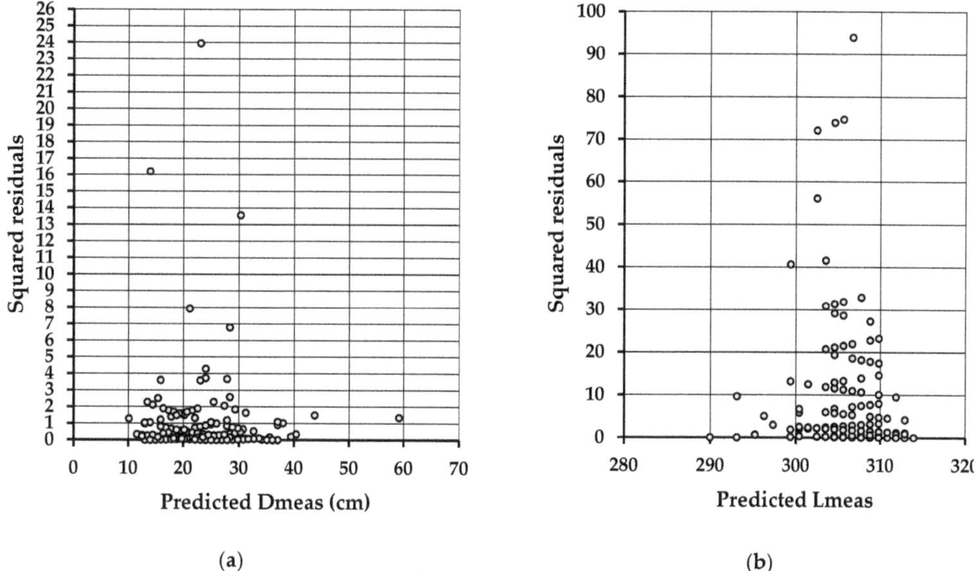

(a)

(b)

Figure A1. Results of the test for homoskedasticity (E1): (**a**) Plot of squared residuals against the predicted Dmeas (Note: data was homoskedastic by Breusch-Pagan test, $p > 0.05$); (**b**) Plot of squared residuals against the predicted Lmeas (Note: data was homoskedastic by Breusch-Pagan test, $p > 0.05$).

Table A2. Results of normality check for the second experiment (E2).

Statistic	Dman	Dmeas	Lman	Lmeas
W-stat	0.951711	0.83517	0.943546	0.824777
p-value	<0.001	<0.001	<0.001	<0.001
α	0.05	0.05	0.05	0.05
Diagnose	no	no	no	no

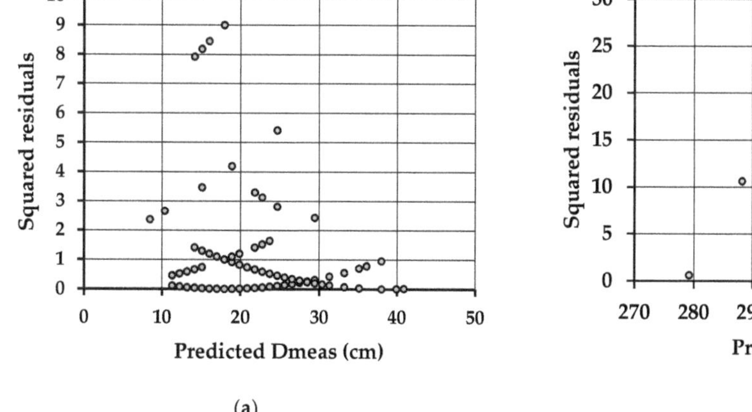

(a)

(b)

Figure A2. Results of the test for homoskedasticity (E2): (**a**) Plot of squared residuals against the predicted Dmeas (Note: data was homoskedastic by Breusch-Pagan test, $p > 0.05$); (**b**) Plot of squared residuals against the predicted Lmeas (Note: data was homoskedastic by Breusch-Pagan test, $p > 0.05$).

References

1. Oprea, I. *Tehnologia Exploatării Lemnului*; Transilvania University Press: Braşov, Romania, 2008; pp. 11–80. ISBN 978-973-598-301-7.
2. Moskalik, T.; Borz, S.A.; Dvorák, J.; Ferencik, M.; Glushkov, S.; Muiste, P.; Lazdiņš, A.; Styranivsky, O. Timber harvesting methods in Eastern European countries: A review. *Croat. J. For. Eng.* **2017**, *38*, 231–241.
3. Mederski, P.; Borz, S.A.; Ðuka, A.; Lazdiņš, A. Challenges in forestry and forest engineering–Case studies from four countries in East Europe. *Croat. J. For. Eng.* **2021**, *42*, 117–134. [CrossRef]
4. Lundbäck, M.; Häggström, C.; Nordfjell, T. Worldwide trends in methods for harvesting and extracting industrial roundwood. *Int. J. For. Eng.* **2021**, *32*, 202–215. [CrossRef]
5. Spinelli, R.; Magagnotti, N.; Visser, R.; O'Neal, B. A survey of skidder fleet of Central, Eastern and Southern Europe. *Eur. J. For. Res.* **2021**, *140*, 901–911. [CrossRef]
6. Müller, F.; Jaeger, D.; Hanewinkel, M. Digitization in wood supply—A review of how Industry 4.0 will change the forest value chain. *Comput. Electron. Agric.* **2019**, *162*, 206–218. [CrossRef]
7. Kemmerer, J.; Labelle, E.R. Using harvester data from on-board computers: A review of key findings, opportunities and challenges. *Eur. J. For. Res.* **2021**, *140*, 206–218. [CrossRef]
8. Hartsch, F.; Kemmerer, J.; Labelle, E.R.; Jaeger, D.; Wagner, T. Integration of harvester production data in German wood supply chains: Legal, social and economic requirements. *Forests* **2021**, *12*, 460. [CrossRef]
9. Rauch, P.; Borz, S.A. Reengineering the Romanian timber supply chain from a process management perspective. *Croat. J. For. Eng.* **2020**, *41*, 85–94. [CrossRef]
10. Government Decision No. 497 of 25 June 2020. Available online: http://www.mmediu.ro/app/webroot/uploads/files/HOT%C4%82R%C3%82RE%20nr.%20497%20din%2025%20iunie%202020.pdf (accessed on 9 April 2020).
11. Acuna, M.; Bigot, M.; Guerra, S.; Hartsough, B.; Kanzian, C.; Kärhä, K.; Lindroos, O.; Magagnotti, N.; Roux, S.; Spinelli, R.; et al. *Good Practice Guidelines for Biomass Production Studies*; CNR IVALSA Sesto Fiorentino (National Research Council of Italy—Trees and Timber Institute): Sesto Fiorentino, Italy, 2012; pp. 1–51. ISBN 978-88-901660-4-4.
12. Björheden, R.; Apel, K.; Shiba, M.; Thompson, M. *IUFRO Forest Work Study Nomenclature*; Swedish University of Agricultural Science: Grapenberg, Sweden, 1995; 16p.
13. Borz, S.A.; Ignea, G.; Popa, B.; Sparchez, G.; Iordache, E. Estimating time consumption and productivity of roundwood skidding in group shelterwood system—A case study in a broadleaved mixed stand located in reduced accessibility conditions. *Croat. J. For. Eng.* **2015**, *36*, 137–146.
14. Borz, S.A.; Ignea, G.; Popa, B. Modelling and comparing timber winching performance in windthrow and uniform selective cuttings for two Romanian skidders. *J. For. Res.* **2014**, *19*, 473–482. [CrossRef]
15. Borz, S.A.; Dinulica, F.; Birda, M.; Ignea, G.; Ciobanu, V.D.; Popa, B. Time consumption and productivity of skidding Silver fir (Abies alba Mill.) round wood in reduced accessibility conditions: A case study in windthorw salvage logging from Romanian Carpathians. *Ann. For. Res.* **2013**, *56*, 363–375.
16. Apafaian, A.I.; Proto, A.R.; Borz, S.A. Performance of a mid-sized harvester-forwarder system in integrated harvesting of sawmill, pulpwood and firewood. *Ann. For. Res.* **2017**, *60*, 227–241.
17. Leahu, I. *Dendrometrie*; "Editura Didactică şi Pedagogică R.A." Press: Bucharest, Romania, 1994; 374p.
18. Hohmann, F.; Ligocki, A.; Frerichs, L. Harvester measuring system for trunk volume determination: Comparison with the real trunk volume and application in the forest industry. *Bull. Transilv. Univ. Bras. Ser. II–For. Wood Ind. Agric. Food Eng.* **2017**, *10*, 27–34.
19. Câmpu, V.R.; Dumitrache, R.; Borz, S.A. The impact of log length on the conversion factor of staked wood to solid content. *Wood Res.* **2015**, *60*, 503–518.
20. Acuna, M.; Sosa, A. Automated volumetric measurements of truckloads through multi-view photogrammetry and 3D reconstruction software. *Croat. J. For. Eng.* **2019**, *40*, 151–162.
21. Tomaštíc, J.; Saloň, S.; Tunák, D.; Chudy, F.; Kardoš, M. Tango in forests—An initial experience of the use of new Google technology in connection with forest inventory tasks. *Comput. Electron. Agric.* **2017**, *141*, 109–117. [CrossRef]
22. Forest Design Scanner (FD Scanner). Available online: https://play.google.com/store/apps/details?id=ro.forestdesign.scanner&gl=RO (accessed on 9 April 2022).
23. Balenović, I.; Liang, X.; Jurjević, L.; Hyyppä, J.; Seletković, A.; Kukko, A. Hand-held personal laser scanning–current status and perspectives for forest inventory application. *Croat. J. For. Eng.* **2021**, *42*, 165–183. [CrossRef]
24. Arboreal Forest—A Real Digital Caliper. Available online: https://www.arboreal.se/en/arboreal-forest/ (accessed on 9 April 2022).
25. Use the Measure App on Your iPhone, iPad of iPod Touch. Available online: https://support.apple.com/en-us/HT208924 (accessed on 9 April 2022).
26. Ucar, Z.; Degermenci, A.S.; Zengin, H.; Bettinger, P. Evaluating the accuracy of remote dendrometers in tree diameter measurements at breast height. *Croat. J. For. Eng.* **2022**, *43*, 185–197. [CrossRef]
27. Şofletea, N.; Curtu, L. *Dendrologie*; "Pentru Viaţă" Publishing House: Braşov, Romania, 2008; 418p.
28. Bland, J.M.; Altman, D.G. Measuring agreement in method comparison studies. *Stat. Methods Med. Res.* **1999**, *8*, 135–160. [CrossRef]
29. Giavarina, D. Lessons in biostatistics. Understanding Bland Altman analysis. *Biochem. Med.* **2015**, *25*, 141–151. [CrossRef]

30. Breusch, T.S.; Pagan, A.R. A simple test for heteroscedasticity and random coefficient variation. *Econometrica* **1979**, *47*, 1287–1294. [CrossRef]
31. Gujarati, D. Econometrics by example. In *Palgrave Macmillan*; Macmillan Publishers Limited: New York, NY, USA, 2011; 371p.
32. Wilmott, C.J.; Matsuura, K. Advantages of the mean absolute error (MAE) over the root mean squared error (RMSE) in assessing the average model performance. *Clim. Res.* **2005**, *30*, 79–82. [CrossRef]
33. Xu, D.; Wang, H.; Xu, W.; Luan, Z.; Xu, X. LiDAR applications to estimate forest biomass at individual tree scale: Opportunities, challenges and future perspectives. *Forests* **2021**, *12*, 550. [CrossRef]
34. Gollob, C.; Ritter, T.; Kraßnitzer, R.; Tockner, A.; Nothdurft, A. Measurement of forest inventory parameters with Apple iPad Pro and integrated LiDAR Technology. *Remote Sens.* **2021**, *13*, 3129. [CrossRef]
35. Marzulli, M.I.; Raumonen, P.; Greco, R.; Persia, M.; Tartarino, P. Estimating tree stem diameters and volume from smartphone photogrammetric point clouds. *Forestry* **2020**, *93*, 411–429. [CrossRef]
36. Borz, S.A.; Proto, A.R. Application and accuracy of smart technologies for measurement of roundwood: Evaluation of time consumption and efficiency. *Comput. Electron. Agric.* **2022**, *197*, 106990. [CrossRef]
37. Hypercube 4.0—Moving Wood Measurement Towards a New Dimension. Project Started on 4 January 2021, Funded by the Executive Unit for Financing Higher Education, Research, Development and Innovation (UEFISCDI). Available online: https://sites.google.com/view/hypercube40/rezumat (accessed on 23 June 2022).
38. Miguel-Díez, F.; Reder, S.; Wallor, E.; Bahr, H.; Blasko, L.; Mund, J.-P.; Cremer, T. Further application of using a personal laser scanner and simultaneous localization and mapping technology to estimate the log's volume and its comparison with traditional methods. *Int. J. Appl. Earth Obs. Geoinf.* **2022**, *190*, 102779. [CrossRef]

Article

Toward a Unified TreeTalker Data Curation Process

Enrico Tomelleri [1,*], Luca Belelli Marchesini [2], Alexey Yaroslavtsev [3], Shahla Asgharinia [4] and Riccardo Valentini [4]

1 Faculty of Science and Technology, Free University of Bolzano, Piazza Università 5, 39100 Bolzano, Italy
2 Forest Ecology Unit, Research and Innovation Centre, Fondazione Edmund Mach, Via E. Mach, 38010 San Michele all'Adige, Italy; luca.belellimarchesini@fmach.it
3 Department of Ecology, Russian State Agrarian University—Moscow Timiryazev Agricultural Academy, Timiryazevskaya St., 49, 127550 Moscow, Russia; yaroslavtsevam@gmail.com
4 Department for Innovations in Biological, Agri-Food and Forest Systems (DIBAF), University of Tuscia, Via San Camillo de Lellis 4, 01100 Viterbo, Italy; asgharinia@unitus.it (S.A.); rik@unitus.it (R.V.)
* Correspondence: etomelleri@unibz.it

Abstract: The Internet of Things (IoT) development is revolutionizing environmental monitoring and research in macroecology. This technology allows for the deployment of sizeable diffuse sensing networks capable of continuous monitoring. Because of this property, the data collected from IoT networks can provide a testbed for scientific hypotheses across large spatial and temporal scales. Nevertheless, data curation is a necessary step to make large and heterogeneous datasets exploitable for synthesis analyses. This process includes data retrieval, quality assurance, standardized formatting, storage, and documentation. TreeTalkers are an excellent example of IoT applied to ecology. These are smart devices for synchronously measuring trees' physiological and environmental parameters. A set of devices can be organized in a mesh and permit data collection from a single tree to plot or transect scale. The deployment of such devices over large-scale networks needs a standardized approach for data curation. For this reason, we developed a unified processing workflow according to the user manual. In this paper, we first introduce the concept of a unified TreeTalker data curation process. The idea was formalized into an R-package, and it is freely available as open software. Secondly, we present the different functions available in "ttalkR", and, lastly, we illustrate the application with a demonstration dataset. With such a unified processing approach, we propose a necessary data curation step to establish a new environmental cyberinfrastructure and allow for synthesis activities across environmental monitoring networks. Our data curation concept is the first step for supporting the TreeTalker data life cycle by improving accessibility and thus creating unprecedented opportunities for TreeTalker-based macroecological analyses.

Keywords: IoT; forest ecology; big data; accessibility

Citation: Tomelleri, E.; Belelli Marchesini, L.; Yaroslavtsev, A.; Asgharinia, S.; Valentini, R. Toward a Unified TreeTalker Data Curation Process. *Forests* **2022**, *13*, 855. https://doi.org/10.3390/f13060855

Academic Editors: Stelian Alexandru Borz, Andrea R. Proto, Robert Keefe, Mihai Nita and Olga Viedma

Received: 24 April 2022
Accepted: 27 May 2022
Published: 30 May 2022

1. Introduction

Technological innovation has frequently been an accelerator for gaining new knowledge in many fields of ecology [1]. The development of Wireless Sensor Network (WSN) technology, combined with the advancements in low-power, high range data transmission, is revolutionizing the approach to environmental monitoring [2]. Such developments, combined with cellular networks capable of supporting massive connectivity with efficient schemes for tethering billions of devices globally, enable Internet of Things (IoT) applications in many fields of science, including ecology [3,4]. This is further advantaged by increasingly ubiquitous connectivity and the extensive coverage provided by the latest generations of communication networks [5]. Thus, such technologies offer new opportunities for collecting continuously consistent environmental and ecophysiological parameters with high temporal frequency across broad spatialized networks [6–8].

The ecological applications of IoT provide different advantages compared to other approaches for environmental monitoring by empowering ecologists to address massive data retrieval in situations where manual collection from multiple and heterogeneous sensors would be time consuming and error-prone. At the same time, diffused networks of environmental sensors allow for conducting measurements with high temporal frequency, even in rough environments or remote experimental sites [9]. Additionally, WSN-based experiments have the possibility for integration with adjacent networks and even with other data streams such as remote sensing [10].

The need for several sensors to be implemented in extended networks requires the usage of low-cost solutions. Such requirement boosts the implementation of open solutions, including open hardware and open software. As a result, IoT technology generally bears unprecedented opportunities for ad hoc customization and community-based development, contributing to the fast diffusion and enhancement of IoT usage in environmental applications [11].

Several smart devices have already supplanted traditional monitoring systems relying on discrete or manual methodologies, giving real-time, continually analyzed, and wirelessly provided data, but different challenges remain unsolved [12]. As a matter of fact, in parallel to increasing ecological data availability, there is a new need to establish and deploy appropriate cyberinfrastructures [8] to enable data science-based research. The first step in designing such an infrastructure is the definition of data curation strategies, including device and data monitoring, quality assurance, storage, analysis, and accessibility. Firstly, a network with many sensors needs the constant monitoring of functionality, especially in harsh environments where an accurate and regular automated control can promptly permit the network manager to act in the case of failure [13]. Additionally, the following crucial step is quality assurance, because the data might suffer from different disturbing factors such as sensor degradation, unstable power availability, and transmission issues [11]. Such factors might generate a disturbed measured signal but could also be the source of missing data or generate duplicates. Secondly, data sourcing from multiple sensors across regional networks needs to include proper and consistent formats [14]. This means adopting self-documenting and recognized standards. This step is of paramount importance for supporting data-sharing and for enabling the final user to proceed with further analyses [15]. Lastly, data are required to be documented, and data access policies should be defined to allow for distribution and accessibility. In detail, accessibility establishes the degree to which researchers can use data. This means that accessible data are not only available but also usable. Therefore, accessibility requires the usage of standard—and possibly self-documenting—formats. In this context, the adoption of a standardized curation approach is fundamental for any Research Data Management plan [16], and it is a milestone for any research hypothesis making use of experimental "big-data" collected across large areas and across different research groups [17]. As mentioned by [18], there is an urgent need for ecologists to establish networks across institutional boundaries to pursue broad scale questions. Thus, in the past few years, the rising data-sharing culture binned uniformed data gathering methods together with data and metadata formats permitted to address regional- to global-scale questions. A good-practice example is the ICOS network, where in situ observations of carbon and other greenhouse gasses (GHG) are continuously collected across 140 measurement stations and 12 European countries according to standardized protocols, processed with a unified processing workflow, and provided to the final user as ICOS data products in a standardized data format [19].

This reasoning showed us the opportunity to develop an open-source toolbox implementing a unified processing workflow for the broad deployment of TreeTalkers (Nature 4.0 SB srl) data within the scientific community. As already demonstrated by the ICOS project in the context of GHG emissions, the presented workflow will support data-sound model development and the inference of forest attributes through the integration of TreeTalker data with complementary data streams [10] and their ultimate feeding into artificial intelligence systems for forestry applications and, eventually, for their contribution to the creation of forest digital twins [20].

2. The TreeTalker

The need to protect forests and technological developments have permitted the development of novel IoT devices for specific applications such as fire and deforestation monitoring. For example, the SeaForest solution is a system for detecting fire, pollution sources, or illegal deforestation based on IoT devices. [21,22] revises the available systems for forest fire detection applications and proposes a novel method based on multi-sensors and cameras. On the other hand, such systems can be deployed to monitor the environment but not multiple parameters related to tree ecophysiology. A more comprehensive approach is required for a better understanding of the resilience of forests to extreme events. The TreeTalker technology provides this opportunity because it is capable of measuring with high frequency and in nearly real-time key processes such as the water consumption, the growth of biomass (diameter), and the health of the leaves. The device consists of a logger enclosed in a plastic case that acquires signals from various sensors (Figure 1). The device is typically installed on a tree trunk at breast height. The current version of the system is described by [6]. It runs on batteries, and it has an average autonomy of up to three months with the default measurement frequency. A separated device equipped solely with a thermohygrometer and a multi-band spectrometer (TT-R) can be placed outside of the canopy for reference. The user can freely program the acquisition rate, which is by default set to hourly, allowing for adaptable and continuous high-frequency environmental monitoring. Raw data are collected via Long Range connection (LoRa) by a master node (the so-called TT-Cloud) and transmitted via the Global System for Mobile Communications (GSM) to a central server. The collected data are centrally saved as digital numbers and require further conversions to be expressed in physical units.

Figure 1. A beech tree monitored with a TreeTalker (power supply not shown).

3. The ttalkR Concept

The ttalkR workflow consists of four subsequential steps (Figure 2). Firstly, it is necessary to retrieve the data relative to a specific mesh of sensors from one or more servers. Secondly, the datasets are merged into four dedicated tables containing detailed information from different classes of devices or specific sensors. One table is for the TT-Cloud, one table is for the data collected from the embedded spectrometers, one table includes all the communication diagnostics, and the last table collects data from all the other sensors. Thirdly, basic quality assurance procedures are applied (i.e., outliers' removal and basic gap filling), followed by the conversions from digital numbers to physical units, which are conducted according to the TreeTalkers' user manual (TT+manual ver. 3.2, September 2020). Lastly, the derived variables are plotted for visualization and saved locally into an SQLite database for further processing.

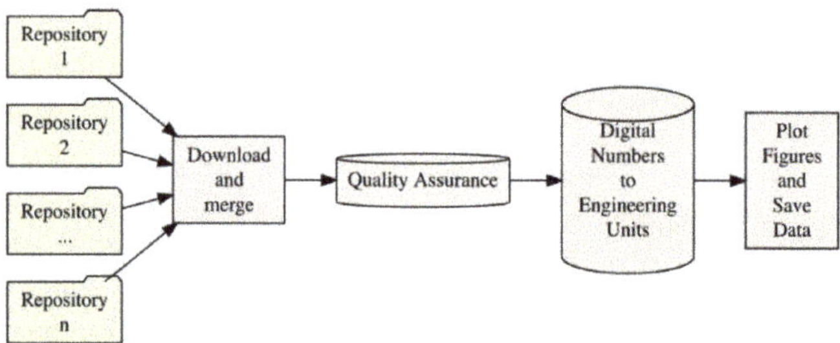

Figure 2. Conceptual workflow of the data curation process within the ttalkR package from the data download to the figures and data saving.

4. Functions of ttalkR

The "ttalkR" package is a collection of functions for curating TreeTalker data. The package is organized in dedicated functions addressing each measured parameter by a TreeTalker device (Figure 3). Such functions are usable between the input function named *ttScrape()* and the output function named *ttOutput()*. The plotting utilities permit the user to visualize the measured parameters' canopy/mesh scale but also the single tree level. While the first approach is useful for gaining an overview at the measurement site, the second approach allows for the identification of anomalous trees and can be used for diagnostic purposes (e.g., for identifying interesting patterns and, eventually, for spotting faulty sensors). Thus, the package can also be deployed for operational site functioning monitoring and maintenance, repair, and operation (MRO).

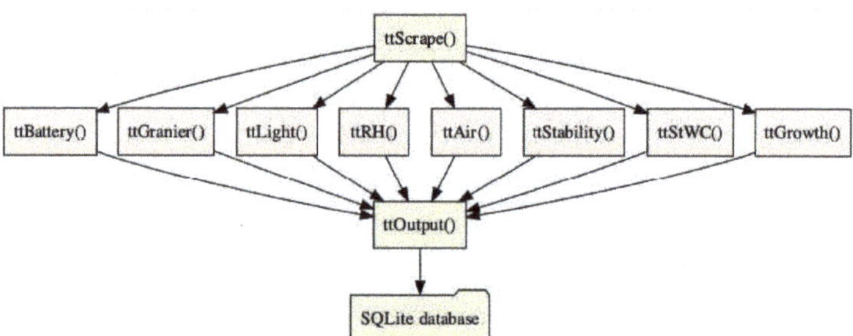

Figure 3. Detailed structure and functions of the ttalkR package. I/O functions are highlighted in green.

4.1. Data Download

The core of the package is formed by the *ttScrape()* function. It downloads TreeTalkers data from specific servers and organizes the different strings into the four aforementioned tables, which are required for further processing. The package was initially developed for the Italian Treetalker network, which also makes use of the standard server initialized by the devices' producer, but it can be deployed for any hosting server by adapting the source URL. The derived tables include information about the (i) status of the master (e.g., battery level and GSM metrics), (ii) communication diagnostics for the connected TreeTalkers (i.e., radio signal strength), (iii) raw data acquired by the spectrometers, and (iv) raw data from all the other attached sensors. The function *ttScrape()* includes a first-tagging of the missing data and removes duplicated fields. Further quality assurance steps are executed in the parameter-specific functions.

4.2. Battery Voltage

The battery voltage of the TreeTalkers within a mesh and the associated master (TTcloud) can be monitored for MRO purposes. Like all the following functions, *ttBattery()* makes use of the data frames created by the *ttScrape()* function. For each TreeTalker, it calculates and plots the battery level consistently with the programmed measurement frequency (Figure 4A). The function considers the bandgap voltage reference of the microcontroller (1.1.v) and the analogue to digital conversion (ADC) values of the bandgap and the battery. No quality assurance is applied to these parameters. As indicated by the producer, the batteries should be recharged at 3500 mV because, below such threshold, the proper functioning of the sensors is not guaranteed (Figure 4A).

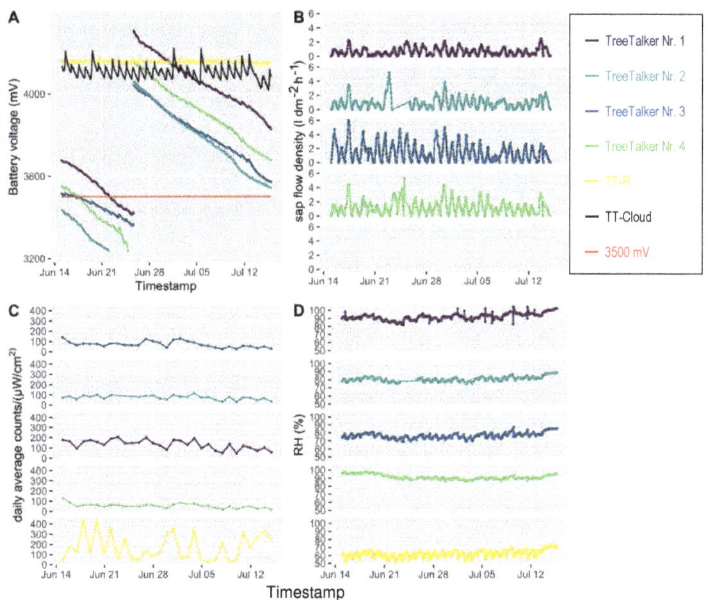

Figure 4. Output of the "ttalkR" plotting utility. (**A**) battery voltage with the indication of the warning threshold (3500 mV); (**B**) sap flow; (**C**) under canopy radiation; (**D**) relative humidity. The example data refer to four TreeTalkers and a TT-R for the period between 14 June 2021 and 14 July 2021. In (**B**), the data from the TT-R are missing because it is a reference device outside of the canopy which is not equipped with the corresponding sensors.

4.3. Sap Flow

Transpiration is a critical process that links the exchange of water, carbon, and energy between the land and the atmosphere, influencing various vegetation–atmosphere feedbacks. Water transfer from the roots to the leaves is driven by transpiration in the form of sap flow through the plant's xylem pathway, and this sap flow influences heat transport in the xylem [23]. The current version (3.2) of the TreeTalker device uses the thermal dissipation method [24] with repeated heating cycles. The default settings foresee 10 min of heating and 50 min of cooling. Probe pairs are inserted in the tree stem with a vertical separation of 10 cm. Normally, the probes are positioned facing north to avoid direct solar heating. Firstly, the function *ttGranier()* converts the voltages from the reference and heat probes into temperatures (Figure 4B). Then, it smooths the time series applying a Savytsky–Golay filter [25] by removing high-frequency components (e.g., electric noise). The function replaces the missing values for the gaps up to 12 h by interpolation. Lately, it estimates the sap flow density for each TreeTalker in a mesh by applying the conversion described by [26], which is assumed to be species independent.

4.4. Under Canopy Radiance

Understanding the spectrum quality of light transmission across the forest canopy can have a significant impact on the design and validation of new forest radiative transfer models. These can be used to better characterize forest–climate interactions or forest production or as a tool to evaluate earth observations [27]. A proper description of the spectral transmittance of forest canopies will enhance our understanding of forest vegetation phenological mechanisms. Each TreeTalker is equipped with two AMS chips (ams-OSRAM AG), the model AS7262 for the visible range and the model AS7263 for the near infra-red range (TT+manual ver. 3.2, September 2020). Each chip can measure six bands. The AS7262 (central wavelengths: 450, 500, 550, 570, 600, 650) has a full width at half maximum (FWHM) of 40 nm, while the with the AS7263 (central wavelengths: 610, 680, 730, 760, 810, 860 nm) has a FWHM of 20nm. The function *ttLight()* makes use of a dedicated data frame created by the *ttScrape()* function. It requires the site coordinates as input arguments, which are used to estimate the sun's position in the sky vault. Firstly, the function smooths the time series for each of the twelve measured spectral bands by applying a Savytsky–Golay filter [25] for removing high-frequency components. Secondly, it filters the spectrometer data according to solar geometry by keeping the measurements acquired with the sun azimuth between +/−30 degrees from the local solar noon. Lastly, it aggregates daily values and plots the spectrometer data from a TreeTalkers mesh for a specific band (Figure 4C). The function does not account for the shading effects related to the site topography.

4.5. Relative Humidity and Air Temperature

Forest canopies act as a thermal insulator, cooling the understory when the weather is hot and warming the understory when the weather is cold [28]. These dynamics affect the relative humidity, and these factors act concurrently with the regulation of the ecological processes occurring below the canopy in natural and urban environments [29]. The function *ttRH()* and the function *ttAir()* make use of the data from the thermohygrometer NXP/Freescale, Model: Si7006 (TT+manual ver. 3.2, September 2020). The relative humidity data (Figure 4D) are filtered for a plausibility range, and the gaps up to 12 h are filled by interpolation, while no quality assurance is applied to the temperature data (Figure 5A).

4.6. Tree Stability

Tree stability is an essential characteristic to be monitored because it can provide information about the resilience of aingle trees as well as of the whole forest ecosystem to abiotic disturbances such as windstorms [30,31]. Accelerometers mounted on a tree trunk can record the sway movement of the tree. The tree sway is affected by tree traits such as the mass, wood density, elasticity, and drag coefficient [32] but also by canopy characteristics such as the closure and roughness. Understanding the behavior of this parameter is crucial in forestry for understanding the response to wind, as storm damage can be a large source of economic loss [33]. TreeTalkers measure the trunk movements by a Silicon Labs MMA8451Q 3-Axis Accelerometer equipped with a very low-power, low-profile capacitive MEMS sensor (TT+manual ver. 3.2, September 2020). The *ttStability()* function processes the oscillation of trees due to gravity with a spherical coordinate system. With basic trigonometry, the angle between the gravity vector and the TreeTalker z-axis are assessed by taking in account variations in the angle of tilt in the xy-plane, as described by [34] (Figure 5B). A positive angle means that the corresponding sensor axis is pointed above the relative horizon (referred to in the standard installation settings), whereas a negative angle indicates that the axis is pointed below the relative horizon.

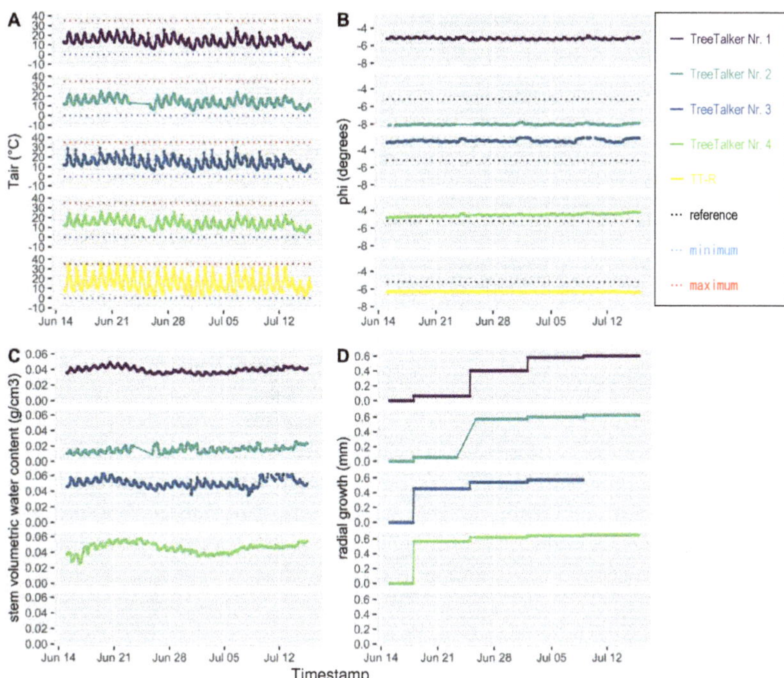

Figure 5. Output of the "ttalkR" plotting utility. (**A**) Air temperature with the reference minimum (dashed blue line) and maximum temperature (dashed red line); (**B**) inclination of the devices; (**C**) stem volumetric content; (**D**) radial growth. The example data refer to four TreeTalkers and a TT-R for the period between 14 June 2021 and 14 July 2021. In (**C,D**), the data from the TT-R are missing because it is a reference device outside of the canopy which is not equipped with the corresponding sensors.

4.7. Stem Volumetric Water Content

The water content in trees varies with diel and seasonal cycles, and it is a reservoir for transpiration [35], with sapwood being the most important storage site [36]. Because of the lack of experimental data, the most often-used models for investigating the water balance of vegetated regions do not take into account differences in plant water storage or their influence on the pathways of the transport in the soil–plant–atmosphere system [37]. TreeTalkers make use of a capacitive sensor (MicroPCB) with copper plates (TT+manual ver. 3.2, September 2020) for measuring stem volumetric water content and its dynamics. The method is based on frequency domain measurements and has been demonstrated to be effective for different tree species, but it requires species-specific calibration [38]. The function *ttStWC()* converts the frequency domain measurements into volumetric water content (Figure 5C) by adopting the calibration functions provided by [39]. Because of the necessary temperature dependence correction, we applied a Savytsky–Golay filter [25] to the temperature data in order to remove high-frequency components, and we used a linear interpolation for gaps up to 12 h.

4.8. Radial Growth

The growth of forests is affected by environmental factors [40], and it is a crucial ecophysiological parameter for quantifying the carbon sink of forests [41]. Radial stem growth occurs based on xylem increments on structures already formed, so trees increase in size with age. The function *ttGrowth()* makes use of data frames created by the *ttScrape()* function. It processes the data from GP2Y0A21 Sharp distance sensors. The sensors are deployed as point dendrometers. The distance sensor is positioned at a few centimeters (typically 3 to 4 cm) away from the tree trunk's surface and is kept in place by a carbon fiber stick anchored in the xylem. The function converts the digital numbers into distance (mm) with a second-degree polynomial regression model provided by the producer (TT+manual ver. 3.2, September 2020), and it applies a temporal averaging (median) on a weekly basis in order to remove the signal noise affecting the hourly measurements (Figure 5D).

4.9. Output

Interacting with databases via scripted languages has advantages over querying databases via a graphical user interface. In fact, data manipulations are preserved in the code, and the aggregates, summaries, and other database operations are not lost. As a result, those pre-analysis data manipulation steps are held and can be replicated. The *ttOutput()* function ingests the output from all the previous functions and creates a new database and an associated structure. In addition to the specific measured variables, each table in the database (Table 1) contains references to time and to a unique TreeTalker identifier (ID).

Table 1. Tables and content of the database created by means of the function *ttOutput()*.

Name	Content
df_ttBattery	timestamp, battery voltage, TreeTalker ID
df_ttGranier	timestamp, sap flow density, TreeTalker ID
df_ttLight	date, daily average counts per band, TreeTalker ID
df_ttRH	timestamp, relative humidity, TreeTalker ID
df_ttAir	timestamp, air temperature, TreeTalker ID
df_ttStability	timestamp, Tree stability, TreeTalker ID
df_ttSTtWC	timestamp, Stem volumetric water content, TreeTalker ID
df_ttGrowth	timestamp, Radial growth, TreeTalker ID

We selected the SQLite format because it is self-contained, stand-alone, and the recognized standard for storage and is therefore suitable for making the data accessible to a broad community for further processing and analysis. Additionally, this format is not software-specific, and it provides the benefits of an easy user setup and the absence of the need to configure or manage a server process.

5. Conclusions

In this article, we proposed and demonstrated the ttalkR package as the first step toward a unified TreeTalker data curation and, therefore, as a crucial advancement toward more formal time series analyses and data interpretation. The ttalkR package was first designed as a toolbox for assisting TreeTalker users in MRO activities and for the unified preprocessing of collected data to allow for cross-site analysis. The toolbox was planned as user-friendly and envisages scientists deploying TreeTalkers and pursuing data formatting for research purposes in a standardized but customizable fashion within the R programming language. The ttalkR package provides an approach to TreeTalker data curation by implementing a workflow for a unified conversion from raw numbers to physical units according to the TreeTalker user manual (TT+manual ver. 3.2, September 2020). The concept behind the package is modular, with I/O functions and parameter dedicated functions. We conceived the general package architecture to be adaptable to further hardware developments (i.e., sensor substitution and addition). At the same time, an open code provides possibilities for implementing advanced quality assurance algorithms and new conversion and calibration procedures. Additionally, we adopted a self-contained and stand-alone relational database as the final output format for facilitating data exchange. In the future, the "ttalkR" package can be extended by including the possibility to curate data from older and newer TreeTalker versions. Furthermore, a finer data elaboration level could be added to provide derived indexes and parameters, which will be helpful for forest modelling and the integration with complementary data streams.

In conclusion, our approach provides new opportunities for synthesis analyses based on the TreeTalker data from large-scale networks and their integration with other data streams such as meteorological information and earth observations. Such an approach is a first step for supporting the TreeTalker data life cycle and is suitable for making the collected data accessible. Ultimately, a unified approach for data curation will enable the exploitation of collected information for data-sound model development, the inference of forest attributes, and for addressing novel and broad research questions in macroecology. Yet, a unified processing of TreeTalker data will be the basis for a new environmental cyberinfrastructure across regional and possibly global research networks.

Author Contributions: Conceptualization, E.T.; methodology, E.T., L.B.M., A.Y., S.A. and R.V.; software, E.T.; funding acquisition, R.V.; drafting the article, E.T.; critical revision of the article, E.T., L.B.M., A.Y., S.A. and R.V. All authors have read and agreed to the published version of the manuscript.

Funding: This research was funded by the Italian Ministry for Instruction, University, and Research—Progetti di ricerca di Rilevante Interesse Nazionale (PRIN 2017), grant number 2017AAA8Z7 (The Italian TREETALKER NETWORK (ITT-Net): continuous large-scale monitoring of tree functional traits and vulnerabilities to climate change).

Data Availability Statement: The code used for this paper and the TreeTalker manual (TT+manual ver. 3.2, September 2020) are freely available at the repository https://github.com/EnricoTomelleri/ttalkR (version 1.0.0) (accessed on 24 April 2022).

Acknowledgments: The authors acknowledge the contribution of Giustino Tonon, who had led the germinal work and deceased on 7 July 2021. The authors thank the Department of Innovation, Research and University of the Autonomous Province of Bozen/Bolzano for covering the Open Access publication costs.

Conflicts of Interest: The authors declare no conflict of interest.

References

1. Allan, B.M.; Nimmo, D.G.; Ierodiaconou, D.; VanDerWal, J.; Koh, L.P.; Ritchie, E.G. Futurecasting Ecological Research: The Rise of Technoecology. *Ecosphere* **2018**, *9*, e02163. [CrossRef]
2. Ali, A.; Ming, Y.; Chakraborty, S.; Iram, S. A Comprehensive Survey on Real-Time Applications of WSN. *Future Internet* **2017**, *9*, 77. [CrossRef]
3. Ibarra-Esquer, J.; González-Navarro, F.; Flores-Rios, B.; Burtseva, L.; Astorga-Vargas, M. Tracking the Evolution of the Internet of Things Concept Across Different Application Domains. *Sensors* **2017**, *17*, 1379. [CrossRef] [PubMed]
4. Li, X.; Zhao, N.; Jin, R.; Liu, S.; Sun, X.; Wen, X.; Wu, D.; Zhou, Y.; Guo, J.; Chen, S.; et al. Internet of Things to Network Smart Devices for Ecosystem Monitoring. *Sci. Bull.* **2019**, *64*, 1234–1245. [CrossRef]
5. Ren, Y.; Zhang, X.; Lu, G. The Wireless Solution to Realize Green IoT: Cellular Networks with Energy Efficient and Energy Harvesting Schemes. *Energies* **2020**, *13*, 5875. [CrossRef]
6. Valentini, R.; Belelli Marchesini, L.; Gianelle, D.; Sala, G.; Yarovslavtsev, A.; Vasenev, V.; Castaldi, S. New Tree Monitoring Systems: From Industry 4.0 to Nature 4.0. *Ann. Silvic. Res.* **2019**, *43*, 84–88. [CrossRef]
7. Zweifel, R.; Etzold, S.; Basler, D.; Bischoff, R.; Braun, S.; Buchmann, N.; Conedera, M.; Fonti, P.; Gessler, A.; Haeni, M.; et al. TreeNet–The Biological Drought and Growth Indicator Network. *Front. For. Glob. Chang.* **2021**, *4*, 776905. [CrossRef]
8. Tognetti, R.; Valentini, R.; Marchesini, L.B.; Gianelle, D.; Panzacchi, P.; Marshall, J.D. Continuous Monitoring of Tree Responses to Climate Change for Smart Forestry: A Cybernetic Web of Trees. In *Climate-Smart Forestry in Mountain Regions*; Tognetti, R., Smith, M., Panzacchi, P., Eds.; Managing Forest Ecosystems; Springer International Publishing: Cham, Switzerlands, 2022; Volume 40, pp. 361–398, ISBN 978-3-030-80766-5. [CrossRef]
9. Porter, J.; Arzberger, P.; Braun, H.-W.; Bryant, P.; Gage, S.; Hansen, T.; Hanson, P.; Lin, C.-C.; Lin, F.-P.; Kratz, T.; et al. Wireless Sensor Networks for Ecology. *BioScience* **2005**, *55*, 561–572. [CrossRef]
10. Tomelleri, E.; Tonon, G. Linking Sap Flow Measurements with Earth Observations. In Proceedings of the 2021 IEEE International Geoscience and Remote Sensing Symposium IGARSS, Brussels, Belgium, 11 July 2021; pp. 6881–6884. [CrossRef]
11. Collins, S.L.; Bettencourt, L.M.; Hagberg, A.; Brown, R.F.; Moore, D.I.; Bonito, G.; Delin, K.A.; Jackson, S.P.; Johnson, D.W.; Burleigh, S.C.; et al. New Opportunities in Ecological Sensing Using Wireless Sensor Networks. *Front. Ecol. Environ.* **2006**, *4*, 402–407. [CrossRef]

12. Chen, C.-P.; Chuang, C.-L.; Jiang, J.-A. Ecological Monitoring Using Wireless Sensor Networks—Overview, Challenges, and Opportunities. In *Advancement in Sensing Technology*; Mukhopadhyay, S.C., Jayasundera, K.P., Fuchs, A., Eds.; Smart Sensors, Measurement and Instrumentation; Springer: Berlin/Heidelberg, Germany, 2013; Volume 1, pp. 1–21, ISBN 978-3-642-32179-5. [CrossRef]

13. Zorzi, I.; Francini, S.; Chirici, G.; Cocozza, C. The TreeTalkersCheck R Package: An Automatic Daily Routine to Check Physiological Traits of Trees in the Forest. *Ecol. Inform.* **2021**, *66*, 101433. [CrossRef]

14. Michener, W.K. Ecological Data Sharing. *Ecol. Inform.* **2015**, *29*, 33–44. [CrossRef]

15. Borgman, C.L.; Wallis, J.C.; Enyedy, N. Little Science Confronts the Data Deluge: Habitat Ecology, Embedded Sensor Networks, and Digital Libraries. *Int. J. Digit. Libr.* **2007**, *7*, 17–30. [CrossRef]

16. Schröder, W.; Nickel, S. Research Data Management as an Integral Part of the Research Process of Empirical Disciplines Using Landscape Ecology as an Example. *Data Sci. J.* **2020**, *19*, 26. [CrossRef]

17. Benedetti-Cecchi, L.; Bulleri, F.; Dal Bello, M.; Maggi, E.; Ravaglioli, C.; Rindi, L. Hybrid Datasets: Integrating Observations with Experiments in the Era of Macroecology and Big Data. *Ecology* **2018**, *99*, 2654–2666. [CrossRef] [PubMed]

18. Peters, D.P.C.; Loescher, H.W.; SanClements, M.D.; Havstad, K.M. Taking the Pulse of a Continent: Expanding Site-Based Research Infrastructure for Regional- to Continental-Scale Ecology. *Ecosphere* **2014**, *5*, art29. [CrossRef]

19. Heiskanen, J.; Brümmer, C.; Buchmann, N.; Calfapietra, C.; Chen, H.; Gielen, B.; Gkritzalis, T.; Hammer, S.; Hartman, S.; Herbst, M.; et al. The Integrated Carbon Observation System in Europe. *Bull. Am. Meteorol. Soc.* **2022**, *103*, E855–E872. [CrossRef]

20. Buonocore, L.; Yates, J.; Valentini, R. A Proposal for a Forest Digital Twin Framework and Its Perspectives. *Forests* **2022**, *13*, 498. [CrossRef]

21. Marcu, A.-E.; Suciu, G.; Olteanu, E.; Miu, D.; Drosu, A.; Marcu, I. IoT System for Forest Monitoring. In Proceedings of the IEEE 2019 42nd International Conference on Telecommunications and Signal Processing (TSP), Budapest, Hungary, 1–3 July 2019; pp. 629–632.

22. Cui, F. Deployment and Integration of Smart Sensors with IoT Devices Detecting Fire Disasters in Huge Forest Environment. *Comput. Commun.* **2020**, *150*, 818–827. [CrossRef]

23. Marshall, D.C. Measurement of Sap Flow in Conifers by Heat Transport. *Plant Physiol.* **1958**, *33*, 385–396. [CrossRef]

24. He, H.; Turner, N.C.; Aogu, K.; Dyck, M.; Feng, H.; Si, B.; Wang, J.; Lv, J. Time and Frequency Domain Reflectometry for the Measurement of Tree Stem Water Content: A Review, Evaluation, and Future Perspectives. *Agric. For. Meteorol.* **2021**, *306*, 108442. [CrossRef]

25. Savitzky, A.; Golay, M.J.E. Smoothing and Differentiation of Data by Simplified Least Squares Procedures. *Anal. Chem.* **1964**, *36*, 1627–1639. [CrossRef]

26. Granier, A. Evaluation of Transpiration in a Douglas-Fir Stand by Means of Sap Flow Measurements. *Tree Physiol.* **1987**, *3*, 309–320. [CrossRef] [PubMed]

27. Hovi, A.; Rautiainen, M. Spectral Composition of Shortwave Radiation Transmitted by Forest Canopies. *Trees* **2020**, *34*, 1499–1506. [CrossRef]

28. De Frenne, P.; Zellweger, F.; Rodríguez-Sánchez, F.; Scheffers, B.R.; Hylander, K.; Luoto, M.; Vellend, M.; Verheyen, K.; Lenoir, J. Global Buffering of Temperatures under Forest Canopies. *Nat. Ecol. Evol.* **2019**, *3*, 744–749. [CrossRef]

29. Matasov, V.; Belelli Marchesini, L.; Yaroslavtsev, A.; Sala, G.; Fareeva, O.; Seregin, I.; Castaldi, S.; Vasenev, V.; Valentini, R. IoT Monitoring of Urban Tree Ecosystem Services: Possibilities and Challenges. *Forests* **2020**, *11*, 775. [CrossRef]

30. Gennari, E.; Latterini, F.; Venanzi, R.; Monaco, A.L.; Picchio, R. Single Tree Stability Assessment in Beech High Forest and Factors That Could Induce Windbreak. *Environ. Sci. Proc.* **2020**, *3*, 60. [CrossRef]

31. Van Haaften, M.; Liu, Y.; Wang, Y.; Zhang, Y.; Gardebroek, C.; Heijman, W.; Meuwissen, M. Understanding Tree Failure—A Systematic Review and Meta-Analysis. *PloS ONE* **2021**, *16*, e0246805. [CrossRef]

32. Van Emmerik, T.; Steele-Dunne, S.; Hut, R.; Gentine, P.; Guerin, M.; Oliveira, R.; Wagner, J.; Selker, J.; van de Giesen, N. Measuring Tree Properties and Responses Using Low-Cost Accelerometers. *Sensors* **2017**, *17*, 1098. [CrossRef]

33. Moore, J.R.; Maguire, D.A. Natural Sway Frequencies and Damping Ratios of Trees: Concepts, Review and Synthesis of Previous Studies. *Trees-Struct. Funct.* **2004**, *18*, 195–203. [CrossRef]

34. Fisher, C.J. *Using an Accelerometer for Inclination Sensing*; Appl. Note Analog Devices Inc.: Norwood, MA, USA, 2010; Volume 1057, pp. 1–8.

35. Waring, R.H.; Running, S.W. Sapwood Water Storage: Its Contribution to Transpiration and Effect upon Water Conductance through the Stems of Old-Growth Douglas-Fir. *Plant Cell Environ.* **1978**, *1*, 131–140. [CrossRef]

36. Cermak, J.; Kucera, J.; Bauerle, W.L.; Phillips, N.; Hinckley, T.M. Tree Water Storage and Its Diurnal Dynamics Related to Sap Flow and Changes in Stem Volume in Old-Growth Douglas-Fir Trees. *Tree Physiol.* **2007**, *27*, 181–198. [CrossRef] [PubMed]

37. Araújo, G.P.; Vellame, L.M.; Costa, J.A.; Costa, C.A.G. A Low-Cost Monitoring System of Stem Water Content: Development and Application to Brazilian Forest Species. *Smart Agric. Technol.* **2021**, *1*, 100012. [CrossRef]

38. Matheny, A.M.; Garrity, S.R.; Bohrer, G. The Calibration and Use of Capacitance Sensors to Monitor Stem Water Content in Trees. *J. Vis. Exp.* **2017**, *130*, 57062. [CrossRef] [PubMed]

39. Asgharinia, S.; Belelli Marchesini, L.; Gianelle, D.; Valentini, R. Design and Performance Evaluation of Internet of Things (IoT) Based Multifunctional Device for Plant Ecophysiology & Hydrology: Toward Stem Water Content & Sap Flow. In Proceedings of the EGU General Assembly Conference Abstracts, Online, 4–8 May 2020. [CrossRef]

40. Zweifel, R. Radial Stem Variations-a Source of Tree Physiological Information Not Fully Exploited Yet. *Plant Cell Environ.* **2016**, *39*, 231–232. [CrossRef] [PubMed]
41. Martinez del Castillo, E.; Zang, C.S.; Buras, A.; Hacket-Pain, A.; Esper, J.; Serrano-Notivoli, R.; Hartl, C.; Weigel, R.; Klesse, S.; Resco de Dios, V.; et al. Climate-Change-Driven Growth Decline of European Beech Forests. *Commun. Biol.* **2022**, *5*, 163. [CrossRef]

Article

Design and Testing of a Novel Unoccupied Aircraft System for the Collection of Forest Canopy Samples

Sean Krisanski [1,2,*], Mohammad Sadegh Taskhiri [1,2,3], James Montgomery [2] and Paul Turner [1,2]

[1] Australian Research Council—Training Centre for Forest Value, University of Tasmania, Tasmania 7001, Australia; mohammadsadegh.taskhiri@utas.edu.au (M.S.T.); paul.turner@utas.edu.au (P.T.)

[2] School of Information and Communication Technology, University of Tasmania, Tasmania 7001, Australia; james.montgomery@utas.edu.au

[3] Institute for Sustainable Industries & Liveable Cities (ISILC), Victoria University, Melbourne, VIC 8001, Australia

[*] Correspondence: sean.krisanski@utas.edu.au

Citation: Krisanski, S.; Taskhiri, M.S.; Montgomery, J.; Turner, P. Design and Testing of a Novel Unoccupied Aircraft System for the Collection of Forest Canopy Samples. *Forests* **2022**, *13*, 153. https://doi.org/10.3390/f13020153

Academic Editors: Stelian Alexandru Borz, Andrea R. Proto, Robert Keefe, Mihai Nita and Nikolay Strigul

Received: 31 October 2021
Accepted: 18 January 2022
Published: 20 January 2022

Publisher's Note: MDPI stays neutral with regard to jurisdictional claims in published maps and institutional affiliations.

Abstract: Unoccupied Aircraft Systems (UAS) are beginning to replace conventional forest plot mensuration through their use as low-cost and powerful remote sensing tools for monitoring growth, estimating biomass, evaluating carbon stocks and detecting weeds; however, physical samples remain mostly collected through time-consuming, expensive and potentially dangerous conventional techniques. Such conventional techniques include the use of arborists to climb the trees to retrieve samples, shooting branches with firearms from the ground, canopy cranes or the use of pole-mounted saws to access lower branches. UAS hold much potential to improve the safety, efficiency, and reduce the cost of acquiring canopy samples. In this work, we describe and demonstrate four iterations of 3D printed canopy sampling UAS. This work includes detailed explanations of designs and how each iteration informed the design decisions in the subsequent iteration. The fourth iteration of the aircraft was tested for the collection of 30 canopy samples from three tree species: eucalyptus pulchella, eucalyptus globulus and acacia dealbata trees. The collection times ranged from 1 min and 23 s, up to 3 min and 41 s for more distant and challenging to capture samples. A vision for the next iteration of this design is also provided. Future work may explore the integration of advanced remote sensing techniques with UAS-based canopy sampling to progress towards a fully-automated and holistic forest information capture system.

Keywords: canopy; drone; leaf; leaves; foliar; samples; sampling; Aerial robotics; UAS; UAV

1. Introduction

Climate change is having a complex variety of effects on our forests from increased atmospheric carbon dioxide levels [1,2], environmental changes such as increasing drought severity and frequency [3–5], and more frequent and severe bushfires [6]. In some cases, local environmental changes are becoming sufficiently persistent and significant enough to shift conditions beyond the tolerable limits of some species, causing the large scale loss of forests and even threatening some species with extinction without assisted migration [7,8]. Scalable and high-fidelity measurements are of considerable importance to furthering our understanding of these changing conditions and their associated impacts on our forests. Forest information will play an important role in enabling evidence-based policy decisions to be made regarding the mitigation of and adaptation to such climate impacts. The enhancement of the tools available for sampling and monitoring our forests will enable larger scale and lower cost collection of forest information.

Unoccupied Aircraft Systems (UAS), remote-sensing and deep-learning technologies have been revolutionising the way we can monitor the structure of forests and quantify carbon stores [9–18] for use in climate models; however, physical samples remain important for calibrating some remote sensing techniques [19,20], directly measuring foliar nutrients, collecting genetic samples, monitoring pests/diseases and studying physical plant traits. Canopy and sub-canopy physical samples can be used to answer questions related to plant, ecosystem, and environmental health. These samples can provide a valuable source of feedback to forest growers and researchers to further optimize and inform the management of our forest resources, but obtaining these samples remains a challenge.

Canopy samples are typically collected with considerable effort through the use of canopy cranes [21–23], arborists, shotguns, crossbows [24,25], line launchers and pole pruners [26]. These techniques can be time-consuming, expensive and in some cases dangerous. Aerial robots/UAS have the potential to help us obtain these samples more safely, cheaply, and rapidly, but to do so, they need to be able to physically interact with trees. Promising research is ongoing to develop platforms such as [27–31] to enable UAS to precisely interact with objects; however, further research is needed to bring this technology to forest sampling. Such robotic systems present an interesting opportunity to improve the way we collect samples and physically interact with forest canopies.

The canopy sampling or pruning designs in the literature fit into three main categories. The most common design category involves a sampling tool that hangs underneath the UAS on a long pole [32–38]. This approach is most suitable for sampling the tops of canopies and is unaffected by a closed canopy, provided that the local UAS regulations around line-of-sight operations and field conditions make this possible. The second most common design category includes those which are lateral reaching [39–42], i.e., the sampling tool protrudes in front of the aircraft rather than hanging beneath it. These systems approach trees from the side rather than from above, so they are able to access locations that the hanging designs cannot reach, such as those which are not at the very top of the tree or locations beneath a closed canopy. Such an approach ideally needs a counterweight to ensure the centre of gravity remains in line with the centre of thrust to avoid wasted performance, instability and other control related issues. Lastly, there is an example of a UAS which approaches from underneath branches, with the cutting tool above it [43]. The latter system is intended for pruning rather than sampling, however, it would still enable leaf sampling and is therefore relevant to this discussion. This design is unable to sample from the top or the side, requiring a vegetation-free section of the branch to fly underneath and hook onto. The tools used to cut the sample or branch by all previous studies identified included circular saws [32–34,41,43], electrically powered [39,42] or spring-loaded secateurs [35] or a simple razor blade [40].

Two studies were particularly noteworthy for their use of computer vision to simplify the task of choosing a branch and making the cut. One demonstration uses machine learning and computer vision to identify stems and branches in real-time and robotically cut them with a secateur style pruning tool [42,44]. Another more recent study used computer vision techniques on a depth image to select a target branch and actively assist the pilot in collecting the sample [37].

With the exception of [42], which had a small propeller guard at the front of the aircraft, and [40], which used off-the-shelf propeller guards, collision tolerance was not present in the existing designs. While this is not a problem for hanging designs, lateral reaching designs are frequently in close proximity to the canopy when cutting/collecting a sample, which puts the aircraft at risk of colliding propellers with branches/leaves and crashing. The two above-mentioned systems with propeller guards may have had some protection from stems and other solid vertical surfaces; however, these systems would not have protection from branches and leaves, which tend to move and do not provide much resistance when flying into them. Collision tolerance which can stop such a lateral reaching UAS from flying too deeply into the canopy would be useful in preventing crashes and would reduce the stress of the sampling operation on the pilot; however, a large surface area is needed to provide sufficient resistance to prevent flying deep into the canopy. Providing such a large surface area for collision presents challenges with regard to aircraft weight and aerodynamics.

In this article, we describe our approach to rapidly prototype a purpose-built UAS for the collection of canopy samples, which was collision tolerant and able to collect samples from the side of trees. Four novel design iterations are presented, with design decisions and justifications described throughout. The designs we present are not intended to replace hanging sampler designs, which are a very logical approach for capturing samples from the top of the canopy. Instead, the goal was to provide the capability of capturing samples from locations from which a hanging design cannot reach, such as from underneath a canopy, from the side of a canopy, inside a forest with a closed canopy or from the side of vertical or overhanging cliff faces.

2. Materials and Methods

The structure of the methodology is as follows: first, the common system components and ideas used in all iterations of the canopy sampling UAS are described. The design process in this study was highly iterative and made extensive use of 3D printing in order to rapidly prototype and test new ideas as different challenges were identified through field testing. In this article, each iteration is presented in order of creation, with lessons learnt from the first design informing the design decisions made for the subsequent design iterations. Lastly, the details of the field testing to evaluate the utility of the approach are described, along with a discussion of relevant safety considerations within the context of existing sampling techniques.

2.1. Overall System Design Considerations and Components

As hanging designs are generally limited to sampling the very tops of trees, and as there are already successful designs for this task such as DeLeaves [32], our design was intended to access samples that would not be accessible with a hanging design, so the first major design goal was to develop a lateral reaching design which was capable of tolerating collisions with the canopy. Further, by having the propellers in line with the sampling tool, the considerable prop-wash does not blow directly into the sample to be captured. If the sample was beneath this aircraft, it would always be a moving target, though a sufficiently long pole can reduce this effect, as demonstrated in other designs.

Lateral reaching designs result in a relatively heavy sampling tool being extended far in front of the centre of gravity, so it is necessary to counterbalance this offset mass to maintain alignment of the centre of gravity with the centre of thrust. It was noted in [32] that there was a need for a counterweight in a lateral reaching design, and this is generally correct; however, it is important to note that the counterweight does not need to be dead weight and does not necessarily reduce the flight time due to weight. Most UAS are already carrying a relatively heavy battery, so it makes sense to use this and/or other components already required by the aircraft as the counterweight for the lateral payload. While this avoids unnecessary weight, it must be acknowledged that a lateral design also means an increase in the moment of inertia about the pitch axis, which does require more energy for a given attitude change. An example of how this counterweight is employed is depicted in Figure 1. Any collisions between the vegetation and the propellers can quickly result in a crash, so a design goal was to keep the sampling tool as far forward as practicable, and therefore keep the vegetation away from the propellers.

Keeping the sampling tool as far in front of the centre of gravity as possible is the first step for avoiding propeller–branch collisions; however, it has limited efficacy on its own. Intelligent sensing of the surrounding environment in order to avoid propeller-branch collisions would be a solution; however, reliable and precise detection and avoidance of small, dynamic obstacles (such as leaves and branches) remains a particularly challenging robotics problem at this time. Consequently, the system was designed to passively tolerate minor collisions with branches rather than attempt to avoid them entirely. The primary risk of these low-speed and minor collisions to the aircraft is the propellers striking branches and being slowed or stopped as a result. This can lead to a loss of thrust/control of that motor, often causing the aircraft to propel itself further into the branch and resulting in the loss of the aircraft. While the UAS designs presented in this work were able to descend safely in the event of losing a single motor, performance is considerably degraded if this occurs.

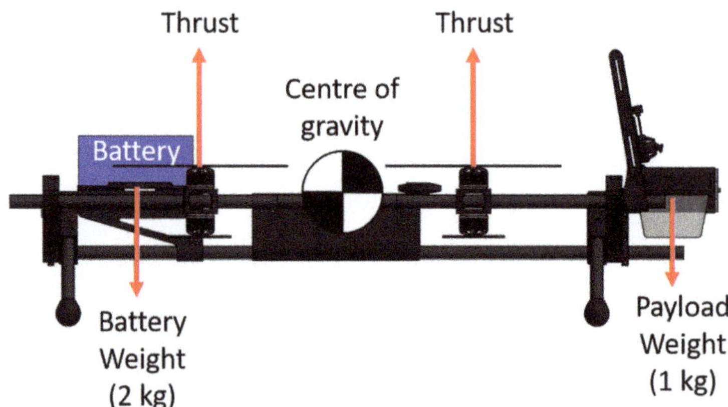

Figure 1. Our designs compensate for the offset payload mass by using the battery as a counterweight. With the battery weighing approximately 2 kg and the payload weighing approximately 1 kg, the payload can be further from the centre of mass than the battery, keeping the sampling operation as far in front of the propellers as practicable.

All iterations of the aircraft used the same base components as described in Table 1, and all used the battery as a counterweight to the sampling payload to ensure that the center of gravity was aligned with the center of thrust. The power components such as the battery, motors and propellers were not our first component choices, as the COVID-19 pandemic had impacted supply chains and greatly limited part availability for this project. A smaller aircraft was preferable, but smaller, suitable combinations of motors, propellers and batteries were unable to be acquired during the project. The resulting airframe was a coaxial octocopter to keep the footprint of the aircraft as small as possible while providing enough thrust to carry the cutting tool payloads.

Table 1. Common UAS components used for all iterations of this system.

Component Type	Model
Flight Controller	Holybro Pixhawk 4—PX4 Firmware
Motors	T-Motor MN4014-11, 330 KV
Propellers	Carbon fibre 431.8 mm diameter, 140 mm pitch
Battery	Turnigy 16 Ah, 6S (Lithium Polymer)
Airframe	Custom 3D printed designs with carbon fibre tubing. Printed using Polylactic Acid (PLA) filament for all components except the motor mounts, which were printed in Polyethylene terephthalate glycol (PETG) for the higher glass transition temperature underneath the motors. Past experience has found that PLA motor mounts can soften and fail catastrophically from motor heat loads.
Companion Computer	Nvidia Jetson Tegra X2 (Nvidia, Santa Clara, California, United States of America) with Auvidea J120 development board (Auvidea, Denklingen, Germany). Note: used from version 3.0 onwards.
Visual Inertial Odometry Sensor	Intel Realsense T265 (Intel, Santa Clara, California, United States of America). Note: used from version 3.0 onwards.

Many prior works made use of entirely custom tooling for the sampling operation; however, our approach was to purchase existing, low-cost power tools and modify them for the task. The penalty of this approach was the additional weight when compared to a custom tool, but the advantages included the ability to rapidly change the approach while also saving considerable engineering effort, time and manufacturing costs during the development of a suitable approach.

2.2. Version 1—Full Collision Box with Electric Secateurs

The first design iteration involved a cage-like structure that was integrated into the airframe, with the intent to cover this in tough plastic meshing. The sampling tool consisted of an electrically powered secateur tool (Ryobi 18 Volt One+ Lopper, https://www.ryobitools.com/outdoor/products/details/18v-one-plus-lopper, last accessed on 10 March 2021) which consisted of sharp jaws driven by a linear actuator and was capable of cutting branches of up to 30 mm in diameter. To adapt the tool for this application, a Pulse Width Modulation (PWM) controlled relay was wired in parallel with the tool's trigger switch, allowing the tool to be controlled using the radio control (RC) transmitter. This version was successful in trimming off sample branches; however, it was difficult to aim due to the small cutting area of the tool. While it was possible to guide the tool into place with rods protruding from the front of the tool [35,39], it was decided that the retrieval of the sample would be greatly preferable to dropping the sample, as it may be challenging to find the trimmed branch, or it may not even fall all the way to the ground. This first version is shown in Figure 2.

Figure 2. Version 1 was most similar to [39]. The first sampling approach used powered secateurs to drop branches of up to 30 mm in diameter to the ground for retrieval. This required precise aiming and flight control during cutting, and while it was functional, a more practical approach that required less precise flight control was sought.

In addition to being challenging to aim, this design was limited to branches with sufficient space for the UAS to fly alongside it with the tool perpendicular to the branch to make the cut. In practice, meeting this requirement was found to be more difficult than initially expected. Further, triggering the cutting mechanism while a branch was in the jaws required both good timing and strong pilot skills; both undesirable traits for a system that needs to be simple to use. Figure 3 show this version trimming a branch sample, though this can be more clearly seen in the provided video in the results section.

Figure 3. Version 1 of the canopy sampling drone trimming off a sample branch. This design was functional but difficult to aim, and it dropped the sample; resulting in the pursuit of a different approach.

While the secateur-based approach did work, and there are other examples of a similar approach being used [35,39,42], it was decided that an alternative approach that required less precise flight control was desirable; especially one with the means to retrieve samples instead of dropping them to the ground (where they may be difficult to identify). To reduce the difficulty of the sample capture process, a simpler approach was sought.

2.3. Version 2—Full Collision Box with Electric Hedge Trimmer

A number of tools were considered, such as small chainsaws and circular saws; however, reacting to the forces required to operate these tools effectively was of concern. Further, it was preferable to avoid the risks associated with getting a saw or secateur-based tool stuck in a branch. While hanging designs such as DeLeaves [32] are able to be released in an emergency, a lateral reaching design does not easily allow for this capability. A single-handed electric hedge trimmer was considered instead, which uses a shearing method of cutting which reacts with its own cutting forces. Further, the blades are designed to limit the size of the branches that can enter the cutting region to a size that should be safely cuttable by the tool: minimising the risks of a stuck tool while sampling. A hedge trimmer also forces the samples to be accumulated slowly and gently into the sample container, minimising the risk of a sample being too heavy (potentially catching the operator by surprise) and shifting the centre of gravity too far from the centre of thrust or overcoming the maximum weight limit of the aircraft.

A low-cost, cordless hedge trimmer (Ryobi 18 Volt "Shrubber" (https://www.ryobitools.com/products/details/18v-one-plus-grass-shear-and-shrubber, last accessed on 10 March 2021), was disassembled and placed into a custom, 3D printed housing with a removable sampling container. As this device did not require any logic-based control, the original electronics were not retained, being replaced with a simple PWM controlled relay for controlling the motor. Figure 4 show this hedge trimmer-based sampling payload. A safety switch was also added to allow the operator to manually prevent the tool from activating unexpectedly, though this aircraft should not be manually handled while powered.

Figure 4. A single-handed hedge trimmer was modified and integrated into a custom 3D printed housing with a removable sample container. A manual safety switch is shown in the top-right of the left image, which allows the operator to prevent the tool from running unexpectedly. Right shows a camera view of the tool while in flight, about to collect a canopy sample.

This payload was designed to easily fit onto the existing frame in place of the previously tested tool. From the first test of this alternative approach, it was clear that this was considerably easier to operate and more effective than the secateur-style approach. The second sample collection test with this system is shown in Figure 5.

Figure 5. Version 2 of the canopy sampling drone made use of a hedge trimmer-based sampling tool. This tool was considerably easier to operate than the previous, secateur-based design. Left and right images show the before and after the collection of a canopy sample, respectively.

The sampling operation requires the front of the aircraft to be flown underneath a branch of interest, then hold a horizontal position while gently raising altitude until the hedge trimmer cuts through the required sample. The collision protecting mesh had not yet been installed on Version 2, so additional care was required not to fly the tool too deep into the canopy, or a crash would be almost guaranteed. The operation of the sampling tool is visualised in Figure 6.

Figure 6. A visualisation and photos showing how canopy samples are collected using this hedge trimmer based approach. Holding altitude, the UAS is piloted gently forward into the vegetation until the collision shield prevents forward motion. The pilot then holds horizontal position while gently raising the altitude until the UAS has cut through to the top of the sample region. The sample will fall into the sample container for retrieval upon landing. Version 3.1 is visualised here due to better photos of operation, but all hedge trimmer based designs in this study used this same sampling approach.

Version 2 has successfully demonstrated the approach was feasible; however, the size of the aircraft required a large vehicle for transportation. Thus a foldable and more portable design was sought.

2.4. Version 3.0—Foldable Airframe with Ducted Fans

Version 3.0 made use of 3D printed ducts to provide protection to the propellers while also allowing the aircraft to fold in half for ease of transport, shown in Figure 7. The eCalc multirotor design tool [45] was used as a tool for estimating the performance of a potential UAS configuration; however, the predictions were found to be too optimistic for this configuration. This version was too heavy for the components used, and while able to hover, was unable to fly out of ground effect, which was approximately 1 m above the ground. The propellers used in this configuration were smaller (407 mm diameter) than the rest of the configurations, as the duct diameter was limited by the print volume of the available 3D printer.

The ducts on this system weighed 600 g each, so these were removed to enable the use of the larger diameter propellers and a considerable mass reduction of 2.4 kg.

Figure 7. Version 3.0 protected the propellers with the use of 3D printed ducts. The design was capable of folding for ease of transport. The calculations used for estimating flight performance were found to be optimistic, with the system unable to fly out of ground effect.

2.5. Version 3.1—Foldable with Collision Shield

Full protection of the propellers remains highly desirable; however, it is the front of the aircraft at the greatest risk of collisions with leaves/branches during the sampling operation. Therefore, a lightweight (approximately 300 g), forward-facing shield was implemented in place of the ducts on the same foldable airframe as shown in Version 3.0. The shield is angled backwards to minimise the risk of the top of the shield catching on a branch while cutting a sample. The offset mass of the shield is counterbalanced by adjusting the battery position to keep the centre of gravity in line with the centre of thrust. This version is shown in Figure 8.

Figure 8. Version 3.1 of the canopy sampling UAS replaced the heavy, ducted fan system with a simple and lightweight, forward-facing shield to prevent collisions between the propellers and the canopy.

The first two iterations required completely manual control of the aircraft during sampling; however, this required considerably more pilot skill and concentration to operate than most modern UAS due to a lack of a precise position holding mode/capability. A means of precise movement control was desired to assist the pilot, so Visual Inertial Odometry (VIO) and a position hold mode which made use of this, was added to the system. VIO was provided by an Intel Realsense T265, which uses a stereo camera and Inertial Measurement Unit (IMU) to precisely track the position and orientation (pose) of the aircraft. An Nvidia Tegra X2 companion computer running Robot Operating System (ROS) [46] provided this information to the Pixhawk 4 flight controller, which used this information for holding a set position.

The VIO sensor was mounted on the rear of the aircraft, as the front of the aircraft was to be intentionally flown into vegetation, which would block the cameras and prevent useful visual tracking. Downward facing VIO was also briefly tested; however, frequently lost position tracking. It was suspected that this was a result of the vegetation beneath this aircraft moving erratically in the propeller wash of the aircraft. The sensor was initially mounted rigidly; however, the frame vibrations from propellers and the hedge trimmer were too severe for the sensor to function correctly, causing a loss of position tracking. This sensor was highly sensitive to vibration and had to be soft mounted with vibration-damping, double-sided tape. The VIO sensor can be seen on the back of the aircraft in Figure 9, with the companion computer hardware inside the centre of the aircraft.

Figure 9. Visual Inertial Odometry (VIO) was used to provide precise flight control to Version 3.1 of the UAS. The VIO sensor was an Intel Realsense T265 and was soft mounted on the rear of the aircraft.

Tuning the flight controller to safely use VIO was particularly challenging for this aircraft, as the large moment of inertia about the pitch axis caused the pitch response to be sluggish. While it could be tuned for stable flight in still conditions, even gentle gusts of wind would cause oscillations about the pitch axis, compromising the efficacy of the sampling operation, even in a gentle breeze. While any breeze is undesirable during a canopy sampling operation due to the canopy becoming a moving target, this design placed too strict of a requirement on the absence of a breeze. To address this, a way of decoupling the pitch of the airframe (for control) from the large moment of inertia about the pitch axis for the payload was sought.

2.6. Version 4—Pitch-Decoupled Hedge Trimmer and Battery

There was no requirement for the battery or sampling payload to be rigidly connected to the airframe, so a novel approach of decoupling the pitch axis of the battery and hedge trimmer from the main airframe was designed. The goal was to allow the aircraft to respond to disturbances (i.e., wind and physical interactions with the canopy) rapidly in the pitch axis while avoiding the need to rotationally accelerate and decelerate the heaviest components of the aircraft, which may also be restrained by the canopy being sampled. This joint should constrain yaw and roll movements; however, the system should also keep a short lever arm for the mass in the roll axis (i.e., keep the roll moment of inertia small). Further, the hedge trimmer should ideally remain level during sampling. The chosen solution was a pinned joint from which the battery and hedge trimmer could hang from, as depicted in Figure 10.

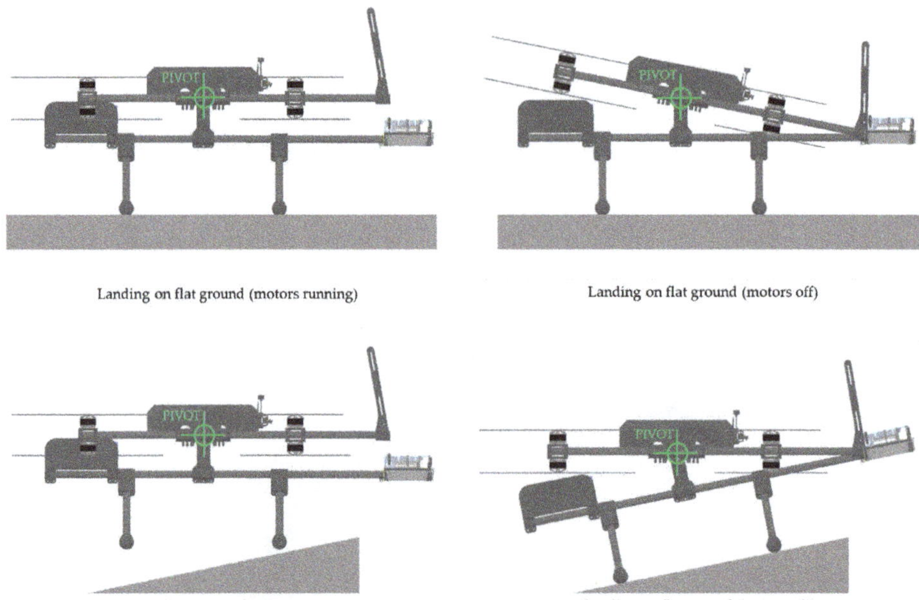

Figure 10. The battery, landing gear and hedge-trimmer sampling system were decoupled in the pitch axis from the flight controller and motors. The result was a highly responsive aircraft with the additional benefit of allowing landing on sloped ground. The prior version required flat ground, which was difficult to come by in the chosen test site.

While designing this aircraft to decouple the main airframe from the payload, an opportunity to provide the aircraft with passive adaptive landing gear was identified and implemented. By putting the landing gear beneath the pivot point, the aircraft was able to safely take-off and land on sloped surfaces, facing either up-slope or down-slope, removing the strict requirements of the previous iterations for a flat landing site. All four legs were able to make contact with a sloped surface while maintaining a vertical thrust vector.

In practice, this idea was found to be highly effective, albeit slightly unusual to pilot, as such a large portion of the aircraft does not respond directly to the control inputs as would typically be expected on a more conventional multirotor UAS. The system was able to land facing upslope on a surface of approximately 25 degrees from the horizontal and downslope to approximately a 20-degree slope while maintaining a vertical thrust vector. The difference is due to the offset mass of the collision shield on the main airframe, which shifts the centre of gravity forward. A sequence of taking off and landing from upward and downward sloped surfaces is shown in Figure 11.

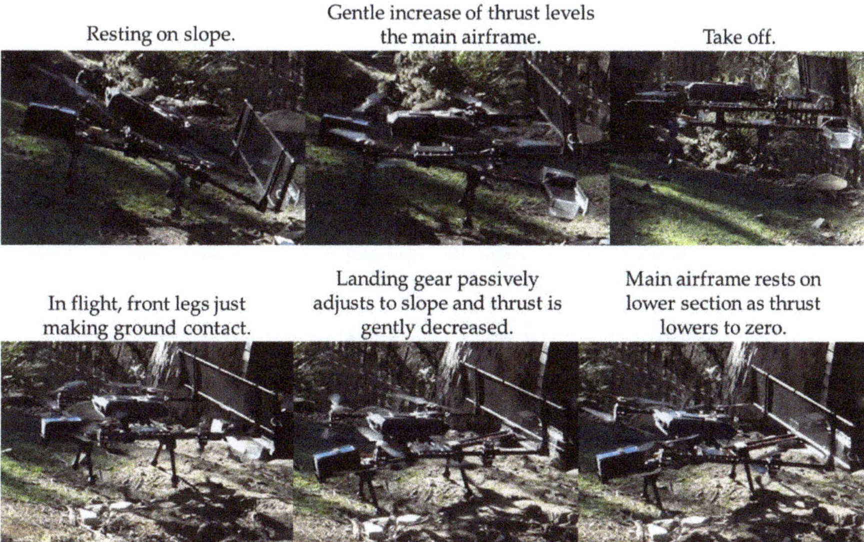

Figure 11. Real-world testing of the landing gear concept found it was highly effective on both upward and downward slopes, albeit unusual to pilot due to the large portion of the aircraft not responding directly to the control input.

While the flight controller assumes rigid body motion, the controller caused no issues in the slow control regime this aircraft was intended for; provided the Proportional, Integral, Derivative (PID) controllers were appropriately tuned. The lower section does act as a pendulum; however, the slow natural frequency meant that it was not an issue in practice. That said, the pilot is required to fly gently, as aggressive flying will cause the lower section to oscillate and potentially reach the angle limit stops (±30 degrees), imparting an undesirable pitching moment on the main airframe.

An additional challenge with lateral reaching canopy sampling UAS which collect samples is that the sample weight shifts the centre of gravity forward, leading to reduced performance and potentially leading to instability. With this pitch-decoupled design, the lower section passively adjusts the weight distribution as samples are captured, keeping the centre of gravity in line with the centre of thrust at all times, an important consideration for any UAS.

2.7. Safety Considerations

Throughout discussion with members of the forestry industry and public, a frequently raised concern about a canopy sampling UAS with a power tool is that it is perceived to be a particularly dangerous creation. These concerns are often raised by people who are not operators of UAS, and perhaps underestimate the severity of the hazard already presented by the propellers on any UAS. While power tools must certainly be respected, we argue that the propellers of any large UAS represent a greater hazard than a hedge trimmer; particularly if someone was particularly cautious of the hedge trimmer but perhaps not so concerned about the propellers. Adult fingers would be unable to fit into the cutting area of this hedge trimmer, and with the exception of a high-energy collision with the tool, the hedge trimmer would be unlikely to cause severe injuries even with skin contact. Large carbon-fibre propellers, on the other hand, contain considerable rotational kinetic energy, are almost invisible when in operation, and may cause large and severe lacerations or even amputations. Thus, we argue that this UAS is no more dangerous than any other similarly sized UAS. That said, any physical interaction with the canopy using a UAS does constitute a greater risk of a crash than one used for remote sensing applications, so it must be operated with care and by an appropriately skilled pilot.

The risks associated with UAS based sampling must also be considered within the context of other, more conventional canopy sampling techniques. The use of pole saws can involve the risk of the cut branch falling on the operator. The use of arborists for sample collection may involve the use of chainsaws at height, while the rope access itself comes with potentially fatal consequences in the event of an accident. Discharging firearms into the air to knock down samples comes with the risk of missing a targeted branch, and while this has a low probability of hitting a bystander, the consequences of which could be fatal. When operated with appropriate skill and caution, and within UAS operational laws, a UAS is highly unlikely to collide with a human, even in the event of a crash since the physical interaction with the canopy, the most likely cause of a crash in the operation, occurs at a considerable distance from the operator.

2.8. Test Site

This study took place in a private forest near Hobart, Tasmania, Australia. The sampled trees consisted of eucalyptus pulchella, eucalyptus globulus and acacia dealbata. The site is steep, with an average slope of approximately 24 degrees.

2.9. Sample Collection Test

To understand the range of sampling times for this system, 30 samples were collected using the final iteration (Version 4) of the aircraft. All samples were within a 50 m horizontal radius from the pilot in these tests, as an unaided line of sight was required to be maintained at all times for both operational reasons and to ensure compliance with Australian UAS regulations. Sampling times were measured using the transmitter/ground station, with a timer automatically starting upon arming and finishing upon disarming the aircraft. An additional 5–10 min was required for both set up and pack down at each field site; however, the typical use case would involve the collection of multiple samples per site.

2.10. Demonstration Video

To demonstrate the effectiveness of the presented systems, a video demonstration of all iterations of the aircraft (excluding Version 3.0) is provided.

3. Results

3.1. Sample Collection Test

Sampling times ranged from 1 min and 23 s, up to 3 min and 41 s, with the main factor being proximity to the pilot. The mean and median sampling times were 2 min, 25 s and 2 min, 20 s, respectively. The distribution of the sampling times is shown in Figure 12.

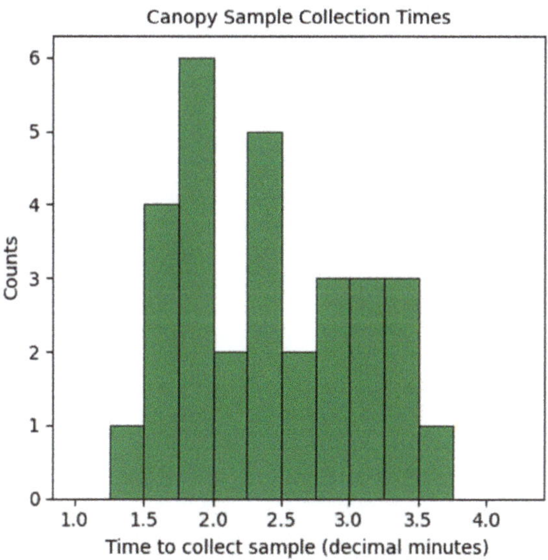

Figure 12. The distribution of the times taken to collect a canopy sample from arming to disarming the aircraft (*n* = 30). An additional 5–10 min is required at the start and end of any sampling session to set up and pack away the equipment.

Ten of the canopy samples collected during this testing are shown in Figure 13.

Figure 13. A collection of samples captured using Version 4 of our canopy sampling UAS. Samples were collected from eucalyptus pulchella, eucalyptus globulus and acacia dealbata species.

The set up and pack down of the system requires an additional 5–10 min; however, multiple flights would typically be performed at each site. The individual flight times are provided in Table A1 in the Appendix A. Version 4 of the aircraft was capable of flying for approximately 12 min, which enables 3–6 samples to be collected per battery, depending on proximity to the take-off/landing site. While the aircraft had a first-person view (FPV) camera on-board, with a live video feed displayed on the ground control station to aid the pilot, a high level of situational awareness was critical during the sampling procedure due to the cluttered environment around the aircraft. Direct line of sight (LOS) is also required by law in Australia (without complex approval processes), so the aircraft was operated by line of sight with only brief checks of the FPV view to assist with aiming of the sampling tool. Most samples were collected by LOS only, without the use of the FPV view; however, this approach becomes increasingly difficult with increasing distance from the pilot, leading to an increased reliance on the FPV system for the sampling action. Stereoscopic depth perception degrades with increasing distance, which makes the precise flying during sample capture considerably more challenging as the distance from the pilot increases. The additional difficulty likely increased the sampling time more than just the added distance to fly to and from the sample location.

3.2. Demonstration Video

An accompanying demonstration video is provided here: https://youtu.be/iM0 RSLVIETY, last accessed on 30 October 2021, which shows all of the designs presented in action.

4. Discussion

As seen in the demonstration video, the described system is capable of rapidly and easily collecting samples from most locations on most trees. By using Visual Inertial Odometry (VIO) and Robotic Operating System (ROS) to provide position control of our aircraft, it was possible to facilitate precise cutting movements with the aircraft, while the simple collision shield reduced the risk of branch-propeller interactions and prevented the aircraft from flying too deeply into the canopy. Precision flight capabilities are not critical for this application, as the hedge trimmer sample collection approach is relatively simple for an adequately skilled pilot to perform in calm conditions; however, VIO based position control makes this operation considerably safer, simpler, and more precise, especially at greater distances from the pilot.

This project has successfully demonstrated an alternative approach to other canopy sampling UAS seen in the literature to date [32–43]. We do not view our approach as a replacement to approaches such as DeLeaves [32], but rather as an alternative tool for forest researchers, which is able to collect samples that tools such as DeLeaves could not reach; notably samples on the side of trees, the side of cliff faces, or those with objects above them which would prevent a hanging pole design from accessing them. Our system is currently limited to sampling from the side of the canopy, which does cover most regions of a tree; however, if the highest tip of the tree is desired, or if the canopy is closed, something such as DeLeaves may be more suitable. On the other hand, if samples beneath a closed canopy are desired, our presented system would be capable of sampling areas which a hanging pole design could not, provided that the UAS can physically fit between the gaps to reach the desired sampling location. Manual sampling of canopies with pole saws would remain more practical when samples are easily reached from the ground. Extremely dense forests, such as unpruned and unthinned plantations, without room for flying a UAS would also be unsuitable for the proposed approach, and hanging sampling systems such as DeLeaves would be necessary.

This study was limited to a single site, with three native Australian tree species sampled; however, as long as a hedge trimmer is capable of cutting the vegetation on a tree, the species should not matter. Further work should explore the effects that differently shaped tree crowns have upon the sampling operation, as the three species this system was tested upon were relatively similar in crown structure. The design presented is suitable for research use, where operators have sufficient expertise with UAS to operate and maintain the system; however, it is not yet sufficiently refined for widespread adoption in forestry. To reach a mature state for industrial adoption, more robust position tracking/holding and a higher quality FPV system would be required for longer-range operations.

There is considerable scope for future work in this space. If this project continues, the next iteration of this system will use an approach inspired by Voliro [27]. It would replace the passive hanging system by actively pivoting the motors about the pitch axis using servos. This would address the issues with the large moment of inertia about the pitch axis, improve yaw responsiveness and still enable safe landing on sloped terrain. As the hedge trimmer based tool was found to be highly effective and practical, we would use this concept to build a dedicated design to reduce weight and reduce non-cutting contact area (i.e., minimise the size of the black box holding the motor or move it out of the way) to minimise drag against the canopy during upward cutting operations. VIO would still be used; however, a pair of VIO sensors (one pointing down and one pointing rearwards) would be used for enhanced robustness, as our current implementation lost track of position occasionally (particularly while flying high up or in a breeze). Our system used the default Intel Realsense VIO package for ROS; however, other VIO packages may be more robust to the conditions. Other aspects of the design would remain similar to Version 4, such as the battery mount with integrated landing gear and the style of collision shield on the front of the aircraft, though landing gear would be integrated into this collision shield to further reduce the part count and weight. This proposed "Version 5" is depicted in Figure 14.

Future Work - Version 5

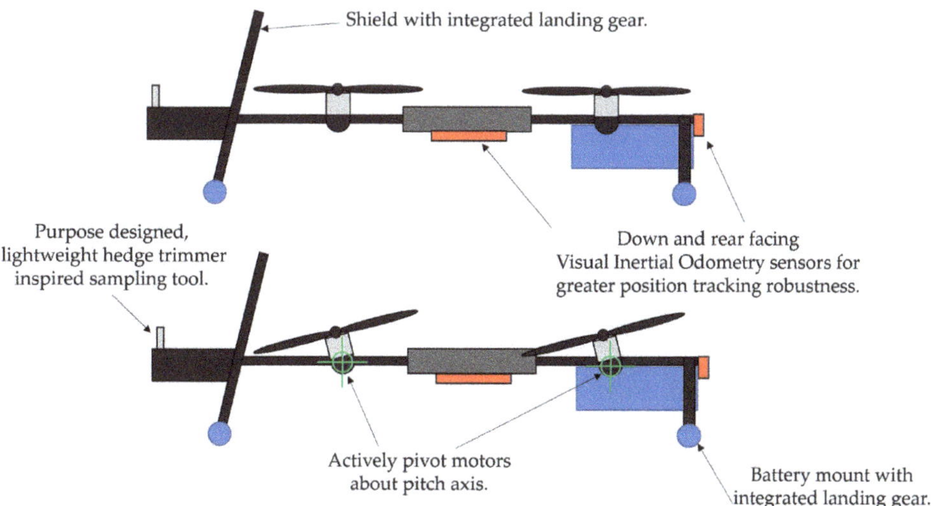

Figure 14. Our vision for the 5th iteration of this aircraft is inspired by Voliro [27], and would actively pivot the motor mounts about the pitch axis to separate forward and rearward movements from the pitch attitude of the entire airframe. This would likely be a better way to address the issues caused by the large moment of inertia about the pitch axis while also enabling take-off and landing on sloped terrain.

Additional, near-future work could see the design of a sample container for collecting multiple samples per flight. This could be achieved by using divided containers, which could rotate into place underneath the hedge trimmer for each sample, analogous to the working mechanism of a revolver. It is also suggested to add the functionality to record the GPS position when the tool is turned on and off to easily provide a GPS position for each sample.

Looking considerably further forward into the future, we could envision a system where remote sensing and physical sampling or other physical interactions could be combined. A UAS with Simultaneous Localisation and Mapping (SLAM) capabilities may not only capture a high-fidelity digital twin of the forest (in the form of a point cloud) but could also collect physical canopy samples during the process. These physical sample locations could then be localised in the captured point cloud, enabling more advanced research of canopy and leaf traits throughout a forest. Alternatively, a sampling location could be selected in a previously captured point cloud, with the UAS able to autonomously go to that position and retrieve a sample. This concept is depicted in Figure 15.

Future Directions – Fully Autonomous Sample Capture

Figure 15. A visual depiction of where this technology may be headed. Advanced remote sensing techniques could be used in conjunction with autonomous sample capture UAS, automating the sample collection process and reducing the human skill required to operate such a system.

Such an approach would need to be able to tolerate the complexities and uncertainties present within point clouds caused by factors such as beam divergence, point cloud registration errors caused by wind during the sensing process, noise and variable scanning resolutions. Deep learning-based approaches such as [47] appear promising for addressing such a challenge.

5. Conclusions

A series of novel canopy sampling UAS were presented, with detailed explanations as to how each iteration informed the design of its successor. These aircraft demonstrated a reliable and rapid method for the capture of canopy samples using a novel hedge trimmer based design not yet seen in the literature. The final prototype was tested for capturing 30 samples, with sample collection times ranging from 1 min and 23 s, up to 3 min and 41 s, depending on the forest conditions and distance from the take-off/landing site to the tree. This design was demonstrated to be capable of rapidly and safely collecting canopy samples that were previously either too difficult, dangerous or expensive to capture or where existing techniques were not suitable. Future work should see to the development of a purpose-built sampling tool based upon a hedge-trimmer-like design, as well as reducing the weight and size of the aircraft carrying it. This approach should also be tested on other types of trees, such as conifers. Looking further forward, fully-autonomous sample capture and simultaneous point cloud capture could be integrated, resulting in a holistic physical and digital sample collection tool for forest research.

Author Contributions: Conceptualisation, S.K. and P.T.; data curation, S.K.; formal analysis, S.K.; funding acquisition, S.K.; investigation, S.K.; methodology, S.K.; project administration, S.K., M.S.T., J.M. and P.T.; resources, S.K., M.S.T. and P.T.; software, S.K.; supervision, M.S.T., J.M. and P.T.;

validation, S.K.; visualisation, S.K.; writing—original draft, S.K.; writing—review and editing, S.K., M.S.T., J.M. and P.T. All authors have read and agreed to the published version of the manuscript.

Funding: This research was supported by Forest and Wood Products Australia (FWPA) and the Department of Agriculture, Water, and the Environment (DAWE), as part of the Science and Innovation Awards grant program (GA75963). This project is also supported by the Australian Research Council, Training Centre for Forest Value (IC150100004).

Data Availability Statement: Not applicable.

Acknowledgments: Thank you to Forest and Wood Products Australia (FWPA) and the Department of Agriculture, Water, and the Environment (DAWE), for awarding this project a "Science and Innovation Award for Young People in Agriculture, Fisheries and Forestry". Thank you to the Australian Research Council, Training Centre for Forest Value for funding my PhD scholarship.

Conflicts of Interest: The authors declare no conflict of interest. The funders had no role in the design of the study; in the collection, analyses, or interpretation of data; in the writing of the manuscript, or in the decision to publish the results.

Appendix A

Table A1. Time to collect canopy samples from arming of the aircraft to disarming.

| | Time to Collect Sample (Arm to Disarm) | |
Sample Number	Minutes:Seconds	Decimal Minutes
1	1:35	1.58
2	1:23	1.38
3	2:12	2.20
4	3:02	3.03
5	2:39	2.65
6	1:41	1.68
7	2:58	2.97
8	2:22	2.37
9	1:41	1.68
10	1:48	1.80
11	1:45	1.75
12	2:28	2.47
13	3:20	3.33
14	1:58	1.97
15	2:24	3.40
16	2:19	2.32
17	1:59	1.98
18	2:41	2.68
19	1:53	1.88
20	1:30	1.50
21	2:01	2.02
22	3:09	3.15
23	3:27	3.45
24	3:41	3.68
25	2:23	2.38
26	2:52	2.87
27	2:49	2.82
28	1:52	1.87
29	3:09	3.15
30	2:19	2.32

References

1. Stinziano, J.R.; Way, D.A. Combined effects of rising [CO$_2$] and temperature on boreal forests: Growth, physiology and limitations. *Botany* **2014**, *92*, 425–436. [CrossRef]
2. Lloyd, J.; Farquhar, G.D. Effects of rising temperatures and [CO$_2$] on the physiology of tropical forest trees. *Philos. Trans. R. Soc. B Biol. Sci.* **2008**, *363*, 1811–1817. [CrossRef]

3. Loukas, A.; Vasiliades, L.; Tzabiras, J. Climate change effects on drought severity. *Adv. Geosci.* **2008**, *17*, 23–29. [CrossRef]
4. Pokhrel, Y.; Felfelani, F.; Satoh, Y.; Boulange, J.; Burek, P.; Gädeke, A.; Gerten, D.; Gosling, S.N.; Grillakis, M.; Gudmundsson, L.; et al. Global terrestrial water storage and drought severity under climate change. *Nat. Clim. Chang.* **2021**, *11*, 226–233. [CrossRef]
5. Trenberth, K.E.; Dai, A.; Van Der Schrier, G.; Jones, P.D.; Barichivich, J.; Briffa, K.R.; Sheffield, J. Global warming and changes in drought. *Nat. Clim. Chang.* **2014**, *4*, 17–22. [CrossRef]
6. Hennessy, K.; Lucas, C.; Nicholls, N.; Bathols, J.; Suppiah, R.; Ricketts, J. *Climate Change Impacts on Fire-Weather in South-East Australia*; Climate Impacts Group, CSIRO Atmospheric Research and the Australian Government Bureau of Meteorology: Aspendale, Australia, 2005.
7. Williams, M.I.; Dumroese, R.K. Preparing for Climate Change: Forestry and Assisted Migration. *J. For.* **2013**, *111*, 287–297. [CrossRef]
8. Sáenz-Romero, C.; Lindig-Cisneros, R.A.; Joyce, D.G.; Beaulieu, J.; St Clair, J.B.; Jaquish, B.C. Assisted migration of forest populations for adapting trees to climate change. *Rev. Chapingo Ser. Cienc. For. Ambiente* **2016**, *22*, 303–323.
9. Kuželka, K.; Surový, P. Mapping Forest Structure Using UAS inside Flight Capabilities. *Sensors* **2018**, *18*, 2245. [CrossRef] [PubMed]
10. Windrim, L.; Bryson, M. Detection, Segmentation, and Model Fitting of Individual Tree Stems from Airborne Laser Scanning of Forests Using Deep Learning. *Remote Sens.* **2020**, *12*, 1469. [CrossRef]
11. Gonzalez de Tanago, J.; Lau, A.; Bartholomeus, H.; Herold, M.; Avitabile, V.; Raumonen, P.; Martius, C.; Goodman, R.C.; Disney, M.; Manuri, S.; et al. Estimation of above-ground biomass of large tropical trees with terrestrial LiDAR. *Methods Ecol. Evol.* **2018**, *9*, 223–234. [CrossRef]
12. Asner, G.P.; Mascaro, J.; Muller-Landau, H.C.; Vieilledent, G.; Vaudry, R.; Rasamoelina, M.; Hall, J.S.; Van Breugel, M. A universal airborne LiDAR approach for tropical forest carbon mapping. *Oecologia* **2012**, *168*, 1147–1160. [CrossRef]
13. Asner, G.; Clark, J.K.; Mascaro, J.; Galindo García, G.A.; Chadwick, K.D.; Navarrete Encinales, D.A.; Paez-Acosta, G.; Cabrera Montenegro, E.; Kennedy-Bowdoin, T.; Duque, Á.; et al. High-resolution mapping of forest carbon stocks in the Colombian Amazon. *Biogeosciences* **2012**, *9*, 2683–2696. [CrossRef]
14. Stephens, P.R.; Kimberley, M.O.; Beets, P.N.; Paul, T.S.; Searles, N.; Bell, A.; Brack, C.; Broadley, J. Airborne scanning LiDAR in a double sampling forest carbon inventory. *Remote Sens. Environ.* **2012**, *117*, 348–357. [CrossRef]
15. Le Toan, T.; Quegan, S.; Davidson MW, J.; Balzter, H.; Paillou, P.; Papathanassiou, K.; Rocca, S.P.; Saatchi, S.; Shugart, H.; Ulander, L. The BIOMASS mission: Mapping global forest biomass to better understand the terrestrial carbon cycle. *Remote Sens. Environ.* **2011**, *115*, 2850–2860. [CrossRef]
16. Shugart, H.H.; Saatchi, S.; Hall, F.G. Importance of structure and its measurement in quantifying function of forest ecosystems. *J. Geophys. Res. Biogeosciences* **2010**, *115*. [CrossRef]
17. Krisanski, S.; Taskhiri, M.S.; Turner, P. Enhancing Methods for Under-Canopy Unmanned Aircraft System Based Photogrammetry in Complex Forests for Tree Diameter Measurement. *Remote Sens.* **2020**, *12*, 1652. [CrossRef]
18. Krisanski, S.; Taskhiri, M.S.; Gonzalez Aracil, S.; Herries, D.; Muneri, A.; Gurung, M.B.; Montgomery, J.; Turner, P. Forest Structural Complexity Tool—An Open Source, Fully-Automated Tool for Measuring Forest Point Clouds. *Remote Sens.* **2021**, *13*, 4677. [CrossRef]
19. Curran, P.J. Remote sensing of foliar chemistry. *Remote Sens. Environ.* **1989**, *30*, 271–278. [CrossRef]
20. Martin, M.E.; Aber, J.D. High spectral resolution remote sensing of forest canopy lignin, nitrogen, and ecosystem processes. *Ecol. Appl.* **1997**, *7*, 431–443. [CrossRef]
21. Stork, N.E. Australian tropical forest canopy crane: New tools for new frontiers. *Austral Ecol.* **2007**, *32*, 4–9. [CrossRef]
22. Gottsberger, G. Canopy Operation Permanent Access System: A novel tool for working in the canopy of tropical forests: History, development, technology and perspectives. *Trees* **2017**, *31*, 791–812. [CrossRef]
23. McCaig, T.; Sam, L.; Nakamura, A.; Stork, N.E. Is insect vertical distribution in rainforests better explained by distance from the canopy top or distance from the ground? *Biodivers. Conserv.* **2020**, *29*, 1081–1103. [CrossRef]
24. Gara, T.W.; Darvishzadeh, R.; Skidmore, A.K.; Wang, T.; Heurich, M. Accurate modelling of canopy traits from seasonal Sentinel-2 imagery based on the vertical distribution of leaf traits. *ISPRS J. Photogramm. Remote Sens.* **2019**, *157*, 108–123. [CrossRef]
25. Gara, T.W.; Darvishzadeh, R.; Skidmore, A.K.; Wang, T.; Heurich, M. Evaluating the performance of PROSPECT in the retrieval of leaf traits across canopy throughout the growing season. *Int. J. Appl. Earth Obs. Geoinf.* **2019**, *83*, 101919. [CrossRef]
26. Kamoske, A.G.; Dahlin, K.M.; Serbin, S.P.; Stark, S.C. Leaf traits and canopy structure together explain canopy functional diversity: An airborne remote sensing approach. *Ecol. Appl.* **2021**, *31*, e02230. [CrossRef] [PubMed]
27. Kamel, M.; Verling, S.; Elkhatib, O.; Sprecher, C.; Wulkop, P.; Taylor, Z.; Siegwart, R.; Gilitschenski, I. The Voliro Omniorientational Hexacopter: An Agile and Maneuverable Tiltable-Rotor Aerial Vehicle. *IEEE Robot. Autom. Mag.* **2018**, *25*, 34–44. [CrossRef]
28. Kim, S.; Choi, S.; Kim, H.J. Aerial manipulation using a quadrotor with a two DOF robotic arm. In Proceedings of the 2013 IEEE/RSJ International Conference on Intelligent Robots and Systems, Tokyo, Japan, 3–7 November 2013.
29. Fumagalli, M.; Naldi, R.; Macchelli, A.; Carloni, R.; Stramigioli, S.; Marconi, L. Modeling and control of a flying robot for contact inspection. In Proceedings of the 2012 IEEE/RSJ International Conference on Intelligent Robots and Systems, Vilamoura-Algarve, Portugal, 7–12 October 2012.

30. Jimenez-Cano, A.E.; Braga, J.; Heredia, G.; Ollero, A. Aerial manipulator for structure inspection by contact from the underside. In Proceedings of the 2015 IEEE/RSJ International Conference on Intelligent Robots and Systems (IROS), Hamburg, Germany, 28 September–2 October 2015.

31. Paul, H.; Ono, K.; Ladig, R.; Shimonomura, K. A multirotor platform employing a three-axis vertical articulated robotic arm for aerial manipulation tasks. In Proceedings of the 2018 IEEE/ASME International Conference on Advanced Intelligent Mechatronics (AIM), Auckland, New Zealand, 9–12 July 2018.

32. Charron, G.; Robichaud-Courteau, T.; La Vigne, H.; Weintraub, S.; Hill, A.; Justice, D.; Bélanger, N.; Lussier Desbiens, A. The DeLeaves: A UAV device for efficient tree canopy sampling. *J. Unmanned Veh. Syst.* **2020**, *8*, 245–264. [CrossRef]

33. Hyneman, J. Jamie Hyneman's 'Arborist' Quadcopter Test. 2015. Available online: https://www.youtube.com/watch?v=1fe9 IDx3vCs (accessed on 1 February 2020).

34. Käslin, F.; Baur, T.; Meier, P.; Koller, P.; Buchmann, N.; D'Odorico, P.; Eugster, W. Novel Twig Sampling Method by Unmanned Aerial Vehicle (UAV). *Front. For. Glob. Chang.* **2018**, *1*, 2. [CrossRef]

35. Finžgar, D.; Bajc, M.; Brezovar, J.; Kladnik, A.; Capuder, R.; Kraigher, H. Development of a patented unmanned aerial vehicle based system for tree canopy sampling. *Folia Biol. Geol.* **2016**, *57*, 35–39. [CrossRef]

36. Schweiger, A.K.; Lussier Desbiens, A.; Charron, G.; La Vigne, H.; Laliberté, E. Foliar sampling with an unmanned aerial system (UAS) reveals spectral and functional trait differences within tree crowns. *Can. J. For. Res.* **2020**, *50*, 966–974. [CrossRef]

37. La Vigne, H.; Charron, G.; Hovington, S.; Desbiens, A.L. Assisted Canopy Sampling Using Unmanned Aerial Vehicles (UAVs). In Proceedings of the 2021 International Conference on Unmanned Aircraft Systems (ICUAS), Athens, Greece, 15–18 June 2021.

38. Bailey, W.; Bryce, M.; Colin, A.; Max, D.; James, F. Sampler Drone for Plant Physiology and Tissue Research. 2018. Available online: https://capstone.engineering.ucsb.edu/projects/dantonio-and-oono-labs-sampler-drone (accessed on 1 February 2020).

39. Hyneman, J.; Colin, C. How Mythbuster Jamie Hyneman Hacked a Drone to Trim His Trees. 2017. Available online: https://www.popularmechanics.com/flight/drones/a26102/jamie-hyneman-drone-plants/ (accessed on 1 February 2020).

40. UC Berkeley Forest Pathology and Mycology Lab. Sampler Drones for Forestry Research. 2015. Available online: https://nature.berkeley.edu/garbelottowp/?p=1801 (accessed on 1 February 2020).

41. Xu, C.; Yang, Z.; Jiang, Y.; Zhang, Q.; Xu, H.; Xu, X. The Design and Control of a Double-saw Cutter on the Aerial Trees-pruning Robot. In Proceedings of the 2018 IEEE International Conference on Robotics and Biomimetics (ROBIO), Kuala Lumpur, Malaysia, 12–15 December 2018.

42. David Lee, W.M.; Beeston, S.; Bates, S.; Schofield, S.; Edwards, M.; Green, R. Autonomous Pruning at Mcleans Island. 2019. Available online: https://www.youtube.com/watch?v=5MERY8vjLqA (accessed on 6 April 2021).

43. Molina, J.; Hirai, S. Aerial pruning mechanism, initial real environment test. In Proceedings of the 2017 IEEE International Conference on Real-Time Computing and Robotics (RCAR), Okinawa, Japan, 14–18 July 2017.

44. Lee, D.; Muir, W.; Beeston, S.; Bates, S.; Schofield, S.D.; Edwards, M.J.; Green, R.D. Analysing Forests Using Dense Point Clouds. In Proceedings of the 2018 International Conference on Image and Vision Computing New Zealand (IVCNZ), Auckland, New Zealand, 19–21 November 2018.

45. Müller, M. eCalc—xcopterCalc—The Most Reliable Multicopter Calculator on the Web. 2020. Available online: https://www.ecalc.ch/xcoptercalc.php (accessed on 14 April 2020).

46. Quigley, M.; Conley, K.; Gerkey, B.; Faust, J.; Foote, T.; Leibs, J.; Berger, E.; Wheeler, R.; Ng, A.Y. ROS: An open-source Robot Operating System. In Proceedings of the ICRA Workshop on Open Source Software, Kobe, Japan, 12–17 May 2009.

47. Chen, X.; Jiang, K.; Zhu, Y.; Wang, X.; Yun, T. Individual Tree Crown Segmentation Directly from UAV-Borne LiDAR Data Using the PointNet of Deep Learning. *Forests* **2021**, *12*, 131. [CrossRef]

Article

Estimating Aboveground Biomass in Dense Hyrcanian Forests by the Use of Sentinel-2 Data

Fardin Moradi [1],* , Ali Asghar Darvishsefat [1] , Manizheh Rajab Pourrahmati [1] , Azade Deljouei [2] and Stelian Alexandru Borz [2],*

1 Department of Forestry and Forest Economics, Faculty of Natural Resources, University of Tehran, Karaj P.O. Box 1417643184, Iran; adarvish@ut.ac.ir (A.A.D.); mrajabpour@ut.ac.ir (M.R.P.)
2 Department of Forest Engineering, Forest Management Planning and Terrestrial Measurements, Faculty of Silviculture and Forest Engineering, Transilvania University of Brasov, Şirul Beethoven 1, 500123 Brasov, Romania; azade.deljouei@unitbv.ro
* Correspondence: moradi.nr@ut.ac.ir (F.M.); stelian.borz@unitbv.ro (S.A.B.); Tel.: +98-910-946-3476 (F.M.); +40-742-042-455 (S.A.B.)

Abstract: Due to the challenges brought by field measurements to estimate the aboveground biomass (AGB), such as the remote locations and difficulties in walking in these areas, more accurate and cost-effective methods are required, by the use of remote sensing. In this study, Sentinel-2 data were used for estimating the AGB in pure stands of *Carpinus betulus* (L., common hornbeam) located in the Hyrcanian forests, northern Iran. For this purpose, the diameter at breast height (DBH) of all trees thicker than 7.5 cm was measured in 55 square plots (45 × 45 m). In situ AGB was estimated using a local volume table and the specific density of wood. To estimate the AGB from remotely sensed data, parametric and nonparametric methods, including Multiple Regression (MR), Artificial Neural Network (ANN), k-Nearest Neighbor (kNN), and Random Forest (RF), were applied to a single image of the Sentinel-2, having as a reference the estimations produced by in situ measurements and their corresponding spectral values of the original spectral (B2, B3, B4, B5, B6, B7, B8, B8a, B11, and B12) and derived synthetic (IPVI, IRECI, GEMI, GNDVI, NDVI, DVI, PSSRA, and RVI) bands. Band 6 located in the red-edge region (0.740 nm) showed the highest correlation with AGB ($r = -0.723$). A comparison of the machine learning methods indicated that the ANN algorithm returned the best ABG-estimating performance (%*RMSE* = 19.9). This study demonstrates that simple vegetation indices extracted from Sentinel-2 multispectral imagery can provide good results in the AGB estimation of *C. betulus* trees of the Hyrcanian forests. The approach used in this study may be extended to similar areas located in temperate forests.

Keywords: aboveground biomass; estimation; remote sensing; Sentinel-2; Iran; multiple regression; artificial neural network; k-nearest neighbor; random forest; performance

Citation: Moradi, F.; Darvishsefat, A.A.; Pourrahmati, M.R.; Deljouei, A.; Borz, S.A. Estimating Aboveground Biomass in Dense Hyrcanian Forests by the Use of Sentinel-2 Data. *Forests* **2022**, *13*, 104. https://doi.org/10.3390/f13010104

Academic Editor: Olga Viedma

Received: 13 December 2021
Accepted: 10 January 2022
Published: 12 January 2022

Publisher's Note: MDPI stays neutral with regard to jurisdictional claims in published maps and institutional affiliations.

1. Introduction

Forests are an essential component of the carbon cycle, as they are both storing and releasing carbon through their biomass into the atmosphere. Globally, forest ecosystems contain approximately 80% of the aboveground and 40% of the underground biomass [1]. Knowledge on the amount of biomass and carbon storage is essential for forest management and planning [2]. Quantifying biomass availability in the forests through field measurements is commonly resource-intensive. Remote sensing techniques integrated with geographic information systems (GISs) provide quick access to useful information, typically available for short cycle times and at lower costs [3]. Combining remotely sensed data with nonspectral ancillary data such as those produced by field sampling has been suggested by many studies as a way to reach better estimates [4]. A variety of remotely sensed data, such as those coming from Landsat, Sentinel, Spot, and ALOS missions, have been used to estimate the volume of wood and biomass stocked in the forests [5–13].

Aboveground biomass (AGB) estimation methods include field measurements and remote sensing approaches [14,15]. There are mainly two methods used in field measurement to estimate the AGB, namely destructive (harvesting) and nondestructive methods. Although the destructive method is useful and accurate in developing equations for the assessment of aboveground biomass over larger areas, it is often constrained to few trees, being time consuming, difficult to implement, and expensive [16]. A nondestructive method is an alternative to estimate the AGB. It is implemented either by climbing to make measurements in different tree parts or, more commonly, by measuring the diameter at the breast height (DBH) and tree height; other options include the estimation of volume and density using allometric equations or remote imagery [17,18]. As a nondestructive method, remote sensing is based on previously developed allometric equations.

The techniques used for estimating the AGB of forests based on remotely sensed data can be divided into two categories, namely those using parametric (statistical regression methods) and nonparametric algorithms, respectively [7]. Nonparametric techniques, including Machine Learning (ML) algorithms such as the k-Nearest Neighbor (kNN), Artificial Neural Networks (ANNs), and Random Forests (RFs), were found to hold a better ability of identifying complex relations between the used predictors and the AGB [7,19]. For instance, ANNs are being considered to be important nonparametric algorithms for estimating forest-related parameters [20]. In addition, the kNN algorithm has received considerable attention because it is easily accessible, and some literature reviews have shown that it holds an excellent capability to increase the precision when estimating vegetation parameters [21–23]. RF regression algorithms have also been widely used for quantifying forest biophysical parameters [5,24–26], standing for an ensemble learning algorithm with applications in classification and regression problems. The RF algorithm was developed by Breiman [27] and can be used to predict continuous and categorical dependent variables. A random subset of observations with replacement, as well as a random set of explanatory variables, are used to build each regression tree [28].

Traditionally, in any part of the world, AGB is estimated by destructive methods, which are used to develop allometric equations based on measured parameters collected from harvested trees (e.g., DBH, tree height, and timber volume) [29]. However, applying allometric equations across a large study area is cumbersome and sometimes impractical as the field measurement input parameters are rare and sometimes unavailable. In comparison, remote sensing techniques can provide large-scale and accurate biophysical information for forest inventory data. Hence, remote sensing data combined with machine learning techniques (i.e., parametric and nonparametric algorithms) have been widely used to estimate forest AGB in the past decade. For example, Muukkonen and Heiskanen [30] predicted the AGB in boreal forests using ANNs applied to ASTER (Advanced Spaceborne Thermal Emission and Reflection Radiometer) data. IRS P6 LISS-III (Indian Remote-Sensing Satellite-P6 Linear Imaging Self-Scanning Sensor-3) data were used by Yadav et al. [31] to estimate the AGB in the Timli forests of India. In their research, the kNN method based on Mahalanobis distance outputted a *RMSE* of 42.25 Mg/ha, while the distance metric used was found to be best, being followed by the fuzzy and Euclidean distances, with *RMSE* of 44.23 Mg/ha and 45.13 Mg/ha, respectively. Lu et al. [32] showed that the estimation of AGB in Amazon forests using Landsat-5 TM data is more accurate in young than in mature stands. Ronoud et al. [33] found that the Landsat-5 TM NIR (near-infrared) band exhibited the highest correlation with AGB ($r = 0.427$). Several studies have used Sentinel-2 data to estimate AGB in various ecosystems, including semiarid [34], Mediterranean [35,36], temperate [7,37,38], tropical [37,39,40], subtropical [41,42] and boreal [43,44] forests, and grasslands [45]. For example, Chrysafis et al. [46] compared Sentinel-2 MSI (MultiSpectral Instrument) and Landsat-8 OLI (Operational Land Imager) imagery for forest growing stock volume (GSV) estimation in a mixed Mediterranean forest in northeastern Greece. GSV was modeled using RF regression based on spectral bands and vegetation indices. They have shown that to estimate the AGB, Sentinel-2 data with an $R^2 = 0.63$ and $RMSE = 63.11$ m^3/ha were better than Landsat-8 OLI data with an $R^2 = 0.62$ and $RMSE = 64.40$ m^3/ha. According to Castillo et al. [37], red and red edge bands produced by Sentinel-2 data combined with elevation data provided the best estimates of AGB in Philippine's mangrove forests when using machine learning methods. Nuthammachot et al. [47] assessed the potential of seven vegetation indices derived from Sentinel-2 images for estimating the AGB in a private forest of Indonesia. They found that, among other indices, including the Normalized Difference Vegetation Index (NDVI), Enhanced Vegetation Index (EVI), Modified Simple Ratio (MSR), Simple Ratio (SR), Sentinel-2 Red-Edge Position (S2REP), and Greenness Normalized Difference Vegetation Index (GNDVI), the Normalized Difference Index (NDI45) exhibited the strongest correlation with AGB ($r = 0.89$, $R^2 = 0.79$). In addition, they found that the NIR spectral band of the Sentinel-2 was the most effective variable in retrieving forest standing volume when using the kNN algorithm. They estimated the standing volume with a relative *RMSE* of 22.94%. Research by Pandit et al. [42] evaluated the usefulness of Sentinel-2 data for estimating the AGB in protected forests from Nepal using the RF algorithm. The effect of the number of input variables, including spectral band values and spectral-derived vegetation indices on the AGB prediction, was also investigated. The model using all spectral bands, in addition to the derived vegetation indices, provided better AGB estimates ($R^2 = 0.81$ and $RMSE = 25.57$ t/ha). Vafaei et al. [48] assessed ALOS-2 (Advanced Land Observing Satellite 2) and Sentinel-2 data for AGB estimation in the Asalem forests of Iran using four machine learning methods, namely the Gaussian process (GP), support vector regression (SVR), RF, and Multi-Layer Perceptron Neural Networks (MLP Neural Nets, MLP NNs). In their study, a SVR model using combined Sentinel-2 spectral information (including blue, green, red, and NIR bands) and six vegetation indices, namely SVI (Simple Vegetation Index), RVI (Ratio Vegetation Index), NDVI (Normalized Difference Vegetation Index), EVI-2 (Enhanced Vegetation Index 2), PVI-2 (Perpendicular Vegetation Index 2), and SAVI (Soil Adjusted Vegetation Index) based on ALOS-2 PALSAR2 (Advanced Land Observing Satellite 2, Phased-Array-type L-band Synthetic Aperture Radar 2) imagery, HH (horizontal transmit and horizontal receive), HV (horizontal transmit and vertical receive), VV (vertical transmit and vertical receive), and VH (vertical and horizontal receive), yielded the best performance to estimate the forest AGB.

Data saturation often causes problems in estimating forest AGB when dealing with high amounts of biomass or high-canopy-density areas [49]. This problem was addressed by combining Sentinel-2 and ALOS2-PALSAR2 data [48]. The studies mentioned above, which evaluated the utility of remotely sensed data for estimating the forest standing volume and AGB, do not show consistency in performance and outcomes, due to the variety of forest conditions, satellite data used, applied methodology, and due to the inherent, specific limitations of each study.

In Iran, an area of ~10.7 million hectares is covered by forests accounting for ca. 7.4% of the country's territory [50]. Hyrcanian forests are the most important forests among the five vegetation regions in Iran due to the density, canopy cover, and diversity in this ecoregion [51,52]. They cover ~2 million hectares and are located on the south coast of the Caspian sea [53]. For these forests, management plans are updated in terms of qualitative and quantitative attributes every ten years, in which collecting data and information are time-consuming and cost-intensive. In contrast, remotely sensed imagery holds a promising potential for monitoring and continuously predicting forest attributes. In conjunction with satellite data, field data can be used to create a continuous map of forest attributes through classification or regression. Therefore, forest attributes have been estimated from remote sensing data with various spatial resolutions, ranging from very high to medium.

To the best of our knowledge, this is the first study attempting to estimate the AGB by the use of remotely sensed data and machine learning algorithms in pure common hornbeam (*Carpinus betulus* L.) forests, as a typical forest type in the temperate forest region of many European and Asian countries. This study was guided by the above mentioned, as well as the fact that pure stands of common hornbeam are distributed from 200 to 1800 m a.s.l., from the western part, characterized by a very humid climate, to the eastern part of the Hyrcanian region, which is characterized by a humid climate [54]. Accordingly, this study aimed to evaluate the usefulness of Sentinel-2 imagery and several machine learning algorithms for estimating the AGB of *C. betulus* forests located in the Patom and Namkhane districts of Kheyrud forest, Northern Iran. The objectives of the study were the following: (i) comparing the performance of different AGB estimation approaches including parametric (i.e., Multiple Regression—MR) and nonparametric algorithms (ANN, kNN, and RF), and (ii) investigating the potential and capability of Sentinel-2 imagery in improving the accuracy of the AGB estimation under the given conditions of the study.

2. Materials and Methods

2.1. Study Site

The study area is located in the Kheyrud forest as part of the mountainous deciduous forests of the Hyrcanian ecoregion, north of Iran (longitude: 51°34′53″ to 51°35′28″ E and latitude: 36°36′14″ to 36°35′28″ N). Kheyrud forest covers a total area of ~8000 ha, and it is a natural and mature forest with uneven-aged and dense to semi-dense stands consisting of seven management districts. Two study sites were selected in Patom and Namkhane districts (Figure 1). The elevation of the selected areas ranges from 480 to 630 m a.s.l. in Patom and from 950 to 1110 m a.s.l. in Namkhane district. According to the Nowshahr synoptic station [51,55], the climate of the area is sub-Mediterranean with an annual temperature averaging 9 °C and a total annual precipitation of 1300 mm. Tilio-buxetum, Querco-carpinetum, Fageto-carpinetum, and Rusco-Fagetum are the main forest communities in the Patom district. Namkhane district contains forest communities of Querco-carpinetum, Fageto-carpinetum, Fagetum mixed, and Fagetum-hyrcanum [34]. Sample plots were selected on flat areas, in pure stands of *C. betulus* to minimize the spectral interference of other species [56]. The stock of *C. betulus* stands based on our plot-level measurements ranged from 174 to 470 m^3 ha^{-1}.

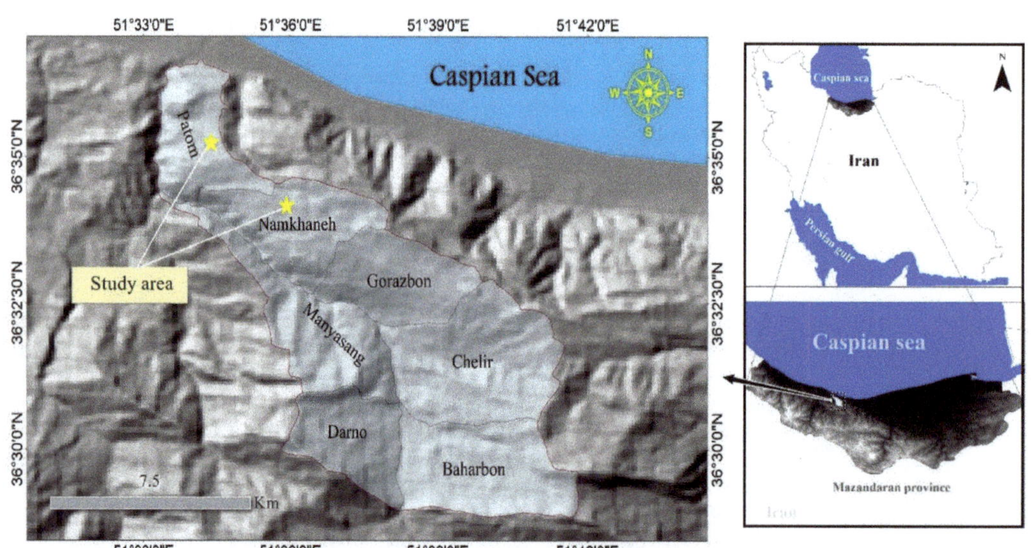

Figure 1. The geographic location of the study sites.

2.2. Remote Sensed Data and Data Preprocessing

Sentinel-2 satellite data (dated 17 July 2016) were obtained from the US Geological Survey (USGS) website (https://earthexplorer.usgs.gov/; accessed on 25 March 2017) and used for AGB estimation. Sentinel-2 carries the Multispectral Imager (MSI) that delivers 13 spectral bands with a spatial resolution ranging from 10 to 60 m. Sentinel-2 10 m spatial resolution bands including B2 (490 nm), B3 (560 nm), B4 (665 nm), and B8 (842 nm), and 20 m spatial resolution bands of B5 (705 nm), B6 (740 nm), B7 (783 nm), B8a (865 nm), B11 (1610 nm), and B12 (2190 nm) were used for analysis. The three 60 m bands (bands 1, 9, and 10), which are mainly focused toward cloud screening and atmospheric correction [57], have not been taken into consideration in this study. The digital topographic maps available at a scale of 1:25,000 and provided by the National Cartographic Center (NCC) of Iran were used to evaluate the geometric accuracy of the satellite image, which was evaluated based on road features extracted from topographic maps.

The sixth version of the Sentinel Application Platform (SNAP) software developed by the European Space Agency (ESA) was used to process the Sentinel-2 data. A visual assessment of radiometric quality was also performed concerning the presence of cloud cover, a scanning line, and duplicated pixels. Then, the two well-known processing methods, namely Principal Component Analysis (PCA) and the spectral band ratio, were applied to all original spectral bands of the images. Table 1 describes the vegetation indices extracted using band rationing.

Table 1. Vegetation indices extracted from Sentinel-2 data.

Index	Equation	Reference
Infrared Percentage Vegetation Index (IPVI)	$\text{NIR}^{1}/(\text{NIR} + \text{RED}^{2})$	[58]
Inverted Red-Edge Chlorophyll Index (IRECI)	$(\text{NIR} - \text{RED})/(\text{RED}/\text{RED})$	[59]
Global Environment Monitoring Index (GEMI)	$n^{3} \times (1 - 0.25 \times n) - (\text{RED} - 0.125)/(1 - \text{RED})$	[60]
Green Normalized Difference Vegetation Index (GNDVI)	$(\text{NIR} - \text{GREEN}^{4})/(\text{NIR} + \text{GREEN})$	[61]
Normalized Difference Vegetation Index (NDVI)	$(\text{NIR} - \text{RED})/(\text{NIR} + \text{RED})$	[62]
Difference Vegetation Index (DVI)	$\text{NIR} - \text{RED}$	[63]
Pigment Specific Simple Ratio (PSSRA)	NIR/RED	[64]
Ratio Vegetation Index (RVI)	NIR/RED	[65]

Note: [1] NIR = near-infrared band, [2] Red = red band, [3] $n = (2 \times (\text{NIR2} - \text{RED2}) + 1.5 \times \text{NIR} + 0.5 \times \text{RED})/(\text{NIR} + \text{RED} + 0.5)$, [4] GREEN = green band.

2.3. In Situ Measurements

Field measurements were conducted to estimate the AGB in August 2016, and in situ data were collected over 55 plots (45 × 45 m; Figure A1) that were navigated by GPS (Garmin Colorado 300; Olathe, KS, USA). Sample plots were distributed selectively to meet the homogeneity of plots in terms of species, terrain slope, and aspect due to the diverse topographic conditions and small extent of pure *C. betulus* stands over the study sites. The DBHs of all trees having a diameter greater than 7.5 cm and species were recorded for each plot. The volume of individual trees was estimated using a local tarif volume table and aggregated at the plot level. Then, AGB (t/ha) was estimated for each plot using Equation (1) [66].

$$AGB = Volume \times WD \tag{1}$$

where *Volume* is the volume per hectare derived from the local tariff table, and WD (t/m^3) is the wood density. The value of 0.68 t/m^3 was used for *C. betulus* as a WD [67].

2.4. Methods

The flowchart of AGB estimation is shown in Figure 2. Pearson's correlation was used to describe the association between AGB and the corresponding spectral values. The AGB (dependent variable) was modeled based on the remote sensing metrics (independent variables) using the parametric method of MR, as well as the well-known nonparametric algorithms of ANNs, kNN [68], and RF [28]. In the MR method, the model was fitted using all variables (main and synthetic spectral imagery). The suitable remote sensing variables that had a strong correlation with AGB were identified by the means of backward elimination and stepwise selection procedures [69]. Before implementing the MR, the normality of the dataset was evaluated using the Kolmogorov—Smirnov test [70].

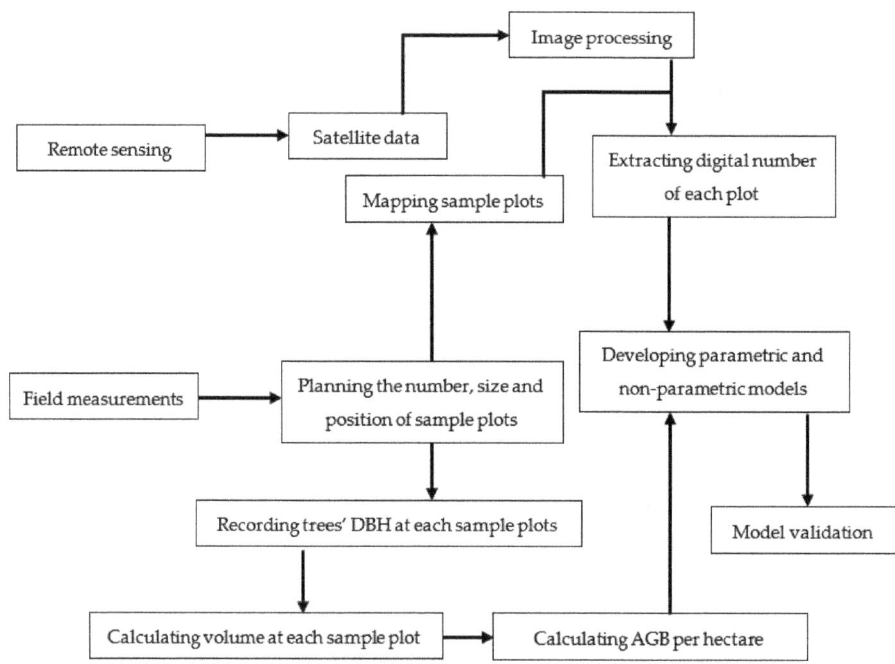

Figure 2. The flowchart of AGB estimation methods used in the study.

Typically, the ANN contains a large number of interconnected nodes and uses mathematical algorithms to model nonlinear problems such as modeling the forest biomass. The MLP (MultiLayer Perceptron) NN model is one of the most commonly used neural network algorithms for environmental modeling, monitoring, mapping of forests, and estimating the forest biomass [69,71,72]. The typical architecture of the MLP NN consists of at least three layers and includes the input, hidden, and output layers. Each layer is composed of several nodes or neurons. The number of neurons used in the input layer was that of the number of input explanatory variables. A significant influence on the performance of the MLP NN model is given by the connection weights between the input and hidden layers, as well as the connection weights between the hidden and output layers. Nonetheless, there is no rule that allows previous decisions to determine the number of neurons in the hidden layer or the number of hidden layers. Some have reported that an insufficient number of hidden neurons made the network learning difficult [73], whereas an excessive number of hidden neurons might lead to unnecessary training time [74]. Therefore, the commonly used strategy to reach the optimum number of neurons in the hidden layer is by trial and error [75]. The output layer contained one neuron and was used to output estimated values of the AGB. The weights assigned at the connections between the input, hidden, and output layers were updated in the training phase and were based on a back-propagation algorithm [76] that minimized the differences between the AGB value estimated by the MLP NN and that produced from AGB in situ inventories. The process was repeated until reaching a predefined accuracy level or the maximum number of iterations.

To develop the architecture of the MLP NN model, in this study, the number of selected hidden neurons was significantly impacting the estimation of AGB [72], as defined by [77]. As a result, by varying the number of neurons against the root-mean-square error (*RMSE*) based on the data contained in the training dataset, the best MLP NN models were reached. These best models were described by the highest R^2 and the lowest *RMSE*. Accordingly, the best MLP NN model was found to be that characterized by two hidden layers containing four neurons in the first and two neurons in the second layer. The model was trained using

70% of the dataset, and the remaining data (30%) were split in half for validation (15%) and testing (15%). The steps described above were implemented in the Statistica software (Ver. 10).

For the kNN algorithm, the choice of the k value, distance metric, and weighting function are critical factors affecting the estimation accuracy [32]. The model performance was tested by the use of k values from 1 to 40 to find the optimum one for implementing the kNN algorithm [23,78]. Moreover, for an efficient comparison of the distance metrics in the kNN implementation, the four distance metrics available in Statistica software (StatSoft. Inc., Tulsa, OK, USA), including the Euclidean, squared Euclidean, Manhattan (city block), and Chebychev distances (Equations (2)–(5)), were used, and their results were compared against each other [78].

The most frequently used distance metric is the Euclidean distance, standing for a simple geometric distance in a multidimensional space [79]. In the case of squared Euclidean distance, the distance between the target and reference units would be squared to give progressively greater weights to data points that are closer or more similar. Absolute distances are considered when using the Manhattan distance metric, although the effect of single large differences (i.e., outlier data) is dampened whether they are not squared [79]. The absolute magnitude of differences between coordinates of a pair of data points was examined by Chebychev's distance metric. This metric can be used for both ordinal and quantitative variables and it is appropriated when one would like to term two data points as "different" if they are different on any one of their dimensions.

$$D(x,p) = \sqrt{(x-p)^2} \tag{2}$$

$$D(x,p) = (x-p)^2 \tag{3}$$

$$D(x,p) = abs(x-p) \tag{4}$$

$$D(x,p) = \max(|x-p|) \tag{5}$$

where D is the distance between the target and reference units, x is the target unit, and p is the reference unit in all equations. The squared Euclidean distance is the most commonly used distance metric among the four mentioned above [78,80–82].

RF is an efficient machine learning algorithm that was developed by Breiman [28], currently being used for classification and regression problems. Typically, its use yields high accuracy, being robust in finding outliers and noise, computes quickly, and shows the relative importance of the input variables [83]. A bagging algorithm [84] is used to generate n sub-datasets (which is called a bootstrap dataset) from the training dataset. By the Classification And Regression Tree (CART) algorithm, each bootstrap dataset is used to construct a base-decision tree [85]. Finally, the RF model is generated by grouping base-decision trees to form a forest. Two-thirds of the total samples from the training dataset, called "in bag" data, should be contained in these bootstrap datasets. Approximately one-third of observations (out-of-bag, OOB) are used to evaluate the RF model [86]. The number of base-decision trees should be selected carefully because the RF model's performance depends on this parameter. In this study, 500 base-decision trees were selected to ensure the stability of the RF model's results, as suggested by Stevens et al. [87], and they were used to produce a graph showing the average squared error rates against each number of trees for training and testing samples, as a robust analytical tool to explore data and to verify the optimal number of trees within RFs. In such graphs, the optimal number of trees is determined based on the number of trees that produces a stable error [55]. Following this, we repeated the RF implementation using this optimal number of trees and other fixed parameters.

2.5. Statistical Analysis and Modeling Performance

PCA analysis was used in this study to identify the main components and to help analyze a subset of features by a dimensionality reduction. PCA is widely used to eliminate waste data in remote sensing studies [88]. In this study, PCA was computed from the bands of the Sentinel-2 image, and it was used for AGB modeling by the means of Statistica (version 10) software. The first component of all bands, except band 10, was included in the PCA analysis. In addition, a sensitivity analysis was used to determine the most effective model parameters [89].

Model testing and validation was performed by using 30% of all observations. The estimated performance metrics of the models were developed in the form of statistics such as the root-mean-square error (*RMSE*), relative *RMSE* (*%RMSE*), which were also used to choose the best model, adjusted coefficient of determination (R^2_{adj}), and standard error of estimates (*SEE*). R^2_{adj} and *SEE* were calculated only for regression models, while *RMSE* and relative *RSME* were used to evaluate the performance of both parametric and nonparametric models (Equations (6)–(9)).

$$RMSE = \sqrt{\frac{\sum_{n-1}^{n}(AGB - AGB_i)^2}{n}} \tag{6}$$

$$\%RMSE = \frac{RMSE \times 100}{\bar{y}} \tag{7}$$

$$R^2_{adj} = 1 - \frac{(1 - R^2)(N - 1)}{N - p - 1} \tag{8}$$

$$SEE = \frac{\sigma}{\sqrt{n}} \tag{9}$$

where *AGB* and AGB_i stand for the estimated and observed *AGB* per plot, respectively, n is the total number of samples, $\bar{y}$ is the average of the testing phase data, R^2 is the coefficient of determination, N is the number of samples, p is the number of predictor variables, and σ is the standard deviation.

3. Results

Based on the in situ measurements, the minimum, maximum, and mean values of the AGB for *C. betulus* stands were estimated at 118, 320, and 210 t/ha, respectively, with a standard deviation of 60 t/ha (Figure 3; Table A1); there was a high variance (3588 t/ha), indicating that the data were spread out from the mean, and from one another (Table A1). The results of the normality test indicated a normal distribution of both in situ and remotely sensed data. Based on Pearson's correlation coefficient, a negative association was found between spectral information and in situ AGB (Table 2). Band 6 of the Sentinel-2 data outputted the highest correlation with in situ AGB ($r = -0.723$; Table 2).

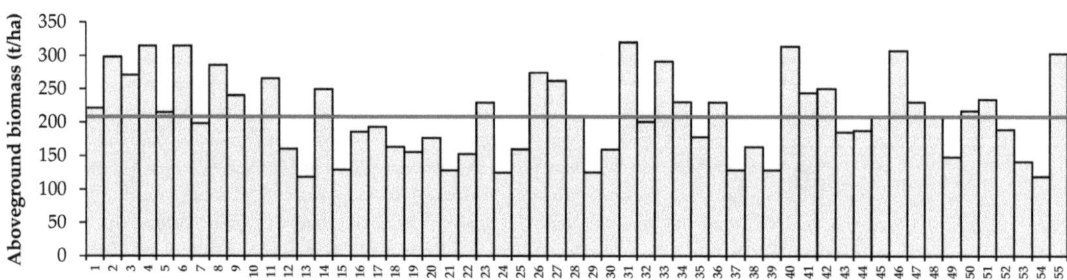

Figure 3. The value of the aboveground biomass for each plot. The red line shows the mean value of aboveground biomass (AGB) at the study level.

Table 2. Pearson's correlation coefficient (r) between spectral information and the aboveground biomass (AGB).

Variable	r	Variable	r
B2	−0.519 **	IPVI	−0.506 **
B3	−0.541 **	IRECI	−0.567 **
B4	−0.580 **	GEMI	−0.666 **
B5	−0.515 **	PC1 [1]	−0.686 **
B6	−0.723 **	GNDVI	−0.322 **
B7	−0.682 **	NDVI	−0.506 **
B8	−0.691 **	DVI	−0.682 **
B8a	−0.674 **	PSSRA	−0.425 **
B11	−0.716 **	RVI	−0.510 **
B12	−0.594 **		

Note: ** Significance level: 0.01, [1] PC1 = first component of PCA.

The result of the AGB prediction using MR indicated that the backward elimination procedure (R^2_{adj} = 0.65, %*RMSE* = 24.72) outperformed the linear regression that used all the variables, as well as the stepwise regression model (Table 3).

Table 3. Performance of the best parametric models for estimating the AGB.

Regression Method	SEE	R^2	R^2_{adj}	%RMSE	Variable
Multiple	40.35	0.757	0.588	29.72	All variables
Stepwise	42.89	0.547	0.535	30.99	B6
Backward	35.73	0.722	0.650	24.72	B2, B4, B5, B6, B7, B11, PCA, GNDVI, NDVI, PSSRA, IRECI, DVI

Table 4 shows the performance of the kNN models that included all the variables and used four distance metrics (Euclidean, Squared Euclidean, Manhattan, and Chebychev). The best distance metric for the kNN algorithm was the Manhattan distance, which returned the lowest %*RMSE* and the highest R^2 (Table 4).

Table 4. Performance of aboveground biomass estimates using the kNN algorithm.

Range of k	Distance Metric	R^2	%RMSE	The Optimal k Value
1–40	Euclidean	0.67	23.90	27
1–40	Squared Euclidean	0.72	22.94	29
1–40	Manhattan	0.73	21.85	25
1–40	Chebychev	0.67	23.87	24

The ANN fitted by a MLP NN model with an input layer containing all variables and two hidden layers produced a relative *RMSE* of 19.93% during the validation phase (Table 5). The sensitivity analysis indicated that PC1 was the most effective variable for estimating AGB.

Table 5. Training and validation results of the aboveground biomass (AGB) using the MLP NN and RF models.

AGB Model	Training Dataset		Validation	
	R^2	%RMSE	R^2	%RMSE
MLP NN	0.89	8.79	0.65	19.93
RF	0.69	19.52	0.60	22.55

As mentioned before, the performance of the RF algorithm depends on choosing the optimal number of trees and numbers of predictors (k) in each node for producing a good response in estimations. For instance, Figure 4 shows the average squared error rates against the number of trees used for AGB estimation when using RF during the training and testing phases. The optimal number of trees is assigned to the point where the error rate does not change by increasing the number of trees (Figure 4). The improvement in accuracy was slow after about 220 trees; therefore, this number was used as a good estimation for an optimum number to use (Figure 4). Based on the variable importance value obtained from the sensitivity analysis, spectral band 6 of Sentinel-2 was the most effective variable. In this study, the best RF model estimated AGB with a relative *RMSE* of 22.55% for k set at 6 (Table 5).

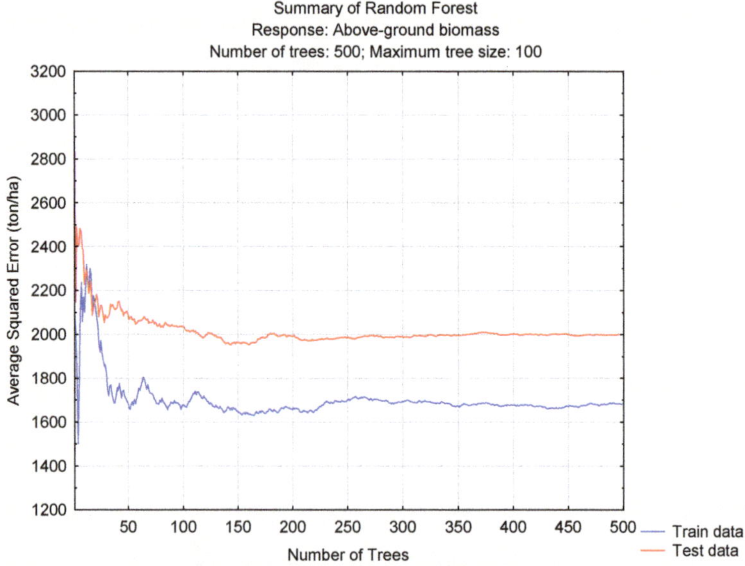

Figure 4. Random Forest error testing graph—the average squared error of aboveground biomass outputted by the Random Forest algorithm, plotted against the number of trees using the training and testing datasets.

4. Discussion

Previous studies have found that remote sensing-based models for AGB estimation are more accurate than empirical-based and GIS-based models [32]. In this study, Sentinel-2 data were used to estimate the AGB in pure stands of *C. betulus* in a part of the Hyrcanian forest, Iran. A total of 19 variables, including original spectral bands, vegetation indices, and the first principal component of PCA (applied to all original bands), were used for estimation. In situ AGB was found to be negatively correlated with all variables. The highest correlation was between the AGB and the two spectral bands located at the red edge (0.731–0.749 nm wavelength) and shortwave infrared (1.539–1.681 nm wavelength), with values of R^2 of −0.723 and −0.716, respectively. The negative correlation between biomass and spectral values has been discussed in many studies [9–11,90], expected to be caused by the canopy shadowing of trees, canopy size, stand volume and density, and consequently, by a more complex vertical structure of the forests. Shadowing is a factor influencing the reduction in spectral reflectance of forests [91]. In addition, the fraction of vegetation cover (FVC) of the ground at the pixel level is another reason affecting the radiation behavior at the canopy level, particularly in taller stands [92,93], which was the case of forests from this study.

The higher spectral radiances of low-density forests characterized by less biomass can be partially explained by a smaller amount of shadows resulting in a higher contribution of the soil to the spectral radiance [12,91]. The age of the studied stands could be another reason for the negative correlation between the amount of AGB and their corresponding spectral values [13,94]. At higher ages, which was the case of this study, the size of the canopy is rising [95], which increases the canopy surface area, size, and number of holes in the canopy [8,94]. Increasing the canopy surface area can reduce the amount of reflection due to the holes created in the tree crowns that is causing the electromagnetic waves to spread through the crown and reduces reflection [94]. In addition, as the age of the trees increases, their requirements for water will increase. As the amount of water increases in the leaves, it will absorb electromagnetic waves and will thus reduce reflection. Furthermore, as the age of the forest stands increases, the number of stories usually develops, causing more propagation of the electromagnetic waves and ultimately a reduction in spectral reflection [10,96]. On the other hand, a positive correlation between biomass and spectral reflectance was reported by different researchers [33,47] and explained by specific characteristics of the study site such as the vertical structure of forest stands, canopy cover percentage, forest health and vitality, species composition, and soil properties. In this study, we found that the relatively strong correlation between AGB and B6, though negative, preserved the presence of this variable in the backward and stepwise regression models (Table 3).

Our results indicated that nonparametric models performed better than MR, and the best result was obtained when using an ANN that outputted a relative *RMSE* of 19.93%. This is in agreement with the findings of Vafaei et al. [48] (relative *RMSE* = 19.17%) and close to those of Gao et al. [19] (relative *RMSE* = 28.8%). The ability to learn during training and to generalize on new datasets makes ANN more powerful and flexible than MR [7,97]. Past research has suggested that whenever an insufficient number of sample plots is available, parametric models can result in a poor performance, while nonparametric models may lead to more accurate predictions [98]. The ANN, as a nonparametric mathematical model, is conceptually similar to biological neural networks and holds excellent linear and nonlinear fitting capabilities [7]. Nevertheless, this is mainly due to the fact that the nonparametric models are able to handle nonlinear relations between variables from multiple sources [34]. By comparing the performance of algorithms for forest AGB estimation on ALOS PALSAR and Landsat data, Gao et al. [19] concluded that ANN performed better than RF. For the temperate forest of China, Chen et al. [7] concluded that ANN was most accurate in assessing the biomass of broadleaved deciduous forests as opposed to regression, SVR, and RF algorithms. As shown by this study, the higher performance of nonparametric algorithms could be due to the complex relations established between AGB and remote sensing variables, which are difficult to understand and explain by parametric algorithms. In addition, nonparametric algorithms are more flexible, by removing some limitations such as the hypotheses on data distribution and the functional form of the mathematical relation between independent and dependent variables. For instance, Lu et al. [32] believed that nonparametric algorithms are more adapted in creating complicated nonlinear biomass models because they do not explicitly predefine the model structure but determine it in a data-driven manner.

As in many other studies, addressing data uncertainty is important. In this study, data uncertainty may be associated with the GPS errors in locating the sample plots, possible errors of the local volume table, the inappropriateness of the available allometric models to calculate the AGB, and spectral interference of other species that existed in the plots. In addition, optical data produced by the Sentinel-2 mission cannot penetrate the forest canopy, preventing it from capturing information about wooden understories. On the one hand, extending the canopy surface will increase the size and number of holes in the canopy. Tree growth will increase in terms of volume, so trees will make a shadow that will cause a reduction in reflection [99]. On the other hand, spreading water on the leaves and increasing the water availability will also reduce the reflectance [99].

Many studies have indicated that integrating multisensor information from optical, radar, and lidar platforms can improve biomass estimation accuracy [32,100]. Furthermore, to improve the estimation of AGB by Sentinel-2 optical data, some points must be considered. Due to the fact that vegetation cover and trees with DBHs less than 7.5 cm are not typically considered in the calculation of the stand volume, studies should be carried out in areas without vegetation cover and small trees, or they should be carried out during the time of year when the vegetation cover is missing. The amount of reflection during the year varies due to the changes in the color of the leaves, water availability, and changes in stand structure; therefore, in situ measurements should be performed close to the time of satellite image acquisition. In addition, further studies should be carried out to clarify the effects of water availability, saturation, canopy cover, vegetation cover, and undergrowth vegetation on the canopy reflectance in a continuum of canopy closure. As one characteristic of our study was the limited number of plots that provided data for modeling and assessment, further studies should be carried out to check the effect of field sampling effort on the improvements in accuracy of the estimates, as one option. Another option would be using a leave-one-out cross-validation (LOOCV) procedure to improve the results [101]. Nevertheless, the approach described herein was commonly used in previous studies [102–105].

5. Conclusions

According to this study, freely available, high-spatial, -temporal, and -resolution multispectral Sentinel-2 data are suitable for estimating *C. betulus* AGB at a small scale over large areas. Our findings showed that in situ AGB is negatively correlated with 19 variables (original spectral bands, vegetation indices, and the first principal component of PCA) extracted from Sentinel-2 data. This negative association was expected to be caused by an increased canopy shadowing of trees, canopy size, stand volume and density, and consequently, a more complex vertical structure. We conclude that nonparametric models (ANN, kNN, and RF) performed slightly better than MR to estimate AGB, because these models are able to account for nonlinear relations between the forest features and AGB. From the group of nonparametric models tested in this study, the use of ANN returned the best result. Therefore, Sentinel-2 data stand as an important information source for assessing and monitoring forest biomass at local and regional scales in complex forest stands. In addition, the efficiency of the models used in this study can inform the selection of predictive mapping techniques for forest AGB modeling.

Author Contributions: Conceptualization, F.M. and A.A.D.; data curation, A.A.D.; formal analysis, F.M., A.A.D. and A.D.; funding acquisition, S.A.B.; methodology, F.M. and A.A.D.; software, F.M. and A.D.; supervision, A.A.D. and S.A.B.; validation, F.M. and A.A.D.; writing—original draft, F.M., A.A.D., M.R.P. and A.D.; writing—review and editing, S.A.B. All authors have read and agreed to the published version of the manuscript.

Funding: This research received no external funding and the APC was funded by the Department of Forest Engineering, Forest Management Planning and Terrestrial Measurements, Faculty of Silviculture and Forest Engineering, Transilvania University of Brasov.

Institutional Review Board Statement: Not applicable.

Informed Consent Statement: Not applicable.

Data Availability Statement: The data that support the findings of this study are available from the first corresponding author (Fardin Moradi), upon reasonable request.

Acknowledgments: We would like to thank Bahram Salehi (State University of New York, USA) for language editing and Ghasem Ronoud (University of Tehran, Iran) for his skilled technical assistance.

Conflicts of Interest: The authors declare that they have no conflict of interest.

Appendix A

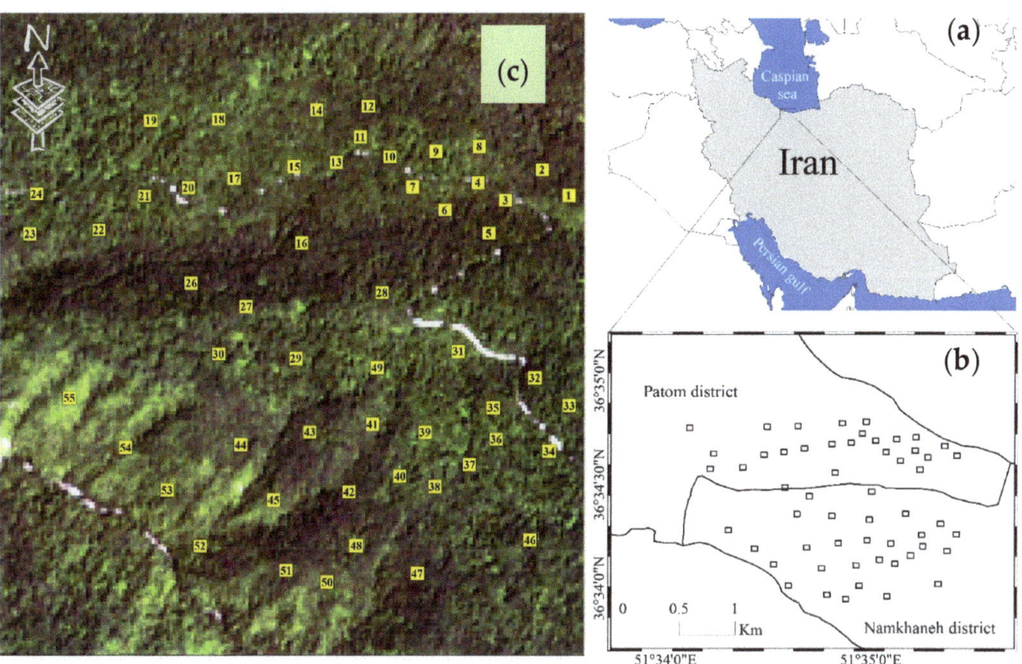

Figure A1. Map of Iran (**a**); location of sample plots in Patom and Namkhaneh district (**b**); location of sample plots over the Sentiel-2 image (**c**).

Table A1. Number of trees, mean value of DBH, volume, and AGB estimation per sample plot.

Plot No.	Number of Trees	Mean Value of DBH (cm)	Volume (m³/ha)	AGB (t/ha)
1	36	39	326	221
2	46	41	439	298
3	46	41	399	271
4	48	39	463	315
5	57	37	316	215
6	70	32	463	315
7	45	35	292	198
8	56	30	420	286
9	88	30	353	240
10	32	50	305	208
11	30	35	390	266
12	42	30	236	160
13	60	22	174	118
14	97	20	366	249
15	88	20	190	129
16	94	24	273	185
17	87	30	284	193
18	60	27	240	163
19	61	26	229	155
20	95	23	260	177
21	81	23	188	128
22	92	26	224	152
23	72	26	337	229

Table A1. *Cont.*

Plot No.	Number of Trees	Mean Value of DBH (cm)	Volume (m^3/ha)	AGB (t/ha)
24	62	35	183	124
25	52	27	234	159
26	61	32	403	274
27	42	46	385	262
28	26	44	304	207
29	32	53	184	125
30	29	41	234	159
31	36	31	470	320
32	30	36	295	201
33	40	33	428	291
34	40	44	338	230
35	27	40	261	178
36	45	34	338	230
37	38	32	189	128
38	48	32	240	163
39	55	38	189	128
40	35	28	461	314
41	50	42	358	244
42	34	30	368	250
43	29	40	272	185
44	36	34	276	188
45	66	47	304	207
46	28	34	452	307
47	57	41	338	230
48	29	36	307	209
49	45	34	218	148
50	50	43	319	217
51	49	43	344	234
52	47	31	278	189
53	55	31	207	141
54	75	29	175	119
55	61	21	446	303
Minimum	26	20	174	118
Maximum	97	53	470	320
Mean	52.58	33.96	308.45	209.73
Variance	397.62	59.85	7752.70	3598.24
Standard Deviation	19.94	7.74	88.05	59.99

References

1. Phat, N.K.; Knorr, W.; Kim, S. Appropriate measures for conservation of terrestrial carbon stocks—Analysis of trends of forest management in Southeast Asia. *For. Ecol. Manag.* **2004**, *191*, 283–299. [CrossRef]
2. Nalaka, G.D.A.; Sivanathawerl, T.; Iqbal, M.C.M. Scaling aboveground biomass from small diameter trees. *Trop. Agric. Res.* **2013**, *24*, 150–162.
3. Darkwah, S.O.; Scoville, M.D.; Wang, L.K. Geographic information systems and remote sensing applications in environmental and water resources. In *Integrated Natural Resources Management*; Springer: Cham, Switzerland, 2021; pp. 197–236.
4. Holmgren, J.; Joyce, S.; Nilsson, M.; Olsson, H. Estimating stem volume and basal area in forest compartments by combining satellite image data with field data. *Scand. J. For. Res.* **2000**, *15*, 103–111. [CrossRef]
5. Agata, H.; Aneta, L.; Dariusz, Z.; Krzysztof, S.; Marek, L.; Christiane, S.; Carsten, P. Forest aboveground biomass estimation using a combination of Sentinel-1 and Sentinel-2 Data. In Proceedings of the IGARSS 2018—2018 IEEE International Geoscience and Remote Sensing Symposium, Valencia, Spain, 22–27 July 2018; pp. 9026–9029.
6. Aranha, J.; Enes, T.; Calvão, A.; Viana, H. Shrub biomass estimates in former burnt areas using Sentinel 2 images processing and classification. *Forests* **2020**, *11*, 555. [CrossRef]
7. Chen, L.; Ren, C.; Zhang, B.; Wang, Z.; Xi, Y. Estimation of forest above-ground biomass by geographically weighted regression and machine learning with sentinel imagery. *Forests* **2018**, *9*, 582. [CrossRef]
8. Heiskanen, J. Estimating aboveground tree biomass and leaf area index in a mountain birch forest using ASTER satellite data. *Int. J. Remote Sens.* **2006**, *27*, 1135–1158. [CrossRef]

9.   Li, L.; Zhou, X.; Chen, L.; Chen, L.; Zhang, Y.; Liu, Y. Estimating urban vegetation biomass from Sentinel-2A image data. *Forests* **2020**, *11*, 125. [CrossRef]

10. Mohammadi, J. Investigating Estimation Some Quantitative Characteristics for Presentation Location Models Using Landsat ETM+ Satellite Data. Master's Thesis, Gorgan University of Agriculture and Natural Sciences, Gorgan, Iran, 2007.

11. Moradi, F.; Darvishsefat, A.A.; Namiranian, M.; Ronoud, G. Investigating the capability of Landsat 8 OLI data for estimation of aboveground woody biomass of common hornbeam (*Carpinus betulus* L.) stands in Khyroud Forest. *Iran. J. For. Pop. Res.* **2018**, *26*, 406–420.

12. Safari, A.; Sohrabi, H. Ability of Landsat-8 OLI derived texture metrics in estimating aboveground carbon stocks of coppice Oak Forests. In Proceedings of the International Archives of the Photogrammetry, Remote Sensing and Spatial Information Sciences, Prague, Czech Republic, 12–19 July 2016; Volume 41.

13. Zheng, D.; Rademacher, J.; Chen, J.; Crow, T.; Bresee, M.; Le Moine, J.; Ryu, S.R. Estimating aboveground biomass using Landsat 7 ETM+ data across a managed landscape in northern Wisconsin, USA. *Remote Sens. Environ.* **2004**, *93*, 402–411. [CrossRef]

14. Gao, L.; Zhang, X. Above-Ground Biomass Estimation of Plantation with Complex Forest Stand Structure Using Multiple Features from Airborne Laser Scanning Point Cloud Data. *Forests* **2021**, *12*, 1713. [CrossRef]

15. Wu, C.; Tao, H.; Zhai, M.; Lin, Y.; Wang, K.; Deng, J.; Shen, A.; Gan, M.; Li, J.; Yang, H. Using nonparametric modeling approaches and remote sensing imagery to estimate ecological welfare forest biomass. *J. For. Res.* **2018**, *29*, 151–161. [CrossRef]

16. Roy, S.; Mudi, S.; Das, P.; Ghosh, S.; Shit, P.K.; Bhunia, G.S.; Kim, J. Estimating Above Ground Biomass (AGB) and Tree Density using Sentinel-1 Data. In *Spatial Modeling in Forest Resources Management*; Springer: Cham, Switzerland, 2021; pp. 259–280.

17. Kumar, L.; Mutanga, O. Remote sensing of above-ground biomass. *Remote Sens.* **2017**, *9*, 935. [CrossRef]

18. Sullivan, M.J.; Lewis, S.L.; Hubau, W.; Qie, L.; Baker, T.R.; Banin, L.F.; Chave, J.; Cuni-Sanchez, A.; Feldpausch, T.R.; Lopez-Gonzalez, G.; et al. Field methods for sampling tree height for tropical forest biomass estimation. *Methods Ecol. Evol.* **2018**, *9*, 1179–1189. [CrossRef] [PubMed]

19. Gao, Y.; Lu, D.; Li, G.; Wang, G.; Chen, Q.; Liu, L.; Li, D. Comparative Analysis of modeling algorithms for forest aboveground biomass estimation in a Subtropical region. *Remote Sens.* **2018**, *10*, 627. [CrossRef]

20. Foody, G.M.; Cutler, M.E.; McMorrow, J.; Pelz, D.; Tangki, H.; Boyd, D.S.; Douglas, I. Mapping the biomass of bornean tropical rain forest from remotely sensed data. *Glob. Ecol. Biogeogr.* **2001**, *10*, 379–387. [CrossRef]

21. Chirici, G.; Barbati, A.; Corona, P.; Marchetti, M.; Travaglini, D.; Maselli, F.; Bertini, R. Non-parametric and parametric methods using satellite images for estimating growing stock volume in alpine and Mediterranean forest ecosystems. *Remote Sens. Environ.* **2008**, *112*, 2686–2700. [CrossRef]

22. Gu, H.; Dai, L.; Wu, G.; Xu, D.; Wang, S.; Wang, H. Estimation of forest volumes by integrating Landsat TM imagery and forest inventory data. *Sci. China Ser. E Technol. Sci.* **2006**, *49*, 54–62. [CrossRef]

23. McRoberts, R.E.; Tomppo, E.O. Remote sensing support for national forest inventories. *Remote Sens. Environ.* **2007**, *110*, 412–419. [CrossRef]

24. Eskelson, B.N.; Barrett, T.M.; Temesgen, H. Imputing mean annual change to estimate current forest attributes. *Silva Fenn.* **2009**, *43*, 649–658. [CrossRef]

25. Hudak, A.T.; Strand, E.K.; Vierling, L.A.; Byrne, J.C.; Eitel, J.U.H.; Martinuzzi, S.; Falkowski, M.J. Quantifying aboveground forest carbon pools and fluxes from repeat LiDAR surveys. *Remote Sens. Environ.* **2012**, *123*, 25–40. [CrossRef]

26. Rajab Pourrahmati, M.; Baghdadi, N.; Darvishsefat, A.; Namiranian, M.; Gond, V.; Bailly, J.S.; Zargham, N. Mapping Lorey's height over Hyrcanian forests of Iran using synergy of ICESat/GLAS and optical images. *Eur. J. Remote Sens.* **2018**, *51*, 100–115. [CrossRef]

27. Breiman, L. Bagging predictors. *Mach. Learn.* **1996**, *24*, 123–140. [CrossRef]

28. Breiman, L. Random forests. *Mach. Learn.* **2001**, *45*, 5–32. [CrossRef]

29. Shahrokhzadeh, U.; Sohrabi, H.; Copenheaver, C.A. Aboveground biomass and leaf area equations for three common tree species of Hyrcanian temperate forests in northern Iran. *Botany* **2015**, *93*, 663–670. [CrossRef]

30. Muukkonen, P.; Heiskanen, J. Estimating biomass for boreal forests using ASTER satellite data combined with standwise forest inventory data. *Remote Sens. Environ.* **2005**, *99*, 434–447. [CrossRef]

31. Yadav, B.K.; Nandy, S. Mapping aboveground woody biomass using forest inventory, remote sensing and geostatistical techniques. *Environ. Monit. Assess.* **2015**, *187*, 308. [CrossRef]

32. Lu, D.; Chen, Q.; Wang, G.; Liu, L.; Li, G.; Moran, E. A survey of remote sensing-based aboveground biomass estimation methods in forest ecosystems. *Int. J. Dig. Earth* **2016**, *9*, 63–105. [CrossRef]

33. Ronoud, G.; Darvishsefat, A. Estimating aboveground woody biomass of *Fagus orientalis* stands in Hyrcanian forest of Iran using Landsat 5 satellite data (Case study: Khyroud Forest). *Geogr. Space* **2018**, *17*, 117–129.

34. Forkuor, G.; Zoungrana, J.B.B.; Dimobe, K.; Ouattara, B.; Vadrevu, K.P.; Tondoh, J.E. Above-ground biomass mapping in West African dryland forest using Sentinel-1 and 2 datasets-A case study. *Remote Sens. Environ.* **2020**, *236*, 111496. [CrossRef]

35. Laurin, G.V.; Balling, J.; Corona, P.; Mattioli, W.; Papale, D.; Puletti, N.; Rizzo, M.; Truckenbrodt, J.; Urban, M. Above-ground biomass prediction by Sentinel-1 multitemporal data in central Italy with integration of ALOS2 and Sentinel-2 data. *J. Appl. Remote Sens.* **2018**, *12*, 016008. [CrossRef]

36. Theofanous, N.; Chrysafis, I.; Mallinis, G.; Domakinis, C.; Verde, N.; Siahalou, S. Aboveground Biomass Estimation in Short Rotation Forest Plantations in Northern Greece Using ESA's Sentinel Medium-High Resolution Multispectral and Radar Imaging Missions. *Forests* **2021**, *12*, 902. [CrossRef]
37. Castillo, J.A.A.; Apan, A.A.; Maraseni, T.N.; Salmo, S.G., III. Estimation and mapping of above-ground biomass of mangrove forests and their replacement land uses in the Philippines using Sentinel imagery. *ISPRS J. Photogram. Remote Sens.* **2017**, *134*, 70–85. [CrossRef]
38. Chen, L.; Wang, Y.; Ren, C.; Zhang, B.; Wang, Z. Assessment of multi-wavelength SAR and multispectral instrument data for forest aboveground biomass mapping using random forest kriging. *For. Ecol. Manag.* **2019**, *447*, 12–25. [CrossRef]
39. Ghosh, S.M.; Behera, M.D. Aboveground biomass estimation using multi-sensor data synergy and machine learning algorithms in a dense tropical forest. *Appl. Geogr.* **2018**, *96*, 29–40. [CrossRef]
40. Hernández-Stefanoni, J.L.; Castillo-Santiago, M.Á.; Mas, J.F.; Wheeler, C.E.; Andres-Mauricio, J.; Tun-Dzul, F.; George-Chacón, S.P.; Reyes-Palomeque, G.; Castellanos-Basto, B.; Vaca, R.; et al. Improving aboveground biomass maps of tropical dry forests by integrating LiDAR, ALOS PALSAR, climate and field data. *Carbon Balance Manag.* **2020**, *15*, 15. [CrossRef]
41. Khan, M.R.; Khan, I.A.; Baig, M.H.A.; Liu, Z.J.; Ashraf, M.I. Exploring the potential of Sentinel-2A satellite data for aboveground biomass estimation in fragmented Himalayan subtropical pine forest. *J. Mt. Sci.* **2020**, *17*, 2880–2896. [CrossRef]
42. Pandit, S.; Tsuyuki, S.; Dube, T. Estimating above-ground biomass in sub-tropical buffer zone community forests, Nepal, using Sentinel 2 data. *Remote Sens.* **2018**, *10*, 601. [CrossRef]
43. Majasalmi, T.; Rautiainen, M. The potential of Sentinel-2 data for estimating biophysical variables in a boreal forest: A simulation study. *Remote Sens. Lett.* **2016**, *7*, 427–436. [CrossRef]
44. Puliti, S.; Hauglin, M.; Breidenbach, J.; Montesano, P.; Neigh, C.S.R.; Rahlf, J.; Solberg, S.; Klingenberg, T.F.; Astrup, R. Modelling above-ground biomass stock over Norway using national forest inventory data with ArcticDEM and Sentinel-2 data. *Remote Sens. Environ.* **2020**, *236*, 111501. [CrossRef]
45. Wang, J.; Xiao, X.; Bajgain, R.; Starks, P.; Steiner, J.; Doughty, R.B.; Chang, Q. Estimating leaf area index and aboveground biomass of grazing pastures using Sentinel-1, Sentinel-2 and Landsat images. *ISPRS J. Photogram. Remote Sens.* **2019**, *154*, 189–201. [CrossRef]
46. Chrysafis, I.; Mallinis, G.; Siachalou, S.; Patias, P. Assessing the relationships between growing stock volume and Sentinel-2 imagery in a Mediterranean forest ecosystem. *Remote Sens. Lett.* **2017**, *8*, 508–517. [CrossRef]
47. Nuthammachot, N.; Phairuang, W.; Wicaksono, P.; Sayektiningsih, T. Estimating aboveground biomass on private forest using Sentinel-2 imagery. *J. Sens.* **2018**, *2018*, 6745629.
48. Vafaei, S.; Soosani, J.; Adeli, K.; Fadaei, H.; Naghavi, H.; Pham, T.D.; Tien Bui, D. Improving accuracy estimation of Forest Aboveground Biomass based on incorporation of ALOS-2 PALSAR-2 and Sentinel-2A imagery and machine learning: A case study of the Hyrcanian forest area (Iran). *Remote Sens.* **2018**, *10*, 172. [CrossRef]
49. Wernick, I.K.; Ciais, P.; Fridman, J.; Högberg, P.; Korhonen, K.T.; Nordin, A.; Kauppi, P.E. Quantifying forest change in the European Union. *Nature* **2021**, *592*, E13–E14. [CrossRef]
50. Food and Agriculture Organization (FAO). *Global Forest Resources Assessment (FRA)*; Report, Iran (Islamic Republic of); FAO: Rome, Italy, 2020.
51. Deljouei, A.; Sadeghi, S.M.M.; Abdi, E.; Bernhardt-Römermann, M.; Pascoe, E.L.; Marcantonio, M. The impact of road disturbance on vegetation and soil properties in a beech stand, Hyrcanian forest. *Eur. J. For. Res.* **2018**, *137*, 759–770. [CrossRef]
52. Rahbarisisakht, S.; Moayeri, M.H.; Hayati, E.; Sadeghi, S.M.M.; Kepfer-Rojas, S.; Pahlavani, M.H.; Kappel Schmidt, I.; Borz, S.A. Changes in Soil's Chemical and Biochemical Properties Induced by Road Geometry in the Hyrcanian Temperate Forests. *Forests* **2021**, *12*, 1805. [CrossRef]
53. Abbasian, P.; Attarod, P.; Sadeghi, S.M.M.; Van Stan, J.T.; Hojjati, S.M. Throughfall nutrients in a degraded indigenous *Fagus orientalis* forest and a *Picea abies* plantation in the of North of Iran. *For. Syst.* **2015**, *24*, 1. [CrossRef]
54. Marvie-Mohadjer, M.R. *Silviculture*, 5th ed.; University of Tehran Press: Tehran, Iran, 2019; p. 418.
55. Haghshenas, M.; Mohadjer, M.R.M.; Attarod, P.; Pourtahmasi, K.; Feldhaus, J.; Sadeghi, S.M.M. Climate effect on tree-ring widths of *Fagus orientalis* in the Caspian forests, northern Iran. *For. Sci. Technol.* **2016**, *12*, 176–182.
56. Lu, D.; Batistella, M.; Moran, E. Satellite estimation of aboveground biomass and impacts of forest stand structure. *Photogramm. Eng. Remote Sens.* **2005**, *71*, 967–974. [CrossRef]
57. Drusch, M.; Del Bello, U.; Carlier, S.; Colin, O.; Fernandez, V.; Gascon, F.; Hoersch, B.; Isola, C.; Laberinti, P.; Martimort, P.; et al. Sentinel-2: ESA's optical high-resolution mission for GMES operational services. *Remote Sens. Environ.* **2012**, *120*, 25–36. [CrossRef]
58. Crippen, R.E. Calculating the vegetation index faster. *Remote Sens. Environ.* **1990**, *34*, 71–73. [CrossRef]
59. Frampton, W.J.; Dash, J.; Watmough, G.; Milton, E.J. Evaluating the capabilities of Sentinel-2 for quantitative estimation of biophysical variables in vegetation. *ISPRS J. Photogram. Remote Sens.* **2013**, *82*, 83–92. [CrossRef]
60. Pinty, B.; Verstraete, M. GEMI: A non-linear index to monitor global vegetation from satellites. *Vegetatio* **1992**, *101*, 15–20. [CrossRef]
61. Gitelson, A.A.; Kaufman, Y.J.; Merzlyak, M.N. Use of a green channel in remote sensing of global vegetation from EOS-MODIS. *Remote Sens. Environ.* **1996**, *58*, 289. [CrossRef]

62. Rouse, J.W.; Haas, R.H.; Schell, J.A.; Deering, D.W. Monitoring vegetation systems in the Great Plains with ERTS. *NASA Spec. Publ.* **1974**, *351*, 309.
63. Tucker, C.J.; Elgin, J., Jr.; McMurtrey, J., III; Fan, C. Monitoring corn and soybean crop development with hand-held radiometer spectral data. *Remote Sens. Environ.* **1979**, *8*, 237–248. [CrossRef]
64. Roujean, J.L.; Breon, F.M. Estimating PAR absorbed by vegetation from bidirectional reflectance measurements. *Remote Sens. Environ.* **1995**, *51*, 375–384. [CrossRef]
65. Jordan, C.F. Derivation of leaf-area index from quality of light on the forest floor. *Ecology* **1969**, *50*, 663–666. [CrossRef]
66. Brown, S. *Estimating Biomass and Biomass Change of Tropical Forests: A Primer*; FAO: Rome, Italy, 1997.
67. Safdari, V.; Ahmed, M.; Palmer, J.; Baig, M.B. Identification of Iranian commercial wood with hand lens. *Pak. J. Bot.* **2008**, *40*, 1851–1864.
68. Tomppo, E.; Halme, M. Using coarse scale forest variables as ancillary information and weighting of variables in k-NN estimation: A genetic algorithm approach. *Remote Sens. Environ.* **2004**, *92*, 1–20. [CrossRef]
69. Tian, X.; Su, Z.; Chen, E.; Li, Z.; van der Tol, C.; Guo, J.; He, Q. Estimation of forest above-ground biomass using multi-parameter remote sensing data over a cold and arid area. *Int. J. Appl. Earth Obs. Geoinf.* **2012**, *14*, 160–168. [CrossRef]
70. Tojal, L.T.; Bastarrika, A.; Barrett, B.; Sanchez Espeso, J.M.; Lopez-Guede, J.M.; Graña, M. Prediction of Aboveground Biomass from Low-Density LiDAR Data: Validation over *P. radiata* Data from a Region North of Spain. *Forests* **2019**, *10*, 819. [CrossRef]
71. Civco, D.L. Artificial neural networks for land-cover classification and mapping. *Int. J. Geogr. Inf. Sci.* **1993**, *7*, 173–186. [CrossRef]
72. Mas, J. Mapping land use/cover in a tropical coastal area using satellite sensor data, GIS and artificial neural networks. *Estuar. Coast. Shelf Sci.* **2004**, *59*, 219–230. [CrossRef]
73. Cheţa, M.; Marcu, M.V.; Borz, S.A. Effect of training parameters on the ability of artificial neural networks to learn: A Simulation on accelerometer data for task recognition in motormanual felling and processing. *Bull. Transilv. Univ. Bras. Ser. II For. Wood Ind. Agric. Food Eng.* **2020**, *13*, 19–36. [CrossRef]
74. Zealand, C.M.; Burn, D.H.; Simonovic, S.P. Short term streamflow forecasting using artificial neural networks. *J. Hydrol.* **1999**, *214*, 32–48. [CrossRef]
75. Tiryaki, S.; Aydın, A. An artificial neural network model for predicting compression strength of heat treated woods and comparison with a multiple linear regression model. *Constr. Build. Mater.* **2014**, *62*, 102–108. [CrossRef]
76. Pham, T.D.; Yoshino, K.; Bui, D.T. Biomass estimation of *Sonneratia caseolaris* (L.) Engler at a coastal area of Hai Phong city (Vietnam) using ALOS-2 PALSAR imagery and GIS-based multi-layer perceptron neural networks. *GISci. Remote Sens.* **2017**, *54*, 329–353. [CrossRef]
77. Bui, D.T.; Tuan, T.A.; Klempe, H.; Pradhan, B.; Revhaug, I. Spatial prediction models for shallow landslide hazards: A comparative assessment of the efficacy of support vector machines, artificial neural networks, kernel logistic regression, and logistic model tree. *Landslides* **2016**, *13*, 361–378.
78. Shataee, S.; Kalbi, S.; Fallah, A.; Pelz, D. Forest attribute imputation using machine-learning methods and ASTER data: Comparison of k-NN, SVR and random forest regression algorithms. *Int. J. Remote Sens.* **2012**, *33*, 6254–6280. [CrossRef]
79. Statistica. *Electronic Text Book*; Stat Soft Inc.: Tulsa, OK, USA, 2010; Available online: www.Statsoft.com (accessed on 22 October 2020).
80. Franco-Lopez, H.; Ek, A.R.; Bauer, M.E. Estimation and mapping of forest stand density, volume, and cover type using the k-nearest neighbors method. *Remote Sens. Environ.* **2001**, *77*, 251–274. [CrossRef]
81. Reese, H.; Nilsson, M.; Sandström, P.; Olsson, H. Applications using estimates of forest parameters derived from satellite and forest inventory data. *Comput. Electron. Agric.* **2002**, *37*, 37–55. [CrossRef]
82. Sironen, S.; Kangas, A.; Maltamo, M. Comparison of different non-parametric growth imputation methods in the presence of correlated observations. *Forestry* **2010**, *83*, 39–51. [CrossRef]
83. Pelletier, C.; Valero, S.; Inglada, J.; Champion, N.; Dedieu, G. Assessing the robustness of Random Forests to map land cover with high resolution satellite image time series over large areas. *Remote Sens. Environ.* **2016**, *187*, 156–168. [CrossRef]
84. Friedman, J.H. *The Elements of Statistical Learning: Data Mining, Inference, and Prediction*; Springer: New York, NY, USA, 2017.
85. Gislason, P.O.; Benediktsson, J.A.; Sveinsson, J.R. Random forests for land cover classification. *Pattern Recognit. Lett.* **2006**, *27*, 294–300. [CrossRef]
86. Rodriguez-Galiano, V.F.; Ghimire, B.; Rogan, J.; Chica-Olmo, M.; Rigol-Sanchez, J.P. An assessment of the effectiveness of a random forest classifier for land-cover classification. *ISPRS J. Photogramm. Remote Sens.* **2012**, *67*, 93–104. [CrossRef]
87. Stevens, F.R.; Gaughan, A.E.; Linard, C.; Tatem, A.J. Disaggregating census data for population mapping using random forests with remotely-sensed and ancillary data. *PLoS ONE* **2015**, *10*, e0107042. [CrossRef]
88. Tavasoli, N.; Arefi, H. Comparison of capability of SAR and optical data in mapping forest above ground biomass based on machine learning. *Environ. Sci. Proc.* **2020**, *5*, 13. [CrossRef]
89. Baloloy, A.B.; Blanco, A.C.; Candido, C.G.; Argamosa, R.J.L.; Dumalag, J.B.L.C.; Dimapilis, L.L.C.; Paringit, E.C. Estimation of mangrove forest aboveground biomass using multispectral bands, vegetation indices and biophysical variables derived from optical satellite imageries: Rapideye, planetscope and sentinel-2. *ISPRS Ann. Photogramm. Remote Sens. Spat. Inf. Sci.* **2018**, *4*, 29–36. [CrossRef]
90. Lu, D. Aboveground biomass estimation using Landsat TM data in the Brazilian Amazon. *Int. J. Remote Sens.* **2005**, *26*, 2509–2525. [CrossRef]

91. Roy, P.; Ravan, S.A. Biomass estimation using satellite remote sensing data—An investigation on possible approaches for natural forest. *J. Biosci.* **1996**, *21*, 535–561. [CrossRef]
92. Liu, Y.; Gong, W.; Xing, Y.; Hu, X.; Gong, J. Estimation of the forest stand mean height and aboveground biomass in Northeast China using SAR Sentinel-1B, multispectral Sentinel-2A, and DEM imagery. *ISPRS J. Photogramm. Remote Sens.* **2019**, *151*, 277–289. [CrossRef]
93. Rudnicki, M.; Silins, U.; Lieffers, V.J. Crown cover is correlated with relative density, tree slenderness, and tree height in lodgepole pine. *For. Sci.* **2004**, *50*, 356–363.
94. Butera, M.K. A correlation and regression analysis of percent canopy closure versus TMS spectral response for selected forest sites in the San Juan National Forest, Colorado. *IEEE Trans. Geosci. Remote Sens.* **1986**, *1*, 122–129. [CrossRef]
95. Sadeghi, S.M.M.; Van Stan, J.T., II; Pypker, T.G.; Friesen, J. Canopy hydrometeorological dynamics across a chronosequence of a globally invasive species, *Ailanthus altissima* (Mill., tree of heaven). *Agric. For. Meteorol.* **2017**, *240*, 10–17. [CrossRef]
96. Khorrami, R.; Darvishsefat, A.; Namiranian, M. Investigation on the capability of Landsat7 ETM+ data for standing volume estimation of beech stands (Case study: Sangdeh forests). *Iran. J. Nat. Res.* **2008**, *60*, 1281–1289.
97. Tóth, T.; Schaap, M.G.; Molnár, Z. Utilization of soil-plant interrelations through the use of multiple regression and artificial neural network in order to predict soil properties in Hungarian solonetzic grasslands. *Cereal Res. Commun.* **2008**, *36*, 1447–1450.
98. Lu, D. The potential and challenge of remote sensing-based biomass estimation. *Int. J. Remote Sens.* **2006**, *27*, 1297–1328. [CrossRef]
99. Fatehi, P.; Damm, A.; Schweiger, A.K.; Schaepman, M.E.; Kneubühler, M. Mapping alpine aboveground biomass from imaging spectrometer data: A Comparison of two approaches. *IEEE J. Sel. Top. Appl. Earth Obs. Remote Sens.* **2015**, *8*, 3123–3139. [CrossRef]
100. Kellndorfer, J.; Walker, W.; LaPoint, E.; Kirsch, K.; Bishop, J.; Fiske, G. Statistical fusion of Lidar, InSAR, and optical remote sensing data for forest stand height characterization: A regional-scale method based on LVIS, SRTM, Landsat ETM+, and ancillary data sets. *J. Geophys. Res. Biogeosci.* **2010**, *115*, G00E08. [CrossRef]
101. Ji, L.; Wylie, B.K.; Nossov, D.R.; Peterson, B.; Waldrop, M.P.; McFarland, J.W.; Rover, J.; Hollingsworth, T.N. Estimating aboveground biomass in interior Alaska with Landsat data and field measurements. *Int. J. Appl. Earth Obs. Geoinf.* **2012**, *18*, 451–461. [CrossRef]
102. Chen, Y.; Li, L.; Lu, D.; Li, D. Exploring bamboo forest aboveground biomass estimation using Sentinel-2 data. *Remote Sens.* **2019**, *11*, 7. [CrossRef]
103. Jiang, X.; Li, G.; Lu, D.; Moran, E.; Batistella, M. Modeling Forest Aboveground Carbon Density in the Brazilian Amazon with Integration of MODIS and Airborne LiDAR Data. *Remote Sens.* **2020**, *12*, 3330. [CrossRef]
104. Persson, H.J.; Jonzén, J.; Nilsson, M. Combining TanDEM-X and Sentinel-2 for large-area species-wise prediction of forest biomass and volume. *Int. J. Appl. Earth Obs. Geoinf.* **2021**, *96*, 102275. [CrossRef]
105. Ronoud, G.; Fatehi, P.; Darvishsefat, A.A.; Tomppo, E.; Praks, J.; Schaepman, M.E. Multi-Sensor Aboveground Biomass Estimation in the Broadleaved Hyrcanian Forest of Iran. *Can. J. Remote Sens.* **2021**, *47*, 818–834. [CrossRef]

Article

Analysis of Factors Influencing Forest Loss in South Korea: Statistical Models and Machine-Learning Model

Jeongmook Park [1], Byeoungmin Lim [2] and Jungsoo Lee [3,*]

[1] Forest ICT Research Center, National Institute of Forest Science, Seoul 02455, Korea; doradoman@naver.com
[2] Department of Forest Management, Kangwon National University, Chuncheon 24341, Korea; bmhanlim@gmail.com
[3] Division of Forest Sciences, Kangwon National University, Chuncheon 24341, Korea
* Correspondence: jslee72@kangwon.ac.kr

Abstract: Analyzing the current status of forest loss and its causes is crucial for understanding and preparing for future forest changes and the spatial pattern of forest loss. We investigated spatial patterns of forest loss in South Korea and assessed the effects of various factors on forest loss based on spatial heterogeneity. We used the local Moran's I to classify forest loss spatial patterns as high–high clusters, low–low clusters, high–low outliers, and high–low outliers. Additionally, to assess the effect of factors on forest loss, two statistical models (i.e., ordinary least squares regression (OLS) and geographically weighted regression (GWR) models) and one machine-learning model (i.e., random forest (RF) model) were used. The accuracy of each model was determined using the R^2, RMSE, MAE, and AICc. Across South Korea, the forest loss rate was highest in the Seoul–Incheon–Gyeonggi region. Moreover, high–high spatial clusters were found in the Seoul–Incheon–Gyeonggi and Daejeon–Chungnam regions. Among the models, the GWR model was the most accurate. Notably, according to the GWR model, the main factors driving forest loss were road density, cropland area, number of households, and number of tertiary industry establishments. However, the factors driving forest loss had varying degrees of influence depending on the location. Therefore, our findings suggest that spatial heterogeneity should be considered when developing policies to reduce forest loss.

Keywords: forest loss; land-cover change; machine learning; spatial heterogeneity; random forest model; geographically weighted regression

Citation: Park, J.; Lim, B.; Lee, J. Analysis of Factors Influencing Forest Loss in South Korea: Statistical Models and Machine-Learning Model. *Forests* **2021**, *12*, 1636. https://doi.org/10.3390/f12121636

Academic Editor: Olga Viedma

Received: 1 October 2021
Accepted: 23 November 2021
Published: 25 November 2021

Publisher's Note: MDPI stays neutral with regard to jurisdictional claims in published maps and institutional affiliations.

1. Introduction

The global forested area is 4.06 billion ha, which accounts for approximately 31% of the total land area; global forest loss since the 1990s has reached approximately 0.42 billion ha [1]. Forest loss increases ground surface temperatures, reduces ecosystem services, and exacerbates climate change [2]. Climate change is caused by factors such as construction and transportation [3,4]. Forest loss can be driven by human activity and biophysical characteristics (i.e., roads, construction, expansion of settlements, industry, wildfires, agricultural activities, mining, industrial logging, etc.) that directly affect forests and cause canopy loss [5]. In particular, the expansion of urban infrastructures, such as roads, transportation, and settlements, causes permanent forest loss [6,7]. Additionally, demand for forest products and the conversion of native forests into commercial forests can simplify forest vegetation structure and reduce biodiversity [8,9]. Therefore, reducing forest loss is necessary to restore and improve the function of forests [10].

In South Korea, the ratio of forest area is about 63%, which is the fourth highest among OECD countries, following Finland, Sweden, and Japan, with a high forest area ratio compared to the global average forest area ratio [11]. However, the forest cover decreased by approximately 3% in 2019 compared to in 1990, with a mean annual decline of 0.1% [12]. This is a higher figure than the 1.7% decrease in the global forest area ratio over the past 30 years, so it is necessary to reduce it by analyzing the causes of forest loss [1]. According to Kim and Hwang, continuous damage to the forest in South Korea has been reported due to tourist sites, golf courses, industrial complexes, housing areas, road construction, and various other factors [13]. To decrease the rate of forest loss, it is necessary to quantitatively analyze the area of forest loss. Additionally, human socioeconomic factors associated with forest loss need to be determined [14]. Recent improvements to geographic information system (GIS) and remote sensing (RS) tools have enabled the rapid collection of data regarding regional forest conversion and loss [15]. The collected data can be analyzed using various techniques, including statistical approaches and machine-learning models, to examine the spatial distribution characteristics of forest loss and the causes of forest loss [16]. Forest loss can occur due to the conversion of forest to many different land uses, and this process is affected by various spatial and socioeconomic factors. Verburg et al. [17] showed that road construction increases human movement and economic activities, which increases the conversion of forest to croplands and grasslands. Damnyag et al. [18] found that, in Ghana, croplands affected forest loss. Scullion et al. [19] pointed out that pasture expansion is the direct cause of forest loss worldwide, with the causes varying among each continent. Echeverria et al. [20] showed that forests closer to rivers were more likely to be lost. Forest in lower altitudes is less accessible; therefore, forest loss is less likely to occur. Similarly, Gayen and Saha [21] showed that forest with a higher slope is less accessible and less likely to experience forest loss. Sharma et al. [22] showed that commercial land use (mining and transportation development) and infrastructure development increased forest loss due to the expansion of surrounding urban areas.

Given that the factors mentioned above vary spatially [23–25], their spatial heterogeneity should be considered when determining their impact on forest loss [26,27]. Therefore, the spatial distribution of forest loss and the relationship between forest loss and its occurrence factors should be analyzed. Regional spatial patterns of forest cover can be quantitatively analyzed using the local Moran's I, first proposed by Anselin [28]. This technique enables statistically significant spatial clusters and outliers to be measured according to characteristics of the forest loss rate of a given area to quantitatively determine the forest loss rate [29]. Correlations between forest loss and various factors have been conducted using statistical models (e.g., ordinary least squares regression (OLS) and geographically weighted regression (GWR) models) and machine-learning model (e.g., random forest (RF) model) [30–32]. The OLS model does not consider the spatial heterogeneity of the area when analyzing correlations among factors, whereas GWR incorporates spatial heterogeneity and, therefore, can provide useful visual information to identify factors impacting forest loss [33]. The GWR model estimates discrete parameters by providing the higher weighted value closer to the observation location [34]. The RF model does not consider spatial heterogeneity; they are similar to the OLS model that provides a single result for the entire range of the research area with high predictive accuracy and efficiency [35,36]. However, the RF model can be used for both classification and regression, which is advantageous for obtaining results very quickly [37]. Nevertheless, the OLS and RF models have rarely been applied to analyze the factors affecting forest loss in South Korea.

In this study, we analyzed the areas of forest loss in South Korea and the factors driving this forest loss. The specific goals were as follows: (1) The distribution of the forest loss area was analyzed using local Moran's I. (2) The suitability of models (OLS, GWR, and RF) to evaluate factors affecting the forest loss rate was compared. (3) Factors affecting forest loss in each region were analyzed. Understanding the causes of forest loss and forest distribution status may contribute to the development of measures that prevent forest loss. In the future, this study can be used to establish forest management policies to prevent forest loss.

2. Materials and Methods

2.1. Study Site

The study was conducted in South Korea at 125°–131° longitude and 33°–38° latitude and included administrative districts, as well as one special city, one special self-governing city, six metropolitan cities, eight provinces, and one special self-governing province. The total area of the study site comprised approximately 10.04 million ha with 63% forest. The highest forest area % of land area was in the Gangwon region (81%), followed by the Daegu–Gyeongbuk (69%) and Busan–Ulsan–Gyeongnam (65%) regions. Forests in the north and east were generally at higher altitudes, and those in the west and south were generally at lower altitudes; however, there were substantial variations in the mean altitude and slope [38]. The study site was divided into eight "spatial regions" containing 152 "spatial areas", based on eight provinces to which each of the seven metropolitan cities belonged (Table 1 and Figure 1). To analyze forest loss and the factors impacting forest loss from a macroscopic perspective, spatial regions were defined by classifying metropolitan cities and provinces by region. Then, for a more detailed analysis, spatial areas were defined to analyze Seoul and other metropolitan cities and the special self-governing city under the same parameters as those used for general cities. Each spatial area was quantitatively analyzed as an independent unit, irrespective of the size of the cities or provinces. Additionally, Seogwipo-si, Jeju-si in the Jeju Special Self-Governing Province, and Ulleung-gun in Gyeongsangbuk-do, which are geographically remote islands, were excluded from analysis because they are distant from other regions, limiting the weighting in spatial pattern analysis [39]. Furthermore, Sejong Special Autonomous City, an administrative district designated in 2012, was excluded from the analysis due to a lack of statistical data from 2005 [40,41].

Table 1. Number of spatial areas and forest rate in each spatial region. For the study, "spatial regions" were defined and split into "spatial areas" (see Figure 1).

Spatial Region	Seoul–Incheon–Gyeonggi Region	Gangwon Region	Busan–Ulsan–Gyeongnam Region	Daegu Gyeongbuk Region	Gwangju–Jeonnam Region	Jeonbuk Region	Daejeon–Chungnam Region	Chungbuk Region
Spatial area (n)	32	15	20	22	22	14	16	11
Forest rate (%)	46	81	65	69	54	51	51	64

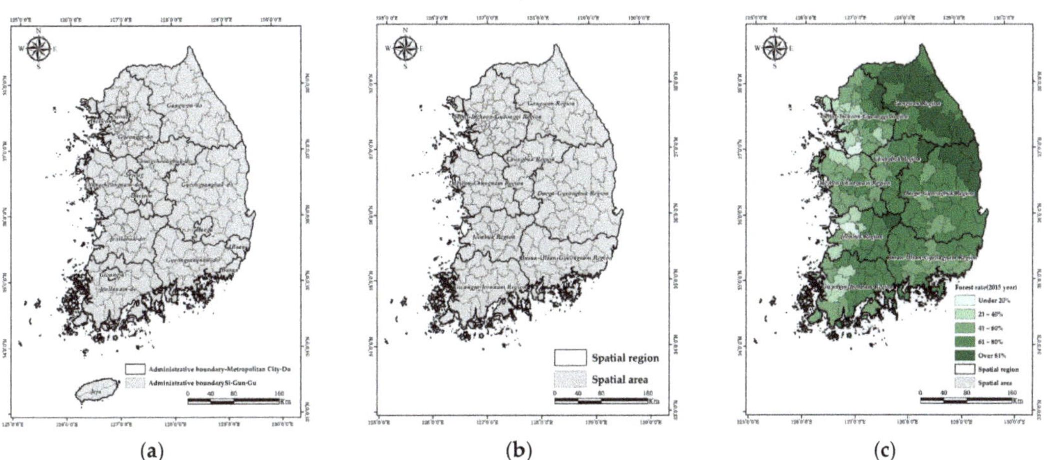

(a)   (b)   (c)

Figure 1. Study area location in South Korea. (**a**) Administrative boundaries at the Metropolitan City·Do level and Si·Gun·Gu level; (**b**) boundaries of spatial regions and spatial areas defined for the study; and (**c**) forest rate in each spatial area in 2015.

2.2. Data Collection

The data used in the status analysis of the spatial distribution of forest loss areas were obtained from the Forest Basic Statistics (FBS), which provides statistics on the current status of national forests in South Korea [42,43]. The FBS data included information regarding the forest type (coniferous forest, deciduous forest, or mixed forest) and age [44,45]. The

FBS data are published every five years; the 2015 and 2005 data were used to analyze changes in forest cover over a ten-year period. Census data and spatial data were used to determine factors affecting forest loss. Census data were obtained from the Cadastral Statistics Chronology [46,47]; the Agricultural Area Survey and the Agriculture, Forestry and Fishery Survey [48–51]; and the Survey of Establishments [52,53]. Spatial data were obtained from the road network map and the railway network map, which were produced in 2005 and 2015 by the Ministry of Land, Infrastructure and Transport (MOLIT). The Cadastral Statistics Chronology includes the area of 28 land categories, including forests, crop fields, paddies, and house sites, for each administrative district [54]. The Agricultural Area Survey provides data on the current status of agricultural land and cultivation within a selected sample area. The Agriculture, Forestry and Fishery Survey analyzes the distribution of agriculture, forestry, and fishery households, number of household members, and farms to construct the data in a cycle of one year and five years [55–57]. The Survey of Establishments collects annual data of each region's establishments, such as size, distribution, industry type, and employees [58]. The road and railway network maps provide the current status of roads and railways across the nation [59] (Table 2).

Table 2. Sources of data for factors affecting forest loss.

Category	Data	Institution	Year of Data Collection	Detailed Data
Census Data	Forest Basic Statistics	Korea Forest Service (KFS)	2005, 2015	Forested area, accumulation of standing
	Cadastral Statistics Chronology	Ministry of Land, Infrastructure and Transport (MOLIT)		28 land categories, including Forestry, Dry paddy-field, Paddy-field, and Building site
	Agricultural Area Survey	Statistics Korea (KOSTAT)		Current status of agricultural land
	Agriculture, Forestry and Fishery Survey			Number of household members, and farms
	Survey of Establishments	Ministry of Employment and Labor (MOEL)		Size, distribution of industry, industry type,
Spatial Data	Road network map	MOLIT		Spatial data
	Railway network map			

2.3. Study Method

In this study, a spatial database was constructed to analyze the forest loss rate for 2005–2015 and the spatial patterns of forest loss. Then, a spatial database for the factors potentially influencing forest loss was constructed. The impact of these factors on forest loss was then analyzed using statistical and machine-learning models (Figure 2).

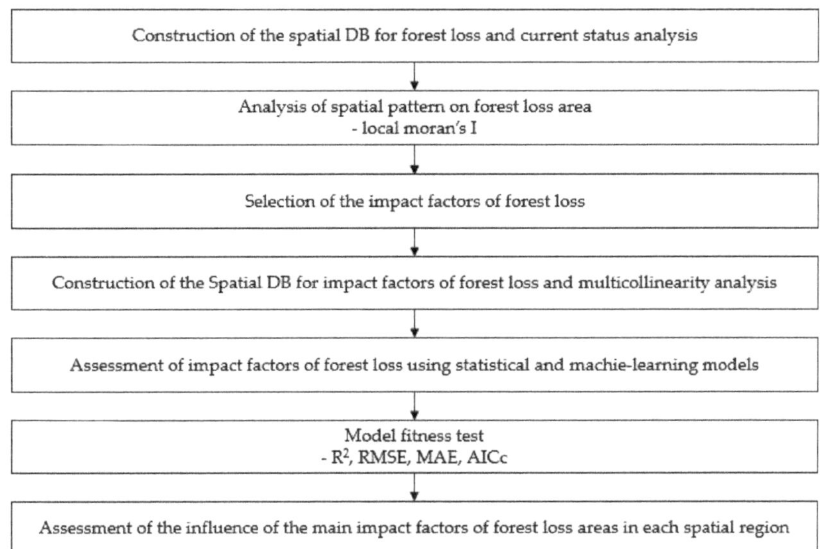

Figure 2. Workflow for analyzing the effects of factors on forest loss.

2.3.1. Construction of the Spatial DB for Forest Loss and Current Status Analysis

For the forest loss area, the data of forest area per spatial area were extracted from the 2005 and 2015 FBS data, and using Equation (1), the rate of change in the forest area was estimated. The rate of change in the forest area was negative (−) if the forest area had decreased, and the negative values were converted into positive (+) values to clearly identify the characteristics of forest loss area, while the spatial areas without a decrease in forest area were excluded from the analysis (Equation (1)).

$$\Delta v_t = \frac{(v_t - v_{t-1})}{v_{t-1}} \tag{1}$$

Δv_t: Forest loss rate at t time; v_t: Forest rate at t time;
v_{t-1} : Forest rate at $t-1$ time; v: Forest rate.

2.3.2. Analysis of Spatial Pattern on Forest Loss Area

The spatial autocorrelation patterns of the forest loss in each area were analyzed using global Moran's I. A Moran's I value >0 indicates that the forest loss area is clustered, and a value <0 indicates that the forest loss area is dispersed [60,61] (Equation (2)). The local Moran's I identifies spatial clusters and outliers based on proximity, which is fundamentally different from the hotspot method but may contribute as a complementary concept because the Moran's I is a high or low level of similarity to the spatial area in the vicinity [62,63]. In the case of proximity, the Euclidean distance was used to measure the distance between features, and similarity with the neighboring area was analyzed based on the result. Based on this, spatial areas were categorized into the following four spatial patterns: high–high (HH) spatial clusters, high–low (HL) spatial outliers, low–low (LL) spatial clusters, and low–high (LH) spatial outliers (Figure 3) [64]. The local Moran's I assigns a weight to a given area based on the spatial proximity among the areas in a cluster, and to analyze the consequent patterns, the range and distance of the weight should be determined. We used a fixed bandwidth to determine the weight range of the local Moran's I in this study [65], and the Euclidean distance was applied for the distance between areas [66,67] (Equation (3)).

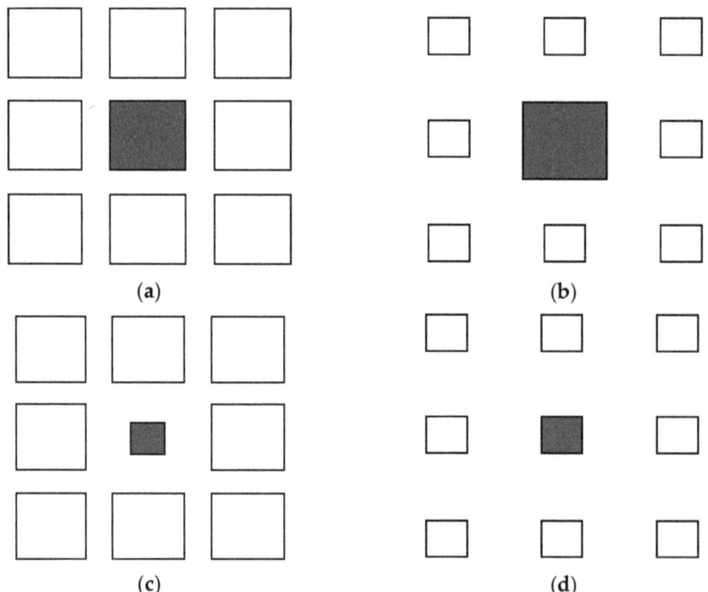

Figure 3. Local Moran's I for spatial clusters and spatial outliers. (**a**) High–high (HH) spatial cluster, (**b**) high–low (HL) spatial outlier, (**c**) low–high (LH) spatial outlier, and (**d**) low–low (LL) spatial cluster.

$$I = \frac{n \sum_{i=1}^{n} \sum_{j=1}^{n} W_{ij}(d)(x_i - \overline{x})(x_j - \overline{x})}{\sum_{i=1}^{n} \sum_{j=1}^{n} W_{ij}(d) \sum_{i}^{n}(x_i - \overline{x})^2} \tag{2}$$

$$I_i = \frac{(x_i - \overline{x}) \sum_j W_{ij}(x_j - \overline{x})}{S^2} \tag{3}$$

x_i: forest loss rate at ith area; x_j : forest loss rate at jth area; $\overline{x}$: the mean of the forest loss rate; W_{ij} : weight index for the location of i relative to j; S^2: variance; n: number of areas.

2.3.3. Selection of Impact Factors

For the variables that influence forest loss, a total of 11 variables were selected in reference to previous studies conducted in South Korea and overseas (Table 3). The selected variables were as follows: road density [68], cropland area [69], grassland area [70], settlement area [71], number of households [72], population [73], and industry employees and establishments [74]. The "industry employees and establishments" variable was analyzed by first recategorizing the industry types in the Survey of Establishments [52,53] that follows the Fisher–Clark categorization of industry (agriculture, forestry, fishery, mining, manufacturing, electricity, gas and waterwork, transportation, and communication) into primary, secondary, and tertiary industries, then counting the employees and establishments in each class of industry [75]. For each variable, the rate of change for each spatial area was estimated following the same method used to determine changes in forest rate. Next, the multicollinearity of variables was determined, and the variables were excluded from further analyses if the variance inflation factor (VIF) was ≥10 [76].

Table 3. Impact factors of forest loss.

Factor	Description	Unit	Data Source
Road density (Rd)	Forest loss is caused by the increased accessibility due to high road density	m/ha	Mena et al., 2006 [68]
Cropland area (Ca)	Forest loss is caused by the expansion of farms leading to increase in population and poverty		Ngwira and Watanbe, 2019 [69]
Grassland area (Ga)	Forest loss is caused by the expansion of industry to meet the increased demand for livestock animals	%	Walker et al., 2013 [70]
Settlement area (Sa)	Forest loss is caused by the settlement and construction of infrastructure		Jayathilake et al., 2020 [71]
Number of households (Nh)	Forest loss is caused by increase in number of households leading to expansion of croplands	n	Godoy et al., 1997 [72]
Population (P)	Forest loss is caused by increase in population leading to increased demand for food and housing	n	Biswas et al., 2012 [73]
Primary industry number of Employees (Pm)			
Primary industry Establishments (Pe)			
Secondary industry number of Employees (Sm)	Forest loss is caused by industrialization and subsequent population pressure	n	Lata et al., 2018 [74]
Secondary industry number of Establishments (Se)			
Tertiary industry Employees (Tm)			
Tertiary industry Establishments (Te)			

2.3.4. Concept of Statistical Learning (OLS and GWR Models)

The OLS and GWR models were used to analyze the spatial correlation between forest loss and human activity. The OLS model is a global model that estimates the influence of a given variable as identical across all study areas, based on the assumption that the variables would have identical correlations in any space. Therefore, the OLS model can be used to confirm the influence of variables on the whole study area [77]. Meanwhile, the global correlation determined through the OLS model may deviate from the locally analyzed correlation so that the estimated correlation may differ from the actual correlation [78]. The estimation equations of OLS are shown in Equations (4) and (5). Thus, the OLS model was used to analyze the global impact factors of forest loss.

$$y = \beta_0 + \sum_{k=1}^{n} \beta_k x_k + \varepsilon \tag{4}$$

$$\hat{\beta} = \left(X'X\right)^{-1}X'Y \tag{5}$$

y: dependent variable; β_0: intercept; β_k: regression coefficient; x_k: kth independent variable; ε: error; $\hat{\beta}$: estimated regression coefficient; X': transpose matrix of variable; X: matrix of variable; Y: vector of the dependent variable.

The GWR model allows for estimation of local parameters as a regional model, enabling estimation of the influence of variables by region. In contrast to the OLS model, the influence of the variable within the study area was estimated per area, and the results of the model were applied to each area [79]. This suggests that the GWR produces a more reliable performance than the OLS because local influences are analyzed to allow the study of spatial migration of variables and as the influence is analyzed per area [80,81]. The weight in the GWR model is assigned through kernel functions based on distance [82]. The bandwidth is divided into fixed and adaptive kernels based on how the bandwidth is set as the weight range. For a fixed kernel, the distribution shows bandwidths of consistent size. For adaptive kernels, the distribution varies according to the data density [83]. The weights were assigned via an adaptive kernel. The estimation equations for the GWR are shown in Equations (6)–(8). Thus, the GWR model was used to analyze the impact factors of forest loss in the dimension of areas (units of spatial areas).

$$y_i = \beta_0(u_i, v_i) + \sum_{k=1}^{n} \beta_k(u_i, v_i)x_{ki} + \varepsilon_i \tag{6}$$

$$\hat{\beta}_i = \left(X'W_iX\right)^{-1}X'W_iY \tag{7}$$

$$w(u_i, v_i) = \begin{pmatrix} w_{i1} & \cdots & 0 \\ \vdots & \ddots & \vdots \\ 0 & \cdots & w_{ik} \end{pmatrix} \tag{8}$$

β_0: intercept; β_k: estimate coefficient for independent variable; y_i: dependent variable; x_k: kth independent variable; (u_i, v_i): longitude and latitude coordinates of ith area; $\beta_k(u_i, v_i)$: estimate coefficient for the location of ith area; ε_i: error; $\hat{\beta}_i$: estimated coefficient for the location of ith area; X': transpose matrix of variable; X: matrix of variable; Y: vector of the dependent variable; W_i: weighted matrix for the location of ith area

2.3.5. Machine-Learning Model (RF Model)

The performance of a machine-learning model and importance of each impact factor were estimated with respect to forest loss for a comparative analysis concept of statistical learning. The RF model was used because it is a representative machine-learning model which is well-known for its simplicity and efficiency [84]. The RF model was implemented using Python's scikit-learn library. A variable is selected at each node, and randomness is exhibited by the learning data at each tree to create an ensemble model of myriads of decision trees [85]. In general, the prediction accuracy and efficiency of the RF model are high, with a low probability of overfitting for learning data [35,86]. The RF model was analyzed using *n_estimators*, *max_depth*, *min_samples_split*, and *min_samples_leaf* as hyperparameters, as shown in Table 4. *n_estimators* is the number of regression trees in the model. As *n_estimators* increases, the fitting effect decreases; therefore, *n_estimators* is often set to 100 [87]. In addition, in reference to previous studies which reported the use of the basic hyperparameter values leading to a high level of accuracy, *n_estimators* was set as 100, *min_samples_split* as 2, *max_depth* as 0, and *min_samples_leaf* as 1 [88,89] (Figure 4). We analyzed the hyperparameter as the default value by referring to previous research when evaluating the relationship between dependent and independent variables as well as statistical models [88] (Table 4).

Table 4. Hyperparameters of the random forest (RF) model.

Hyperparameter	Value
n_estimators	100
Max_depth	None
Min_sample_split	2
Min_samples_leaf	1

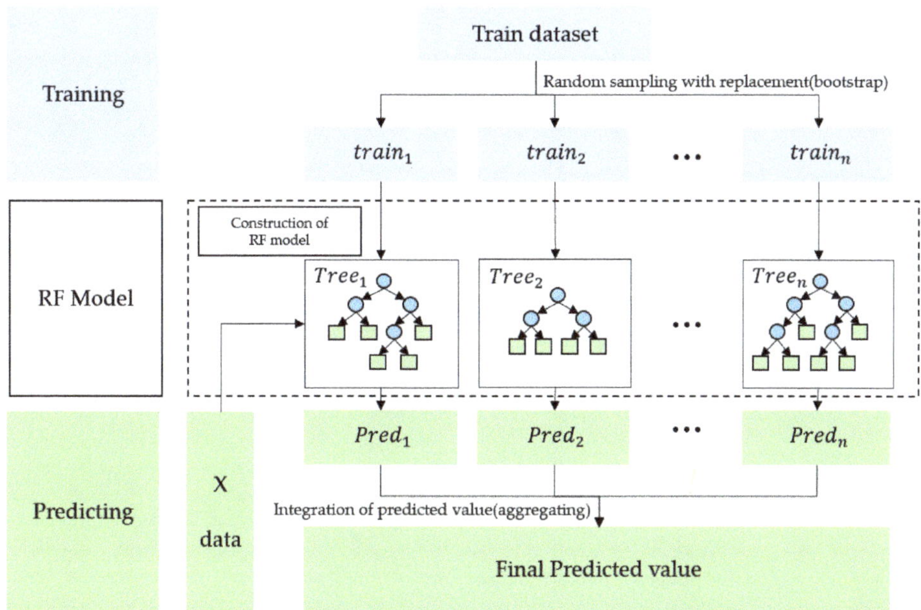

Figure 4. Regression tree node based on RF model.

2.3.6. Model Fitness

The coefficient of determination (R^2), root mean squared error (RMSE), mean absolute error (MAE), and Akaike's information corrected criterion (AICc) were used to test the performance of the statistical and machine-learning models. The R^2 was used to analyze the predictive power of the models. RMSE is a scale that represents the differences between the model-predicted values and the actual observed values and is used to evaluate the accuracy of spatial analyses and remote sensing with error distributions [90,91]. MAE is the mean value for the absolute difference between the model-predicted value and actual value, which indicates the mean error size. As in the RMSE, smaller estimates indicate smaller errors, which verifies a higher prediction accuracy [92,93]. AICc allows the estimation of the relative quantity of data lost in the statistical model. Smaller values indicate a higher model fitness. In general, AICc provides the solution to overfitting when the sample size is small; thus, it is more useful than AIC [94]. The R^2, RMSE, MAE, and AICc values were obtained using Equations (9)–(12), respectively. Additionally, the influence and importance of variables were analyzed based on the regression coefficients from the statistical models and the IncMSE from the machine-learning model. The regression coefficient is indicative of the influence of the impact of factors on forest loss, and a positive or negative value indicates a positive or negative, respectively [95]. The % IncMSE is indicative of an increase in the mean squared error, and a higher value indicates a more critical variable within the RF model [96].

$$R^2 = 1 - \frac{\sum_{i=1}^{n}(y_i - \hat{y}_i)}{\sum_{i=1}^{n}(y_i - \overline{y})^2} \tag{9}$$

$$RMSE = \sqrt{\frac{1}{n}\sum_{i=1}^{n}(y_i - \hat{y}_i)^2} \tag{10}$$

$$MAE = \frac{1}{n}\sum_{i=1}^{n}|y_i - \hat{y}_i| \tag{11}$$

$$\text{AICc} = 2\text{nlog}_e(\hat{\sigma}) + \text{nlog}_e(2\pi) + n\left(\frac{n + tr(s)}{n - 2 - tr(s)}\right)(S = \frac{\hat{y}_i}{y_i}) \qquad (12)$$

y_i: dependent variable; $\hat{y}_i$: estimated value of dependent variables; $\overline{y}_i$: mean of dependent variables; σ: residual standard error ; $\hat{\sigma}$: estimated value of residual standard error; n: number of variables; tr(s): trace of the hat matrix.

3. Results and Discussion

3.1. Spatial Distribution of Forest Loss during 2005–2015

The forest rate in South Korea decreased by approximately 1% in 2015 compared to 2005, with significant differences among spatial regions. The highest forest loss rate was observed in the Seoul–Incheon–Gyeonggi region, and the lowest forest loss rate was observed in the Gangwon region (Table 5). The mean forest loss rate in the Seoul–Incheon–Gyeonggi region was 3.3%, which was 1.8-fold higher than the mean national rate of forest loss and approximately 5-fold higher than the Gangwon region with the lowest forest loss rate. In particular, the Seoul–Incheon–Gyeonggi region exhibited a 14.4% maximum forest loss rate, a level far higher than other spatial regions. This result indicated that forest loss caused by forest conversion and land use change was concentrated in the Seoul region over the past decade (Figure 5a). Such changes in forest conversion and land use seem to have occurred due to urbanization such as the expansion of road network and the construction of various infrastructure facilities centered on the metropolitan area where the altitude of the sea level is relatively low due to the cancellation policy of the development restriction area in the metropolitan area [97]. This is similar to the case of China, which is geographically neighboring. In order to analyze the impact of development due to urban expansion on forest loss, Zhou et al. [98] analyzed the impact of urbanization on forest loss in six major urban megaregions of China, including Beijing–Tianjin–Hebei (BTH), Yangtze River Delta (YRD), Pearl River Delta (PRD), Wuhan (WH), and Chengdu–Chongqing (CY). As a result, forest loss was slightly different in each region, but urban expansion showed a major impact on forest loss [98]. Conversely, the standard deviation was 3.3% for the Seoul region, which was higher than all other spatial regions, with 14.4% maximum and 0.1% minimum forest loss rates, indicating that forest loss occurred intensively in the Seoul–Incheon–Gyeonggi region and its surrounding regions. However, the deviation in forest loss among spatial areas was substantially higher than other spatial regions. On the other hand, one area in the Seoul–Incheon–Gyeonggi region, three in Gangwon region, one in the Daegu–Gyeongsangbuk region, and one in the Gwangju–Jeonnam region showed increases in forest area during 2005–2015. These areas were excluded from the analyses of the spatial clusters and outliers of forest loss areas and the factors influencing forest loss (Figure 5b).

Table 5. Distribution of forest loss rate across spatial regions.

Category	Minimum Forest Loss Rate	Mean Forest Loss Rate	Maximum Forest Loss Rate	Standard Deviation
Seoul–Incheon–Gyeonggi region	0.1%	3.3%	14.4%	3.3%
Gangwon region	0.2%	0.6%	1.5%	0.4%
Busan–Ulsan–Gyeongnam region	0.2%	1.1%	2.6%	0.8%
Daegu–Gyeongbuk region	0.1%	0.8%	2.8%	0.6%
Gwangju–Jeonnam region	0.1%	1.6%	7.7%	1.9%
Jeonbuk region	0.3%	1.7%	3.8%	1.0%
Daejeon–Chungnam region	0.3%	2.7%	8.1%	2.0%
Chungbuk region	0.1%	1.5%	5.3%	1.5%

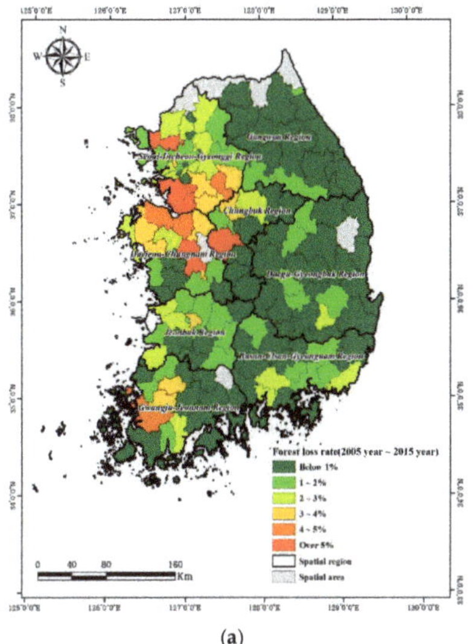
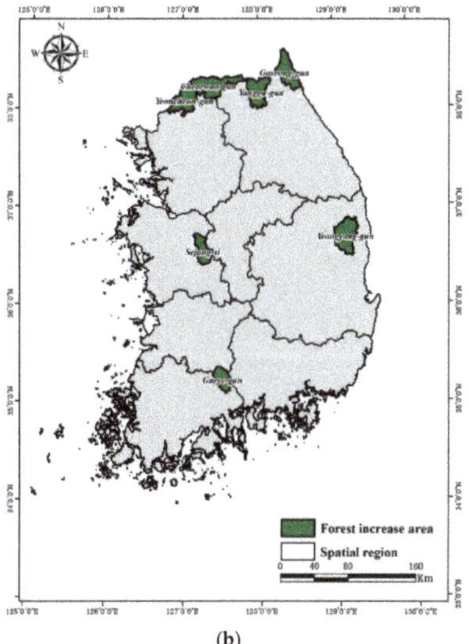

(a)   (b)

Figure 5. Forest loss area and forest increase area by spatial region and spatial area: (**a**) forest loss areas per spatial region and area; (**b**) forest increase areas per spatial region and area.

3.2. Spatial Patterns of Forest Loss

The spatial distribution characteristics of forest loss are shown in Figure 6. Forest loss showed significant positive spatial autocorrelation (global Moran's I = 0.29, $p < 0.01$), indicating that forest loss was clustered. A high number of HH clusters occurred in the Seoul–Incheon–Gyeonggi region, approximately 77% of all HH clusters. Seoul-si, Incheon-si, Gimpo-si, Hwaseong-si, and Pyeongtaek-si, which are in the capital region, had many large-scale development projects (e.g., for housing, urban development, multicomplexes, free economic zones, etc.), which were either ongoing or completed in 2021. This is presumed to have led to the higher rate of forest loss in this region compared with other spatial regions. The possibility of continuous forest loss is also predicted to be high in this region [99]. Approximately 42% of clusters in the Gangwon region were LL clusters, and the mean forest loss rate was 0.6%, which was 1.2% lower than the mean forest loss rate across all spatial areas (1.8%). HL spatial outliers were mostly found close to LL clusters, whereas LH spatial outliers were mostly found close to HH clusters. The forest loss rate in the areas with HL outliers was 1.9%, which was 0.1% higher than the mean forest loss rate across all spatial areas (1.8%). This is presumably because the distribution of forest loss rate in the neighboring regions (i.e., Daegu-si, Gumi-si, Gimcheon-si, Gunwui-gun, and Seongju-gun; 0.6%) was 1.2% lower than the mean forest loss rate across all areas. Most LH spatial outliers were found in the Seoul–Incheon–Gyeonggi region, presumably due to the presence of HH clusters in the surrounding areas (Figure 6).

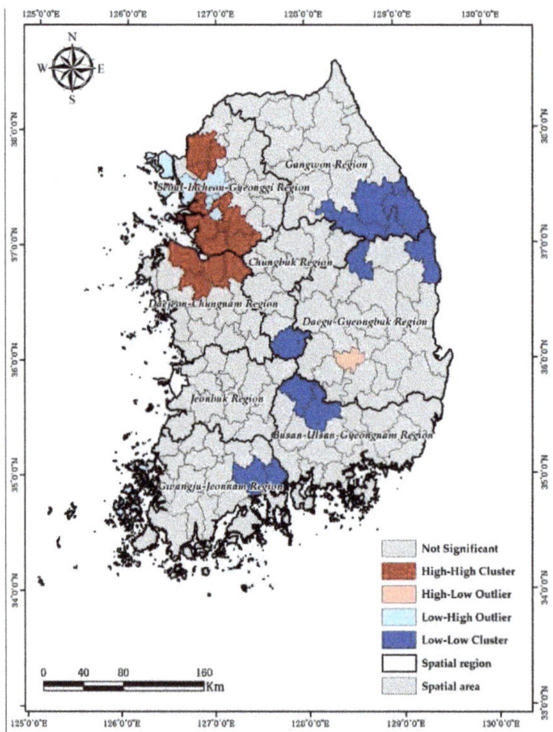

Figure 6. Map of spatial clusters and spatial outliers in forest loss area in South Korea.

3.3. Assessment of Factors Impacting Forest Loss

3.3.1. Selection of Variables Related to the Factors Impacting Forest Loss

Prior to the selection of variables related to the impact factors of forest loss, multicollinearity and correlation analyses were performed for the variables. As shown in Table 5, the population and number of households showed VIFs of approximately 28 and 26, respectively; therefore, reanalysis was conducted after excluding the population. The results of the reanalysis showed that the multicollinearity was reduced to ≤10, and the remaining variables were selected as the final variables. (Table 6).

Table 6. Correlation coefficients and variance inflation factors (VIFs) among variables potentially influencing forest loss. See Table 3 for variable abbreviations.

Category	Rd	Ca	Ga	Sa	P	Nh	Pm	Pe	Sm	Se	Tm	Te	VIF
Rd	1	-	-	-	-	-	-						1.3
Ca	−0.312 **	1	-	-	-	-	-	-	-	-	-	-	1.5
Ga	−0.059	0.143	1	-	-	-	-	-	-	-	-	-	1.0
Sa	0.047	0.145	−0.044	1	-	-	-	-	-	-	-	-	1.6
P	0.333 **	−0.267 **	−0.166 *	0.448 **	1	-	-	-	-	-	-	-	28.7
Nh	0.347 **	−0.328 **	−0.168 *	0.455 **	0.976 **	1	-	-	-	-	-	-	26.5
Pm	−0.039	0.093	0.025	0.089	0.011	−0.012	1	-	-	-	-	-	3.1
Pe	−0.121	0.174	0.025	0.097	−0.036	−0.067	0.812 **	1	-	-	-	-	3.2
Sm	−0.033	0.138	−0.033	0.340 **	0.206 *	0.177 *	0.058	0.049	1	-	-	-	1.8
Se	−0.171 *	0.186 *	−0.011	0.246 **	0.186 *	0.153	0.092	0.176 *	0.624 **	1	-	-	2.0
Tm	0.340 **	−0.183 *	−0.171*	0.373 **	0.811 **	0.769 **	0.049	−0.032	0.245 **	0.292 **	1	-	6.7
Te	0.351 **	−0.288 **	−0.197*	0.321 **	0.855 **	0.830 **	0.050	−0.036	0.243 **	0.278 **	0.910 **	1	8.6

$* \, p < 0.10, \, ** \, p < 0.05.$

3.3.2. Model Fitness Test

The fitness of each of the three models is presented in Table 7. The GWR model showed better performance than the OLS model and the RF model in explaining the correlation between forest loss and its impact factors. The R2 of the GWR model was 0.69, which was 1.4-fold higher than the OLS model and equivalent to the RF model. The RMSE of the GWR model was 1.17, which was 0.37 lower than the OLS model but equivalent to the RF model. The MAE of the GWR model was 0.85, which was 0.2 and 0.03 lower than that of the OLS and RF models, respectively. The AICc of the GWR was lower than that of the OLS model by approximately 48. The GWR model was similar to the RF model with respect to R^2 and RMSE; however, a lower MAE meant that the GWR model most accurately explained the relationship between the variables and forest loss (Table 7).

Table 7. Model fitness test of the statistical models and machine-learning model.

Model	R^2	RMSE	MAE	AICc
OLS	0.48	1.54	1.05	591.4
GWR	0.69	1.17	0.85	543.9
RF	0.69	1.17	0.88	-

Our results suggest that the GWR model is more suitable than the OLS model, and unlike the OLS model, it is possible to emphasize the relationship between forest loss and impact factors by deriving results according to geographical location and regional characteristics [100,101]. In addition, as with the OLS model, the RF model analyzes the relationship between forest loss and impact factors across the entire range of research areas, so it is limited to analyze the effects of regional characteristics [36]. The GWR model is considered to be the best explanation for the relationship between forest loss and impact factors.

Table 8 shows the influence of the OLS, GWR, and RF models. The influence and importance of the models reflect the quantitative degree of the effect of independent variables on forest loss in each model. In the OLS model, the variables with the highest influence ($\geq$0.01) on forest loss were the number of households, number of tertiary industry establishments, grassland area, and road density, whereas the variables with the lowest influence (<0.001) were the number of secondary industry establishments, number of primary industry establishments, number of secondary industry employees, and number of tertiary industry employees. In the GWR model, the variables with the highest influence ($\geq$0.01) on forest loss were road density, number of households, cropland area, and number of tertiary industry establishments, whereas the variables with the lowest influence (<0.001) were grassland area, number of primary industry employees, and number of tertiary industry employees. In the RF model, the variables with the highest influence ($\geq$0.01) on forest loss were road density, number of households, and number of tertiary industry establishments, whereas the variables with the lowest influence (<0.03) were cropland area, grassland area, number of primary industry employees, and number of secondary industry employees. Therefore, three variables (road density, number of households, and number of tertiary industry establishments) were the most influential variables across the three models.

Table 8. The influence of each variable on forest loss in the statistical (OLS and GWR) and machine-learning (RF) models. See Table 3 for variable abbreviations.

	OLS Model	GWR Model	RF Model
Rd	0.016	0.0400	0.278
Ca	0.009	0.0245	0.012
Ga	0.019	0.0009	0.025
Sa	0.008	−0.0013	0.057
Nh	0.033	0.0253	0.258
Pm	−0.000	−0.0002	0.020
Pe	−0.000	−0.0009	0.039
Sm	0.000	0.0056	0.022
Se	−0.001	−0.0042	0.046
Tm	0.000	0.0008	0.097
Te	0.023	0.0167	0.141

3.3.3. Assessment of Factors Impacting Forest Loss Areas in Each Spatial Region

Since the OLS model and the RF model are global models that cannot deal with spatial heterogeneity, the GWR model was used to determine the influence of factors on forest loss in each spatial region [36,90]. Figure 7 shows the results of the GWR model for each spatial area. The factors that affect the forest loss by each spatial region through the GWR model show the following characteristics.

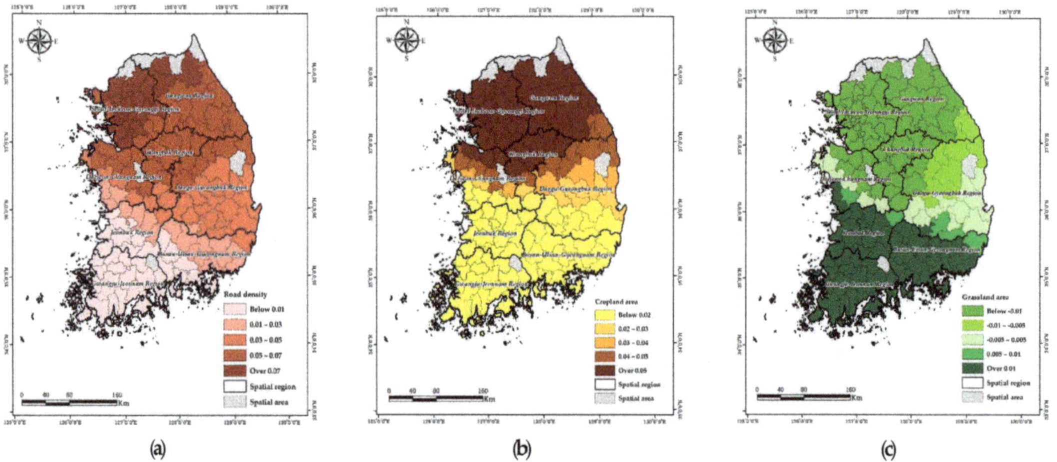

Figure 7. *Cont.*

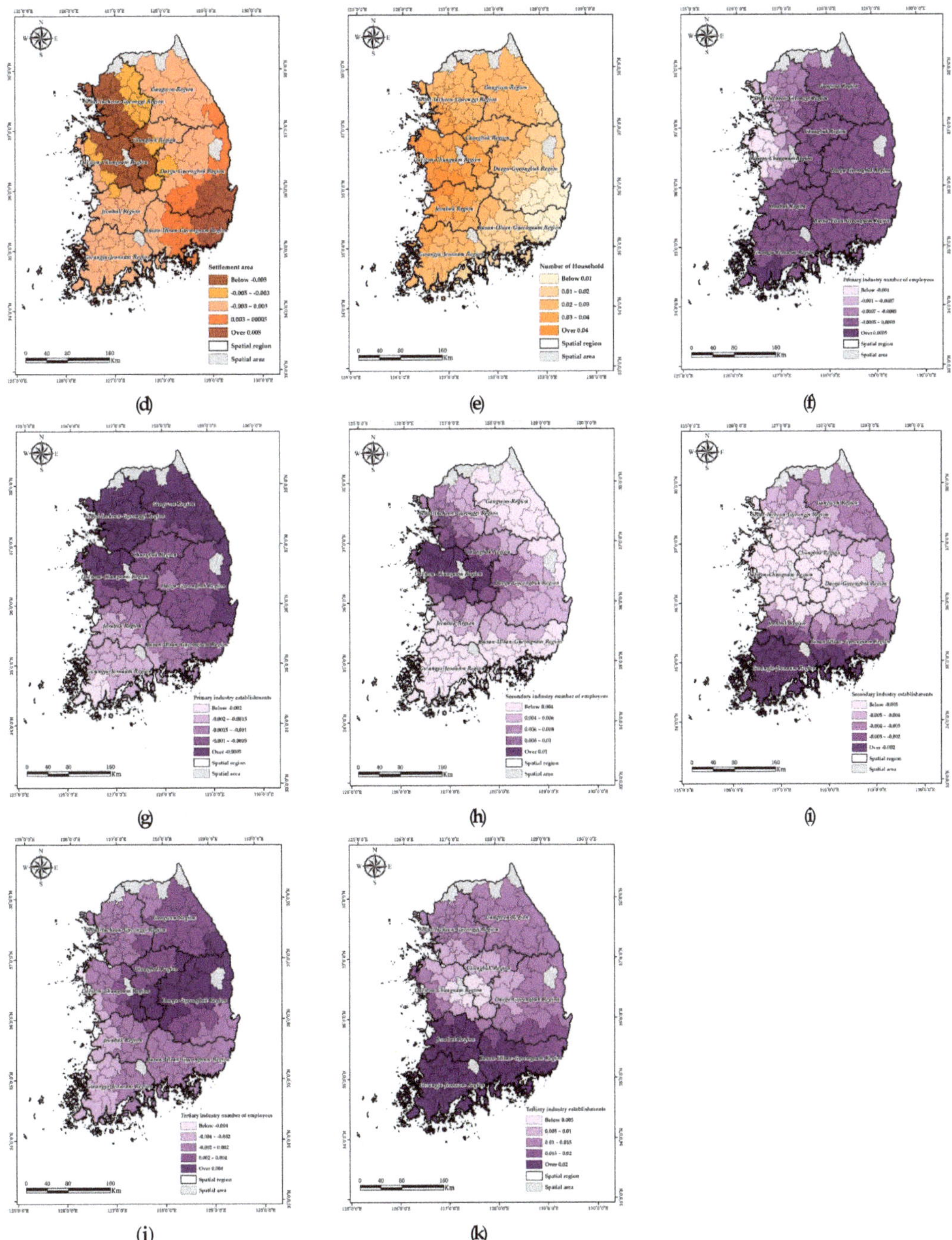

Figure 7. Distribution map of regression coefficients for the forest loss impact factors in the GWR model: (**a**) road density, (**b**) cropland area, (**c**) grassland area, (**d**) settlement area, (**e**) number of households, (**f**) primary industry number of employees, (**g**) primary industry establishment, (**h**) secondary industry number of employees, (**i**) secondary industry establishment, (**j**) tertiary industry number of employees, and (**k**) tertiary industry establishment.

Road density showed a major influence on forest loss in the Seoul–Incheon–Gyeonggi region, but on the contrary, a low influence in the Gwangju–Jeonnam region. The rate of increase in road density and mean rate of forest loss during 2005–2015 in the Seoul–Incheon–Gyeonggi region were 54.1% and 3.2%, respectively, both of which were the highest across the nation. The higher rate of forest loss in this region compared with other spatial regions may be attributed to the high road density in Seoul-si, Incheon-si, and Gyeonggi-do, which are categorized as the capital regions of South Korea within the Seoul–Incheon—yeonggi region, which accounted for 60% of the top 22 spatial areas with reported high road densities in 2010 [102]. The cropland area was the main cause of forest loss in the Seoul–Incheon–Gyeonggi region, but not the main cause of forest loss in the Gwangju–Jeonnam and Jeonbuk regions. The grassland area was closely related to forest loss in Gwangju–Jeonnam and Jeonbuk regions, but the relationship between forest loss and the grassland area was weak in the Seoul–Incheon–Gyeonggi and Gangwon regions. The settlement area showed a high influence on forest loss in the Busan–Ulsan–Gyeongnam region, but it showed a low influence in the Daejeon–Chungnam region. The number of households was the main factor of forest loss compared to other regions in the Daejeon–Chungnam region, whereas the number of households in the Busan–Ulsan–Gyeongnam region was not enough to factor for forest loss. Of all the metropolitan cities and provinces, Daejeon-si and Chungcheongnam-do, which belong to the Daejeon–Chungnam region, had the highest rate of increase in the number of households during 2000–2010, followed by Gyeonggi-do. This is presumed to be the reason for the high rate of forest loss in the Daejeon–Chungnam region [103].

Regarding the number of industry employees and establishments, the number of primary industry employees had a high positive effect on forest loss in the Gwangju–Jeonnam region but a negative effect in certain spatial areas in the Daejeon–Chungnam region. The number of primary industry establishments did not have much effect on the forest loss in the Gwangju–Jeonnam region. The number of secondary industry employees was the main cause of forest loss in the Daejeon–Chungnam region, but not in Gwangju–Jeonnam region. In the Chungcheongnam-do area of the Daejeon–Chungnam region, the manufacturing industry contributed to 46.9% of the gross domestic regional production in 2006, which was higher than the national average (28.2%) [104]. Therefore, it is presumed that the increase in secondary industry employees had an impact on forest loss. The number of secondary industry establishments showed a high influence on forest loss in certain spatial areas of the Gwangju–Jeonnam region, but a low influence in the Daejeon–Chungnam region. The number of tertiary industry employees was analyzed as the main factor causing forest loss in the Daegu–Gyeongbuk region, but the influence was relatively insufficient in the Gwangju–Jeonnam region. The number of tertiary industry establishments had a high influence on forest loss in the Gwangju–Jeonnam region, but a low influence in the Daejeon–Chungnam area. Mo and Lee [105] reported that in 2015, Gwangju-si in the Gwangju–Jeonnam region specialized in tertiary industries, with a high number of wholesale, retail, accommodation, food service, banking, insurance, real estate, and lease service establishments, among other tertiary industries. This is thought to have resulted in the high positive effect of tertiary industry establishments in the Gwangju–Jeonnam region. The effects of variables on forest loss differed among the spatial regions. For example, in the Seoul–Incheon–Gyeonggi region, in which there was a higher rate of forest loss, road density and number of households had a strong effect on forest loss. Conversely, in the Daejeon–Chungnam region, the number of secondary industry employees had a strong effect on forest loss (Figure 7).

The effect of these factors on forest loss is due to urbanization [106,107]. Urbanization causes forest loss by generating high demand for residential facilities and infrastructure facilities in neighboring areas, especially in the development area, as the population and the number of households increase beyond simple regional development [108,109]. This is consistent with the results of Chen et al. that forest loss occurred due to urban development including the increase of roads and residential areas [110]. Urbanization and development are likely to continue in the future, so it is necessary to prepare measures to maintain the balance of forest conservation and forest loss between urbanization and regional development.

On the other hand, the causes of forest loss are different according to the region, as in our research [98]. In developing countries, rapid agricultural expansion and excessive use of forest resources are the main causes of forest loss, resulting in forest loss due to food demand problems and agricultural investment [111,112]. For example, in Malawi, the expansion of agricultural land such as corn farm expansion, tobacco cultivation, and brick production is one of the main causes of forest loss [113]. Myanmar's expansion of agricultural land following rapid agricultural investment and expansion of the city is among the main causes of forest loss. Then, the factors for forest loss differ according to the region [69,114]. In addition, in countries such as Bhutan, Laos, Nepal, Sri Lanka, and Vietnam, topographical conditions (altitude, slope) and biophysical requirements such as temperature and precipitation were also among the main factors in the loss of forest area [115]. In the United States, factories, houses, and roads have had a great impact on forest loss, which is also the difference in accessibility, lifestyle, and institution [116]. As the relationship between forest loss and impact factors differs by country and region, studies are being conducted on various regions. Mwangi et al. analyzed the relationship between forest loss and impact factors on randomly selected sites using land coverage maps in the Central Region and analyzed that topographic factors (altitude, slope), distance from roads and distance from rivers are the main causes of forest loss [117]. This study analyzed the influence of topographic factors and forest loss, unlike our research, which showed that the closer the distance from the road and the closer the river, the easier the transportation, resulting in forest loss. Santos et al. [118] analyzed the relationship between forest loss and impact factors in the Amazon region of Brazil and confirmed that the rapid expansion of roads, ranches, and agricultural products affected the loss of forest. This means that the increase in roads increases accessibility, which is believed to have promoted the change of forest into cropland and pasture [9,118]. In addition, Mas and Cuevas analyzed the forest loss status based on the municipality, and then analyzed the effect on forest loss using the GWR model, and confirmed that the same factors could have different effects depending on the region. On the other hand, the forest loss and its impact factors were also conducted through preceding research. Geist and Lambin analyzed the causes of forest loss by dividing them into proximate causes and underlying driving forces through preceding research review and analyzed that the impact of the forest loss was on agricultural expansion, the use of timber and infrastructure expansion, economic and commercialization, and institutional and demographic factors [23]. In addition, Armenteras et al. [119] analyzed the previous studies conducted on Latin American countries to analyze the factors affecting forest loss and its impacts and confirmed that access to markets and agricultural and forest activities had a major impact on forest loss. As the factors of forest loss and its impact differ by region, studies are being conducted on various continents and regions. Forests are decreased by the above-mentioned factors, and this is affected by regional socioeconomic factors, institutional factors, and topographic factors, so they should be analyzed considering these factors. Therefore, causes of forest loss are different in each region, which is judged to be due to the differences in socioeconomic, biophysical characteristics, policies, and institutions of each region [19,22,100,120].

3.4. Limitations of the Study

Meanwhile, this study has certain limitations. First, we analyzed forest loss and factors at the administrative district level. However, spatial analyses using grids or micropolygon units can provide more details regarding the effects of factors on forests [121,122]. Another limitation was the selection of factors influencing forest loss. We did not include several variables that have recently been found to affect forest loss in South Korea, such as altitude, slope, and photovoltaic solar plants [123,124]. The lower the altitude and slope, the easier the accessibility, so the agricultural forest clearing is advantageous, and the forest loss appears. However, the altitude and slope were not used in this study because securing the time-series data was limited compared to other forest loss factors [123]. In the case of the photovoltaic solar plants, according to Mori and Tabata [125], there are benefits such as mitigation of climate change and economic benefits, but it can cause biodiversity due to forest loss, loss of carbon sinks, and risk of landslides. However, this study did not utilize it due to the limitation of securing time-series data. Therefore, future studies need to discuss the effects of geographical factors (high altitude, slope), photovoltaic solar plants, etc., on forest loss and the problems that can be caused.

Therefore, considerable efforts are required to more clearly predict factors affecting forest loss by including suitable factors in each spatial area. Results of this study may contribute to the development of policies for reducing forest loss and provide valuable data on the correlation between forest loss and the factors impacting this process. Further studies are needed to address the limitations of this study to enhance the applicability of the results.

4. Conclusions

We analyzed the spatial distribution of forest loss in Korea and the factors affecting forest loss. The results of this study showed that forest loss occurred in large quantities mainly in the Seoul–Incheon–Gyeonggi region and was 1.8 times higher than the average forest loss in South Korea. As a result of Moran's I analysis, HH clusters occurred mainly in the Seoul–Incheon–Gyeonggi region, which shows that forest loss occurred mainly in the Seoul–Incheon–Gyeonggi region. The forest loss and its impact factors were analyzed using OLS, GWR, and RF models. The GWR model had a 1.4-fold higher R^2 than the OLS model, and the AICc was about 48 less. In addition, the MAE was lower than the RF model, showing the highest model suitability. This means that the GWR model can perform a better regional approach to forest loss and its impact compared to the OLS model and the RF model, and it suggests that the GWR model is easy to analyze according to regional differences. The most frequent forest loss in the Seoul–Incheon–Gyeonggi region was found to have a strong impact on road density and number of households. This is due to the progress of road construction and infrastructure installation as urbanization progresses mainly in the Seoul–Incheon–Gyeonggi region between 2005 and 2015. In particular, according to Liu et al. [126], infrastructure construction and economic growth are the main causes of forest loss, and forest loss appears to have occurred as developments have progressed around the region.

On the other hand, since forest loss varies according to regional characteristics, research needs to be conducted based on background knowledge of the region [101]. Therefore, the analysis of factors affecting forest loss should be carried out in consideration of the situation of each country and region as in the previous studies, and both biophysical and socioeconomic factors should be considered as much as possible. The GWR model is useful for quantitative analysis of forest loss factors by region, and it is expected to be useful for policy design and evaluation of forest loss by using it together with qualitative analysis. In addition, if we analyze the changes in the forest loss and its impact factors, which were mentioned earlier, it will be useful data for policy setting. Therefore, in future studies, it is necessary to analyze the changes and causes of forest loss over time using the local Moran's I, time-series and hotspot analysis, and GWR model. Clearing the factors affecting forest loss will be useful for establishing forest management plans and improving forest protection systems.

Author Contributions: Conceptualization, J.P., B.L. and J.L.; methodology, J.P. and J.L.; software, J.P. and B.L.; validation, J.P., B.L. and J.L.; formal analysis, J.P.; investigation, B.L.; resources, J.P. and B.L.; data curation, J.P. and B.L.; writing—original draft preparation, J.P., B.L. and J.L.; writing—review and editing, J.P. and J.L.; visualization, B.L.; supervision, J.P. and J.L.; project administration, J.L.; funding acquisition, J.L. All authors have read and agreed to the published version of the manuscript.

Funding: This study was carried out with the support of the 'R&D Program for Forest Science Technology (2019151D10-2123-0301)' provided by Korea Forest Service (Korea Forestry Promotion Institute).

Institutional Review Board Statement: Not applicable.

Informed Consent Statement: Not applicable.

Data Availability Statement: Not applicable.

Conflicts of Interest: The authors declare no conflict of interest.

References

1. Food and Agriculture Organization of the United Nations (FAO). *Global Forest Resources Assessment 2020*; Main Report; FAO: Rome, Italy, 2020.
2. Prevedello, J.A.; Winck, G.R.; Weber, M.M.; Nichols, E.; Sinervo, B. Impacts of Forestation and Deforestation on Local Temperature Across the Globe. *PLoS ONE* **2019**, *14*, e0213368. [CrossRef]
3. Intergovernmental Panel on Climate Change (IPCC). *Climate Change 2014: Contribution of Working Groups I, II and III to the Fifth Assessment Report of the Intergovernmental Panel on Climate Change*; Synthesis Report; IPCC: Geneva, Switzerland, 2014.
4. Giesekam, J.; Tingley, D.D.; Cotton, I. Aligning carbon targets for construction with (inter) national climate change mitigation commitments. *Energy Build.* **2018**, *165*, 106–117. [CrossRef]
5. Tyukavina, A.; Hansen, M.C.; Potapov, P.; Parker, D.; Okpa, C.; Stehman, S.V.; Kommareddy, I.; Turubanova, S. Congo Basin forest loss dominated by increasing smallholder clearing. *Sci. Adv.* **2018**, *4*, eaat2993. [CrossRef]
6. Cohn, A.S.; Bhattarai, N.; Campolo, J.; Crompton, O.; Dralle, D.; Duncan, J.; Thompson, S. Forest loss in Brazil increases maximum temperatures within 50 km. *Environ. Res. Lett.* **2019**, *14*, 084047. [CrossRef]
7. Zhu, Z.; Zhu, X. Study on Spatiotemporal Characteristic and Mechanism of Forest Loss in Urban Agglomeration in the Middle Reaches of the Yangtze River. *Forests* **2021**, *12*, 1242. [CrossRef]
8. Zheng, Z.; Ma, T.; Roberts, P.; Li, Z.; Yue, Y.; Peng Huang, K.; Han, Z.; Wan, Q.; Zhang, Y.; Zhang, X.; et al. Anthropogenic impacts on Late Holocene land-cover change and floristic biodiversity loss in tropical southeastern Asia. *Proc. Natl. Acad. Sci. USA* **2021**, *118*, e2022210118. [CrossRef]
9. Laurance, W.F.; Sayer, J.; Cassman, K.G. Agricultural expansion and its impacts on tropical nature. *Trends Ecol. Evol.* **2014**, *29*, 107–116. [CrossRef] [PubMed]
10. Nobre, C.A.; Sampaio, G.; Borma, L.S.; Castilla-Rubio, J.C.; Silva, J.S.; Cardoso, M. Land-use and climate change risks in the Amazon and the need of a novel sustainable development paradigm. *Proc. Natl. Acad. Sci. USA* **2016**, *113*, 10759–10768. [CrossRef] [PubMed]
11. Korea Forest Service. *Forest Basic Statistics*; Korea Forest Service: Daejeon, Korea, 2020.
12. Korea Forest Service. *Statistical Yearbook of Forestry*; Korea Forest Service: Daejeon, Korea, 2020.
13. Kim, S.B.; Hwang, S.T. A Study on Extraction and Classification of Unlawful Forest Land Diversion by Using GIS. *Korean Public Adm. Q.* **2015**, *27*, 513–542.
14. Aide, T.M.; Clark, M.L.; Grau, H.R.; López-Carr, D.; Levy, M.A.; Redo, D.; Andrade-Núñez, M.J.; Muñiz, M. Deforestation and Reforestation of Latin America and the Caribbean (2001–2010). *Biotropica* **2012**, *45*, 262–271. [CrossRef]
15. Dlamini, W.M. Analysis of Deforestation Patterns and Drivers in Swaziland using Efficient Bayesian Multivariate Classifiers. *Model. Earth Syst. Environ.* **2016**, *2*, 1–14. [CrossRef]
16. Mayfield, H. Making the Most of Machine Learning and Freely Available Datasets: A Deforestation Case Study. Ph.D. Thesis, The University of Queensland, Queensland, Australia, 2015.
17. Verburg, R.; Filho, S.R.; Lindoso, D.; Debortoli, N.; Litre, G.; Bursztyn, M. The Impact of Commodity Price and Conservation Policy Scenarios on Deforestation and Agricultural Land Use in a Frontier Area within the Amazon. *Land Use Policy* **2014**, *37*, 14–26. [CrossRef]
18. Damnyag, L.; Saastamoinen, O.; Blay, D.; Dwomoh, F.K.; Anglaaere, L.C.; Pappinen, A. Sustaining Protected Areas: Identifying and Controlling Deforestation and Forest Degradation Drivers in the Ankasa Conservation Area, Ghana. *Biol. Conserv.* **2013**, *165*, 86–94. [CrossRef]
19. Scullion, J.J.; Vogt, K.A.; Drahota, B.; Winkler-Schor, S.; Lyons, M. Conserving the Last Great Forests: A Meta-analysis Review of the Drivers of Intact Forest Loss and the Strategies and Policies to Save Them. *Front. For. Glob. Change.* **2019**, *2*, 62. [CrossRef]
20. Echeverria, C.; Coomes, D.A.; Hall, M.; Newton, A.C. Spatially explicit models to analyze forest loss and fragmentation between 1976 and 2020 in southern Chile. *Ecol. Modell.* **2008**, *212*, 439–449. [CrossRef]
21. Gayen, A.; Saha, S. Deforestation probable area predicted by logistic regression in Pathro river basin: A tributary of Ajay River. *Spatial. Inf. Res.* **2018**, *26*, 1–9. [CrossRef]
22. Sharma, P.; Thapa, R.B.; Matin, M.A. Examining Forest cover change and deforestation drivers in Taunggyi District, Shan State, Myanmar. *Environ. Dev. Sustain.* **2020**, *22*, 5521–5538. [CrossRef]
23. Geist, H.J.; Lambin, E.F. Proximate Causes and Underlying Driving Forces of Tropical Deforestation: Tropical Forests are Disappearing as the Result of Many Pressures, both Local and Regional, Acting in Various Combinations in Different Geographical Locations. *Bioscience* **2002**, *52*, 143–150. [CrossRef]

24. Brondizio, E.S.; Moran, E.F. Level-Dependent Deforestation Trajectories in the Brazilian Amazon from 1970 to 2001. *Popul. Environ.* **2012**, *34*, 69–85. [CrossRef]
25. Cochard, R.; Ngo, D.T.; Waeber, P.O.; Kull, C.A. Extent and Causes of Forest Cover Changes in Vietnam's Provinces 1993–2013: A Review and Analysis of Official Data. *Environ. Rev.* **2017**, *25*, 199–217. [CrossRef]
26. Jusys, T. Fundamental Causes and Spatial Heterogeneity of Deforestation in Legal Amazon. *Appl. Geogr.* **2016**, *75*, 188–199. [CrossRef]
27. Trigueiro, W.R.; Nabout, J.C.; Tessarolo, G. Uncovering the Spatial Variability of Recent Deforestation Drivers in the Brazilian Cerrado. *J. Environ. Manag.* **2020**, *275*, 111243. [CrossRef] [PubMed]
28. Anselin, L. Local Indicators of Spatial Association-LISA. *Geogr. Anal.* **1995**, *27*, 93–115. [CrossRef]
29. Harris, N.L.; Goldman, E.; Gabris, C.; Nordling, J.; Minnemeyer, S.; Ansari, S.; Lippmann, M.; Bennett, L.; Raad, M.; Hansen, M.; et al. Using Spatial Statistics to Identify Emerging Hot Spots of Forest Loss. *Environ. Res. Lett.* **2017**, *12*, 024012. [CrossRef]
30. Cushman, S.A.; Macdonald, E.A.; Landguth, E.L.; Malhi, Y.; Macdonald, D.W. Multiple-scale Prediction of Forest Loss Risk Across Borneo. *Landsc. Ecol.* **2017**, *32*, 1581–1598. [CrossRef]
31. Uvsh, D.; Gehlbach, S.; Potapov, P.V.; Munteanu, C.; Bragina, E.V.; Radeloff, V.C. Correlates of Forest-cover Change in European Russia, 1989–2012. *Land Use Policy.* **2020**, *96*, 104648. [CrossRef]
32. Mohamed, M.A. An Assessment of Forest Cover Change and Its Driving Forces in the Syrian Coastal Region during a Period of Conflict, 2010 to 2020. *Land* **2021**, *10*, 191. [CrossRef]
33. Clement, F.; Orange, D.; Williams, M.; Mulley, C.; Epprecht, M. Drivers of Afforestation in Northern Vietnam: Assessing Local Variations using Geographically Weighted Regression. *Appl. Geogr.* **2009**, *29*, 561–576. [CrossRef]
34. Chen, L.; Ren, C.; Zhang, B.; Wang, Z.; Xi, Y. Estimation of forest above-ground biomass by geographically weighted regression and machine learning with sentinel imagery. *Forests* **2018**, *9*, 582. [CrossRef]
35. Breiman, L. Random forests. *Mach. Learn.* **2001**, *45*, 5–32. [CrossRef]
36. Georganos, S.; Grippa, T.; Niang Gadiaga, A.; Linard, C.; Lennert, M.; Vanhuysse, S.; Mboga, N.; Wolff, E.; Kalogirou, S. Geographical Random Forests: A Spatial Extension of the Random Forest Algorithm to Address Spatial Heterogeneity in Remote Sensing and Population Modelling. *Geocarto. Int.* **2021**, *36*, 121–136. [CrossRef]
37. Zamani Joharestani, M.; Cao, C.; Ni, X.; Bashir, B.; Talebiesfandarani, S. PM2. 5 prediction based on random forest, XGBoost, and deep learning using multisource remote sensing data. *Atmosphere* **2019**, *10*, 373. [CrossRef]
38. Kim, C.K.; Jeong, S.Y.; Kwon, S.D.; Kim, W.K. A Study on the Possibility for the Introduction of Forest Land Reverse Mortgage System. *J. KIFR* **2014**, *18*, 39–47.
39. Park, I.S.; Kim, E.J.; Hong, S.O.; Kang, S.H. A Study on Factors Related with Regional Occurrence of Cardiac Arrest Using Geographically Weighted Regression. *Health Soc. Welf. Rev.* **2013**, *33*, 237–257.
40. Sung, J.H.; Chae, G.S. Analysis of Economic Impact in Drought: Focusing on Rice production. *JRD* **2018**, *41*, 1–23.
41. Lee, C.W.; Yoo, D.G. Evaluation of Drought Resilience Reflecting Regional Characteristics: Focused on 160 Local Governments in Korea. *Water* **2021**, *13*, 1873. [CrossRef]
42. Korea Forest Service. *Forest Basic Statistics*; Korea Forest Service: Daejeon, Korea, 2005.
43. Korea Forest Service. *Forest Basic Statistics*; Korea Forest Service: Daejeon, Korea, 2015.
44. Lee, S.J.; Yim, J.S.; Son, Y.M.; Kim, R.H. Recalculation of Forest Growing Stock for National Greenhouse Gas Inventory. *J. Climate. Change. Res.* **2016**, *7*, 485–492. [CrossRef]
45. Park, J.M.; Lee, J.S.; Lee, H.S.; Park, J.W. Study on Timber Yield Regulation Method using Probability Density Function. *J. Korean Soc. For. Sci.* **2020**, *109*, 504–511.
46. Ministry of Land, Infrastructure and Transport. *Cadastral Statistic Annual Report*; Ministry of Land, Infrastructure and Transport: Sejong, Korea, 2005.
47. Ministry of Land, Infrastructure and Transport. *Cadastral Statistic Annual Report*; Ministry of Land, Infrastructure and Transport: Sejong, Korea, 2015.
48. Statistics Korea. *Agricultural Area Survey*; Statistics Korea: Daejeon, Korea, 2005.
49. Statistics Korea. *Agricultural Area Survey*; Statistics Korea: Daejeon, Korea, 2015.
50. Statistics Korea. *Agriculture, Forestry and Fishery Survey*; Statistics Korea: Daejeon, Korea, 2005.
51. Statistics Korea. *Agriculture, Forestry and Fishery Survey*; Statistics Korea: Daejeon, Korea, 2015.
52. Ministry of Employment and Labor. *Survey of Establishments*; Ministry of Employment and Labor: Sejong, Korea, 2005.
53. Ministry of Employment and Labor. *Survey of Establishments*; Ministry of Employment and Labor: Sejong, Korea, 2015.
54. Park, E.B.; Song, C.H.; Ham, B.Y.; Kim, J.W.; Lee, J.Y.; Choi, S.E.; Lee, W.K. Comparison of Sampling and Wall-to-Wall Methodologies for Reporting the GHG Inventory of the LULUCF Sector in Korea. *J. Climate Change Res.* **2018**, *9*, 385–398. [CrossRef]
55. Lee, H.M.; Goh, J.T. Factors Influencing Cultivated Area Decisions of the Rural Area in the Fringe of Small and Medium Sizes City. *J. Korean Soc. Rural Plann.* **2015**, *21*, 1–10. [CrossRef]
56. Sim, S.Y.; Lee, K.W.; Kim, M.G.; Park, J.Y. The Case of Utilization on Regional Agricultural Statistics information. *Korea Rural Econ. Inst.* **2019**, 1–151.
57. Lee, J.M. Analysis of the Status of Agricultural Communities and Location Quotient (LQ) using Regional Survey Data in 2015 Census of Agriculture, Forestry, and Fisheries. *J. Korean Soc. Rural Plan.* **2020**, *26*, 83–93. [CrossRef]

58. Kim, S.Y.; Park, H.; Koo, H.M.; Ryoo, D.K. The Effects of the Port Logistics Industry on Port City's Economy. *J. Navig. Port. Res.* **2015**, *39*, 267–275. [CrossRef]
59. Lee, Y.G.; Jung, C.G.; Kim, W.J.; Kim, S.J. Analysis of National Stream Drying Phenomena using DrySAT-WFT Model: Focusing on Inflow of Dam and Weir Watersheds in 5 River Basins. *J. KAGIS* **2020**, *23*, 53–69.
60. Dai, W.; Li, Y.; Fu, W.; Jiang, P.; Zhao, K.; Li, Y.; Penttinen, P. Spatial Variability of Soil nutrients in Forest Areas: A Case Study from Subtropical China. *J. Plant. Nutr. Soil. Sci.* **2018**, *181*, 827–835. [CrossRef]
61. Pan, P.; Sun, Y.; Ouyang, X.; Zang, H.; Rao, J.; Ning, J. Factors Affecting Spatial Variation in Vegetation Carbon Density in Pinus massoniana Lamb. Forest in Subtropical China. *Forests* **2019**, *10*, 880. [CrossRef]
62. Ma, W.; Chen, J.; Chen, P. Illegal Activities Hotspot Analysis Based on GIS methods. In Proceedings of the 2nd IEEE International Conference on Emergency Management and Management Sciences, Beijing, China, 8–10 August 2011; Institute of Electrical and Electronics Engineers: Piscataway, NJ, USA, 2011; pp. 270–273.
63. Sánchez-Martín, J.M.; Rengifo-Gallego, J.I.; Blas-Morato, R. Hot Spot Analysis Versus Cluster and Outlier Analysis: An Enquiry into the Grouping of Rural Accommodation in Extremadura (Spain). *ISPRS Int. J. GeoInf.* **2019**, *8*, 176. [CrossRef]
64. Islam, A.; Sayeed, M.A.; Rahman, M.K.; Ferdous, J.; Islam, S.; Hassan, M.M. Geospatial Dynamics of COVID-19 Clusters and Hospots in Bangladesh. *Transbound Emerg. Dis.* **2021**, *68*, 3643–3657. [CrossRef] [PubMed]
65. Kim, S.M.; Choi, Y. Assessment of Lead (Pb) and Zinc (Zn) Contamination in Beach Sands by Hot Spot Analysis. *J. Coast. Res.* **2019**, *91*, 321–325. [CrossRef]
66. Zhang, C.; Luo, L.; Xu, W.; Ledwith, V. Use of Local Moran's I and GIS to Identify Pollution Hotspots of Pb in Urban Soils of Galway, Ireland. *Sci. Total. Environ.* **2008**, *398*, 212–221. [CrossRef]
67. Dilig, I.J.A.; San Juan, W.I.M. Geostatistical and Cluster Analysis of Earthquakes in the Philippines. *Int. Arch. Photogramm. Remote. Sens. Spat. Inf. Sci.* **2019**, *42*, 185–192. [CrossRef]
68. Mena, C.F.; Bilsborrow, R.E.; McClain, M.E. Socioeconomic Drivers of Deforestation in the Northern Ecuadorian Amazon. *Environ. Manag.* **2006**, *37*, 802–815. [CrossRef]
69. Ngwira, S.; Watanabe, T. An Analysis of the Causes of Deforestation in Malawi: A Case of Mwazisi. *Land* **2019**, *8*, 48. [CrossRef]
70. Walker, N.F.; Patel, S.A.; Kalif, K.A. From Amazon Pasture to the High Street: Deforestation and the Brazilian Cattle Product Supply Chain. *Trop. Conserv. Sci.* **2013**, *6*, 446–467. [CrossRef]
71. Jayathilake, H.M.; Prescott, G.W.; Carrasco, L.R.; Rao, M.; Symes, W.S. Drivers of Deforestation and Degradation for 28 tropical Conservation Landscapes. *AMBIO* **2021**, *50*, 215–228. [CrossRef] [PubMed]
72. Godoy, R.; Franks, J.R.; Wilkie, D.; Alvarado, M.; Gray-Molina, G.; Roca, R.; Escóbar, J.; Cardenas, M. *The Effects of Economic Development on Neotropical Deforestation: Household and Village evidence from Amerindians in Bolivia*; Development Discussion Papers; Harvard Institute for International Development: Cambridge, MA, USA, 1996.
73. Biswas, S.; Swanson, M.E.; Vacik, H. Natural Resources Depletion in Hill Areas of Bangladesh: A Review. *J. Math. Sci.* **2012**, *9*, 147–156. [CrossRef]
74. Lata, K.; Misra, A.K.; Shukla, J.B. Modeling the Effect of Deforestation Caused by Human Population Pressure on Wildlife Species. *Nonlinear. Anal. Modell. Control.* **2018**, *23*, 303–320. [CrossRef]
75. Fisher, A.G. *Clash of Progress and Security*; Macmillan and Co. Ltd.: London, UK, 1935.
76. Freund, R.J.; Wilson, W.J.; Sa, P. *Regression: Analysis*; Elsevier: Amsterdam, The Netherlands, 2006.
77. Tu, J. Spatial Variations in the Relationships Between Land Use and Water Quality across an Urbanization Gradient in the Watersheds of Northern Georgia, USA. *Environ. Manag.* **2013**, *51*, 1–17. [CrossRef]
78. Foody, G.M. Geographical Weighting as a Further Refinement to Regression Modelling: An Example Focused on the NDVI–Rainfall Relationship. *Remote Sens. Environ.* **2003**, *88*, 283–293. [CrossRef]
79. Xie, W.; Yi, S.; Leng, C.A. Study to Compare Three Different Spatial Downscaling Algorithms of Annual TRMM 3B43 Precipitation. In Proceedings of the 26th International. Conference on Geoinformatics, Kunming, China, 28–30 June 2018; Institute of Electrical and Electronics Engineers: Piscataway, NJ, USA, 2018; pp. 1–6.
80. Wang, Q.; Ni, J.; Tenhunen, J. Application of a Geographically-weighted Regression Analysis to Estimate Net Primary Production of Chinese Forest Ecosystems. *Glob. Ecol. Biogeogr.* **2005**, *14*, 379–393. [CrossRef]
81. Wang, J.; Wang, S.; Li, S. Examining the Spatially Varying Effects of Factors on PM2. 5 Concentrations in Chinese Cities using Geographically Weighted Regression Modeling. *Environ. Pollut.* **2019**, *248*, 792–803. [CrossRef]
82. Kumari, M.; Singh, C.K.; Bakimchandra, O.; Basistha, A. Geographically Weighted Regression Based Quantification of Rainfall–Topography Relationship and Rainfall Gradient in Central Himalayas. *Int. J. Climatol.* **2017**, *37*, 1299–1309. [CrossRef]
83. Tu, J.; Xia, Z.G. Examining spatially varying relationships between land use and water quality using geographically weighted regression I: Model design and evaluation. *Sci. Total. Environ.* **2008**, *407*, 358–378. [CrossRef] [PubMed]
84. Shah, S.H.; Angel, Y.; Houborg, R.; Ali, S.; McCabe, M.F. A Random Forest Machine Learning Approach for the Retrieval of Leaf Chlorophyll Content in Wheat. *Remote. Sens.* **2019**, *11*, 920. [CrossRef]
85. Abdulkareem, N.M.; Abdulazeez, A.M. Machine Learning Classification Based on Radom Forest Algorithm: A Review. *Int. J. Sci. Bus.* **2021**, *5*, 128–142.
86. Bunn, C.; Talsma, T.; Läderach, P.; Castro, F. Climate Change Impacts on Indonesian Cocoa Areas. In Cocoa Climate Suitability Indonesia 2017 CIAT 50 and Mondelez International. Available online: https://www.cocoalife.org/progress/climate-change-impacts-on-indonesian-cocoa-areas-report (accessed on 3 November 2021).

87. Liu, Y.; Wu, H. Prediction of Road Traffic Congestion Based on Random Forest. In Proceedings of the 10th International Symposium on Computational Intelligence and Design (ISCID), Hangzhou, China, 9–10 December 2017; Institute of Electrical and Electronics Engineers: Piscataway, NJ, USA, 2017; Volume 2, pp. 361–364.

88. Krishnamurthy, N.; Maddali, S.; Hawk, J.A.; Romanov, V.N. 9Cr steel visualization and predictive modeling. *Comput. Mater. Sci.* **2019**, *168*, 268–279. [CrossRef]

89. Song, X.; Long, Y.; Zhang, L.; Rossiter, D.G.; Liu, F.; Jiang, W. Spatial Accuracy Evaluation for Mobile Phone Location Data with Consideration of Geographical Context. *IEEE Access.* **2020**, *8*, 221176–221190. [CrossRef]

90. Koh, E.H.; Lee, E.; Lee, K.K. Application of Geographically Weighted Regression Models to Predict Spatial Characteristics of Nitrate Contamination: Implications for an Effective Groundwater Management Strategy. *J. Environ. Manag.* **2020**, *268*, 110646. [CrossRef]

91. Chen, F.W.; Liu, C.W. Estimation of the Spatial Rainfall Distribution using Inverse Distance Weighting (IDW) in the Middle of Taiwan. *Paddy Water Environ.* **2012**, *10*, 209–222. [CrossRef]

92. Willmott, C.J. Some Comments on the Evaluation of Model Performance. *Bull. Am. Meteor. Soc.* **1982**, *63*, 1309–1313. [CrossRef]

93. Khodadoust Siuki, S.; Saghafian, B.; Moazami, S. Comprehensive Evaluation of 3-hourly TRMM and Half-hourly GPM-IMERG Satellite Precipitation Products. *Int. J. Remote Sens.* **2017**, *38*, 558–571. [CrossRef]

94. Burnham, K.P.; Anderson, D.R. Multimodel Inference: Understanding AIC and BIC in Model Selection. *Soc. Methods Res.* **2004**, *33*, 261–304. [CrossRef]

95. Nashwari, I.P.; Rustiadi, E.; Siregar, H.; Juanda, B. Geographically Weighted Regression Model for Poverty Analysis in Jambi Province. *Indones. J. Geogr.* **2017**, *49*, 42. [CrossRef]

96. Meng, D.; Xi, Z.; Zhao, J. Analysis on the Impacting Factors of Hand, Foot and Mouth Disease Incidence Using Random Forest. In Proceedings of the 10th Data Driven Control and Learning Systems Conference (DDCLS), Suzhou, China, 14–16 May 2021; Institute of Electrical and Electronics Engineers: Piscataway, NJ, USA, 2021.

97. Park, S.; Choi, G.Y. Urban Sprawl in the Seoul Metropolitan Region, Korea Since the 1980s Observed in Satellite Imagery. *JAKG* **2016**, *5*, 331–343. [CrossRef]

98. Zhou, W.; Zhang, S.; Yu, W.; Wang, J.; Wang, W. Effects of urban expansion on forest loss and fragmentation in six megaregions, China. *Remote Sens.* **2017**, *9*, 991. [CrossRef]

99. Kim, S.H. The Analysis of Influence Factors on Vitalization of Large Scale Development Projects in Seoul Metropolitan Area. *Korea Real Estate Acad.* **2013**, *55*, 145–155.

100. Jaimes, N.B.P.; Sendra, J.B.; Delgado, M.G.; Plata, R.F. Exploring the driving forces behind deforestation in the state of Mexico (Mexico) using geographically weighted regression. *Appl. Geogr.* **2010**, *30*, 576–591. [CrossRef]

101. Mas, J.F.; Cuevas, G. Identifying Local Deforestation Patterns Using Geographically Weighted Regression Models. In Proceedings of the International Conference on Geographical Information Systems Theory, Applications and Management, Barcelona, Spain, 28–30 April 2015; SCITEPRESS—Science and Technology Publications: Setubal, Portugal, 2015; pp. 36–49.

102. Kim, Y.H.; Jeong, G.O.; Kim, S.G. Comparison of Transport Infrastructure and Its Utilization Features in between Korean and Foreign Cities. *KOTI* **2013**, *1*, 1–158.

103. Choi, E.Y. *Analysis of the Changes in Rural Villages in Chungnam Province based on GIS(2005~2010)*; ChungNam Institute: Gongju-si, Korea, 2014; Volume 21, pp. 1–157.

104. Yeo, C.H.; Kim, D.C.; Pyo, K.J. An Analysis of GRDP Growth Factors and Spatial Locational Characteristics of Growth-Leading Industries in Chungcheongnam-Do. *Chungcheong Reg. Stud.* **2009**, *2*, 25–42.

105. Mo, S.W.; Lee, K.B. Industrial Competitiveness of Gwangju and Jeonnam. *J. Ind. Econ. Bus.* **2017**, *30*, 445–460. [CrossRef]

106. Keleş, S.; Sivrikaya, F.; Çakir, G.; Köse, S. Urbanization and forest cover change in regional directorate of Trabzon forestry from 1975 to 2000 using landsat data. *Environ. Monit. Assess.* **2008**, *140*, 1–14. [CrossRef] [PubMed]

107. Wakeel, A.; Rao, K.S.; Maikhuri, R.K.; Saxena, K.G. Forest management and land use/cover changes in a typical micro watershed in the mid elevation zone of Central Himalaya, India. *For. Ecol. Manag.* **2005**, *213*, 229–242. [CrossRef]

108. Baehr, C.; BenYishay, A.; Parks, B. Linking Local Infrastructure Development and Deforestation: Evidence from Satellite and Administrative Data. *J. Assoc. Environ. Resour. Econ.* **2021**, *8*, 375–409. [CrossRef]

109. BenYishay, A.; Parks, B.; Runfola, D.; Trichler, R. Forest cover impacts of Chinese development projects in ecologically sensitive areas. In Proceedings of the SAIS CARI 2016 Conference, Washington, DC, USA, 13–14 October 2016; China Africa Research Initiative: Washington, DC, USA, 2016; pp. 13–14.

110. Chen, X.; Li, F.; Li, X.; Hu, Y.; Hu, P. Quantifying the Compound Factors of Forest Land Changes in the Pearl River Delta, China. *Remote Sens.* **2021**, *13*, 1911. [CrossRef]

111. Tejaswe, G. *Mannual on Deforestation, Degradation, and Fragmentation Using Remote Sensing and GIS. Strengthening Monitoring, Assessment and Reporting on Sustainable Forest Management in Asia (GCP/INT/988/JPN)*; MAR-SFM Working Paper 5; Food and Agriculture Organization: Rome, Italy, 2007.

112. Angelsen, A. Policies for reduced deforestation and their impact on agricultural production. *Proc. Natl. Acad. Sci. USA* **2010**, *107*, 19639–19644. [CrossRef]

113. Yang, R.; Luo, Y.; Yang, K.; Hong, L.; Zhou, X. Analysis of forest deforestation and its driving factors in Myanmar from 1988 to 2017. *Sustainability* **2019**, *11*, 3047. [CrossRef]

114. Lim, C.L.; Prescott, G.W.; De Alban, J.D.T.; Ziegler, A.D.; Webb, E.L. Untangling the proximate causes and underlying drivers of deforestation and forest degradation in Myanmar. *Conserv. Lett.* **2017**, *31*, 1362–1372. [CrossRef] [PubMed]

115. Xu, X.; Jain, A.K.; Calvin, K.V. Quantifying the biophysical and socioeconomic drivers of changes in forest and agricultural land in South and Southeast Asia. *Glob. Chang. Biol.* **2019**, *25*, 2137–2151. [CrossRef]

116. Li, M.; Mao, L.; Zhou, C.; Vogelmann, J.E.; Zhu, Z. Comparing Forest fragmentation and its drivers in China and the USA with Globcover v2. 2. *J. Environ. Manag.* **2010**, *91*, 2572–2580. [CrossRef]

117. Mwangi, N.; Waithaka, H.; Mundia, C.; Kinyanjui, M.; Mutua, F. Assessment of drivers of forest changes using multi-temporal analysis and boosted regression trees model: A case study of Nyeri County, Central Region of Kenya. *Model Earth Syst. Environ.* **2020**, *6*, 1657–1670. [CrossRef]

118. Dos Santos, A.M.; da Silva, C.F.A.; de Almeida Junior, P.M.; Rudke, A.P.; de Melo, S.N. Deforestation drivers in the Brazilian Amazon: Assessing new spatial predictors. *J. Environ. Manag.* **2021**, *294*, 113020. [CrossRef]

119. Armenteras, D.; Espelta, J.M.; Rodríguez, N.; Retana, J. Deforestation dynamics and drivers in different forest types in Latin America: Three decades of studies (1980–2010). *Glob. Environ. Change* **2017**, *46*, 139–147. [CrossRef]

120. Lambin, E.F.; Geist, H.J.; Lepers, E. Dynamics of land-use and land-cover change in tropical regions. *Annu. Rev. Environ. Resour.* **2003**, *28*, 205–241. [CrossRef]

121. Cai, J.; Huang, B.; Song, Y. Using Multi-source Geospatial Big Data to Identify the Structure of Polycentric Cities. *Remote Sens. Environ.* **2017**, *202*, 210–221. [CrossRef]

122. Li, F.; Zhou, T.; Lan, F. Relationships Between Urban Form and Air Quality at Different Spatial Scales: A Case Study from Northern China. *Ecol. Indic.* **2021**, *121*, 107029. [CrossRef]

123. Park, J.Y.; Lee, Y.J.; Lee, W.S.; Lee, B.K. Status of Photovoltaic Power Plant Installation Projects Proceeded through EIA and its Environmental Discussion. *Korea Environ. Inst.* **2018**, *5*, 1–78.

124. Choi, J.S.; Kwak, D.A.; Kwon, S.D.; Baek, S.A. Study on Applicability of Slope Types to Permission Standard for Forestland Use Conversion. *J. KAGIS* **2018**, *21*, 145–157.

125. Mori, K.; Tabata, T. Comprehensive Evaluation of Photovoltaic Solar Plants vs. Natural Ecosystems in Green Conflict Situations. *Energies* **2020**, *13*, 6224. [CrossRef]

126. Liu, Y.; Feng, Y.; Zhao, Z.; Zhang, Q.; Su, S. Socioeconomic drivers of forest loss and fragmentation: A comparison between different land use planning schemes and policy implications. *Land Use Policy.* **2016**, *54*, 58–68. [CrossRef]

forests

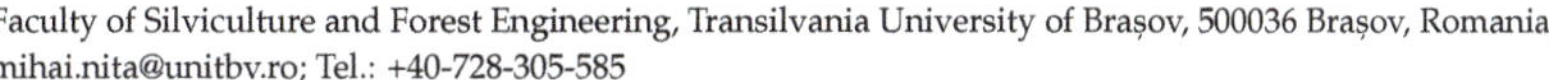

Article

Testing Forestry Digital Twinning Workflow Based on Mobile LiDAR Scanner and AI Platform

Mihai Daniel Niță

Faculty of Silviculture and Forest Engineering, Transilvania University of Brașov, 500036 Brașov, Romania; mihai.nita@unitbv.ro; Tel.: +40-728-305-585

Abstract: Climate-smart forestry is a sustainable forest management approach for increasing positive climate impacts on society. As climate-smart forestry is focusing on more sustainable solutions that are resource-efficient and circular, digitalization plays an important role in its implementation. The article aimed to validate an automatic workflow of processing 3D pointclouds to produce digital twins for every tree on large 1-ha sample plots using a GeoSLAM mobile LiDAR scanner and VirtSilv AI platform. Specific objectives were to test the efficiency of segmentation technique developed in the platform for individual trees from an initial cloud of 3D points observed in the field and to quantify the efficiency of digital twinning by comparing the automatically generated results of (DBH, H, and Volume) with traditional measurements. A number of 1399 trees were scanned with LiDAR to create digital twins and, for validation, were measured with traditional tools such as forest tape and vertex. The segmentation algorithm developed in the platform to extract individual 3D trees recorded an accuracy varying between 95 and 98%. This result was higher in accuracy than reported by other solutions. When compared to traditional measurements the bias for diameter at breast height (DBH) and height was not significant. Digital twinning offers a blockchain solution for digitalization, and AI platforms are able to provide technological advantage in preserving and restoring biodiversity with sustainable forest management.

Keywords: digital twinning; climate smart; LiDAR; artificial intelligence; digitalization

Citation: Niță, M.D. Testing Forestry Digital Twinning Workflow Based on Mobile LiDAR Scanner and AI Platform. *Forests* **2021**, *12*, 1576. https://doi.org/10.3390/f12111576

Academic Editor: Mark Vanderwel

Received: 25 May 2021
Accepted: 13 November 2021
Published: 16 November 2021

1. Introduction

Technology-based on digital twins extends well beyond the initial design to the merging of the world of IoT (Internet of Things), AI (artificial intelligence), and big data analytics [1]. Digitally replicating the real world, as more data becomes available, empowers data scientists and other IT specialists to optimize deployments for peak efficiency, as well as create other potential what-if scenarios [2]. Buildings, factories, and even entire cities are now digitally represented as digital twins [3]. Some have suggested even people, processes, and organizations have digital twins, expanding the concept of digital twins even further [4].

Known as the part of world who harbors the highest biodiversity, forests are one of the most complex systems from a structural and functional point of view. In addition to their role as recreational resources, wood products, and material and energy providers, forests and the forest sector are fundamental in reducing greenhouse gas emissions by capturing carbon dioxide in tree biomass. [5]. Climate-smart forestry is a sustainable forest management approach for increasing these positive climate impacts on society [6]. In response to climate change, the approach intends to reduce greenhouse gas emissions, adapt forest management to create resilient forests and focus on active forest management with the goal of sustainability by increasing productivity while simultaneously offering forest benefits [7,8]. With two big challenges ahead, a green and digital transition, digital twinning in forestry is the next development step [9]. Digitalization plays a key role in climate-smart forestry's focus on sustainable, resource-efficiency, and circular solutions [10].

In forests, we observe extensive vertical stratification, making them among the world's most complex ecosystems [11]. Forests containing conifers are the simplest as they consist of a tree layer reaching about 30 m in height, a shrub layer that often is spotty, and a soil layer covered with mosses and lichens [12]. Forests with deciduous foliage are more complex; a rainforest canopy consists of at least three different layers, while deciduous trees have a separate upper and lower layer [13]. Due to this complexity, accurate characterization of forests using precise inventory remains one of the most challenging activities in the digitalization of forestry [14,15].

Equipment and techniques have become more affordable and accessible in recent years. With the development of technology to generate 3D scenes from measurements, LIDAR has become more portable and more affordable [16]. This has enabled the building of virtual worlds that reflect natural landscapes using precision measurements. Particularly, terrestrial lidar systems collect large amounts of data varying from tens of thousands to billions of 3D points to determine the 3D space surrounding a given point in 3D [17].

In forest inventories, a TLS (terrestrial lidars scanner) can document forests rapidly, automatically, and provide inch-by-inch details in minutes. Early work related to forest inventory estimation via TLS started with the development of Cyra Technologies' TLS system around 2000; and it was later acquired by Leica in 2001 [18–21]. Forest inventories used TLS as a way to improve harvest efficiency by replacing manual measurements with measurements derived from TLS data in the forest plots [22]. As a result, TLS has been used in collecting basic attributes such as DBH (diameter at breast height), tree height, and tree position in forest sample plots [23,24]. A scientifically confirmed fact is that the measuring diameter and height of a tree are affected by an error of at least 5.6% and the measurement bias of DBH and H affects estimation up to 26.4% [25–27]. Therefore, using classical methods for estimating volume and biomass are not suitable for modern needs in the context of a circular economy.

Virtual tree measurements are achieved today by using software applications and allometric approaches [28–33]. However, the quality of results and maturity of these algorithms are still low [34,35]. Furthermore, there is no digitalization workflow on the market that would be able to provide a complete set of solutions to the problem, from the measurements in the forest to creating digital twins of each tree [36]. There are several challenges in the field of measuring trees in the real world, and multiple scans are needed from a variety of angles to capture all trees in the area of interest in their full height if possible [37,38]. Another aspect is the problem of segmentation of individual trees and the delimitation of the soil surface [35,39]. These are crucial for the entire process of forest digital twinning, and current solutions often fail due to certain oblique orientations of the trunks, the presence of shrubs in the soil, and other obstructions present in various cases [40,41]. With regards to biometric data extraction, most known methods use overly simplified models that aspire to approximate trunk geometry through cylinders or cones and excessively complex models that try to model the observational data with as much precision as possible [36,40].

Among other software that provide partial or total solution to digital twinning (e.g., 3D Forest, OPALS, TreeQSM), VirtSilv is a newly developed platform that responds to the realities of the forest and provides industry-specific services in all segments [42]. VirtSilv is an online platform that uses AI customizable algorithms to produce unique shapes of trees as digital support for a fully automated traceability IT circuit between forest management, transport, and the wood industry.

In the current context of software development there is a need to validate an entire workflow starting with data collection and finishing with providing digital twins usable in forestry, easily accessible to decision makers. The article aims to validate an automatic workflow of processing 3D pointclouds to produce digital twins for every tree in a specific forest using GeoSLAM mobile LiDAR scanner and VirtSilv AI platform. The specific goals were: (1) to test the efficiency of segmentation technique developed in the platform for individual trees from an initial cloud of 3D points observed in the field; and (2) to quantify the efficiency of digital twinning by comparing the automatically generated results of (DBH, H and volume) with traditional measurements.

2. Materials and Methods

2.1. Study Area

Several measurement campaigns were carried out, using both a mobile LIDAR device and traditional forest inventory tools (forestry tape for DBH—diameter at breast height and vertex logger IV for tree height), focusing on three plots of 1 ha size in Carpathian Mountains, Ciucas Massif (Figure 1). For this part of the Carpathians, according to the WorldClim global database on weather and climate data [43], elevations range between 800 and 2400 m a.s.l. and the climate is temperate-continental, with wet summers and cold winters. The same source mentions for this area mean annual precipitation from 615.4 mm to 1095 mm (overall annual mean 793.4 mm; standard deviation 84.7) and mean annual temperature between 1.2 and 9.2 °C.

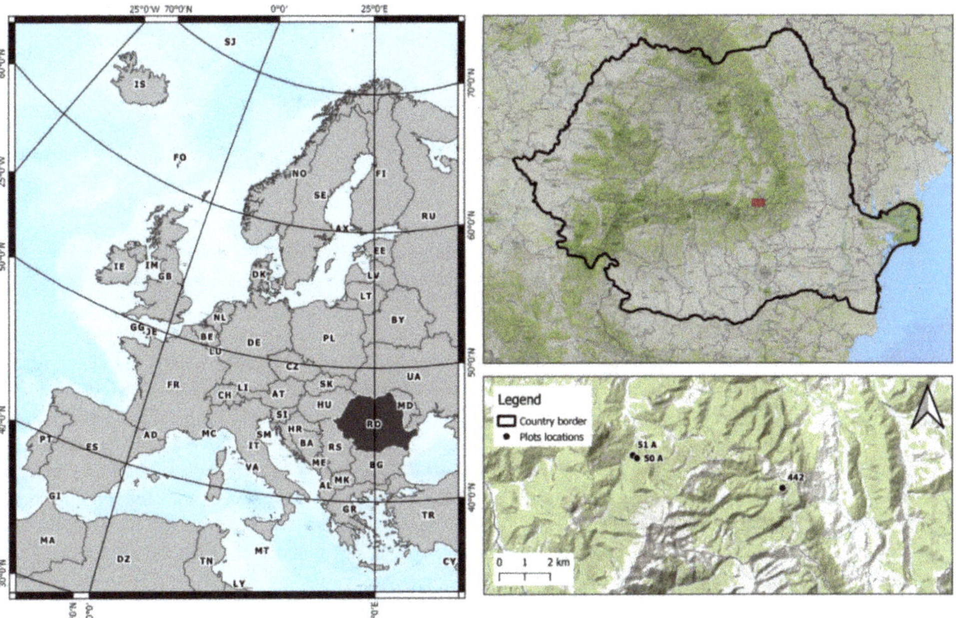

Figure 1. Location of the study.

The 1-ha size plots were selected based on the tree density, geo-spatial distribution, focusing on the forests of economic interest planned for thinnings and selective logging.

2.2. Data

The plots were scanned using ZEB Horizon, a scanner based on LiDAR technology, and included in the category of terrestrial laser scanners (TLS). This is a 3D scanner of high-speed used for measurements that require recording of details. A ZEB Horizon Scanner uses laser technology, weighing 1.3 kg it is designed for outdoor applications that require scanning up to 100 m and at an accuracy of 1–3 cm. The scanner uses a rotating mirror to beam around the area that is scanned. The measurement characteristics consist of up to 300,000 repetitions per second. Data acquired using GEOSLAM Horizon technology is a point cloud in the form of three-dimensional data compiled using SLAM (simultaneous localization and mapping). The scanning time suitable to produce dense pointclouds was on an average of approximately 20 min/hectare for each plot.

Data collection was carried out in mixed forests, predominated by spruce and beech. This results in a point cloud made representing the scanner's environment in a three-dimensional dataset (Figure 2). Later, the point cloud is mentioned as the laser scan (or simply scan).

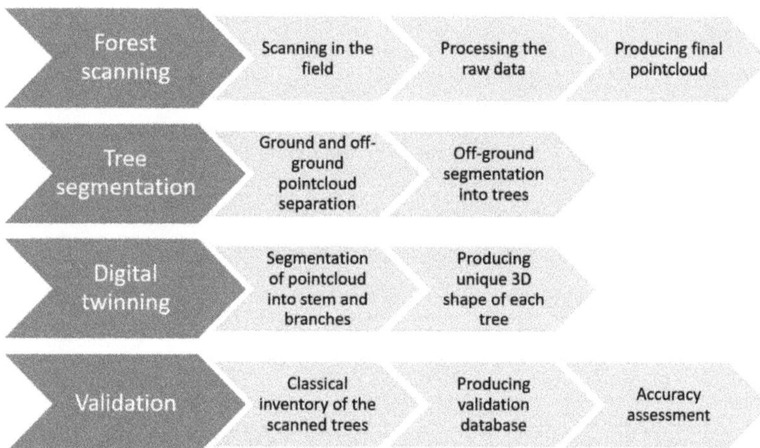

Figure 2. Methodology workflow.

The traditional inventory was made by two teams consisting of two forest engineers. The average processing time per team was on average of 8 h/hectare.

2.3. Tree Segmentation

The raw data generated during the scanning process enables the visual identification of individual tree structures, but they are not yet quantitatively differentiated. To create individual raw material for digital twin, VirtSilv first separates the ground from the trees, and then it reconstructs each tree separately (Figure 2). The algorithm takes three steps to estimate each tree's footprint simultaneously. The algorithm begins at a large nucleus of points with high density and then grows by accretion until it meets neighboring trees. As a result, the operator is given many options to customize the algorithm and is given the option to change data sets according to their needs. The average processing time of segmentation was 30 min for a 1-ha plot.

2.4. Digital Twinning Process

When all of the individual tree segments are identified, the remaining task is to recognize tree trunks and model their numerical dimensions on a simple and flexible basis, thereby giving the potential for the digital twinning process. To overcome the limitations of current techniques, VirtSilv algorithms are designed around the following principles:

- The trunk shape of segments of sufficiently small height can be approximated very well by inclined cone trunks;
- The vertical projection of the data obtained from segments of sufficiently small height can be approximated by a ring of points with relatively high density;
- Generally, the successive segments in the vertical array are very well aligned, in the sense that the angle and bending of each segment, concerning that vertical changes are low.

Thus, the VirtSilv algorithm is focused on extracting chains of cone trunks as a numerical model for trunks. The average time of producing the 3D model of a tree digital twin is less than one minute.

2.5. Validation, Accuracy Assessment, Robustness Check

For validation, a Bland–Altman test was used. These plots are extensively used to evaluate the agreement among two different instruments or two measurement techniques. Bland–Altman plots allow identification of any systematic difference between the measurements (i.e., fixed bias) or possible outliers. This can be carried out by Bland and Altman's approximate method or by more precise methods [44,45].

For detecting heteroskedasticity the Breusch-Pagan test was used. It involved using a variance function and using a $\chi 2$-squared test: the test statistic is distributed $n\chi 2$ with k degrees of freedom. If the test statistic has a p-value below an appropriate threshold (e.g., $p < 0.05$) then the null hypothesis of homoskedasticity is rejected and heteroskedasticity assumed.

The boundaries of Bland and Altman's agreement have traditionally been used to evaluate the agreement between different methods of measuring discrete variables. However, when the variances of the measurement errors of different methods are similar, Bland and Altman's plot can be misleading. Therefore, it was used the R package "MethodCompare" to generate a bias plot and a precision plot based [46].

3. Results

After the traditional inventory was realized, a total number of 1369 trees were sampled in the three forest plots (Figure 3). Plot 442 had the highest tree density of 739 per hectare, average DBH of 32.54 cm, average height 29.8, average tree volume 1.25 m^3, and a total volume of 923.06 m^3/ha. Plot 051A had a tree density of 373 per hectare, average DBH of 52.14 cm, average height 36.4, average tree volume 3.65 m^3, and a total volume of 1363.35 m^3/ha. Plot 050A had the lowest tree density of 258 per hectare, average DBH of 56.13 cm, average height 38.9, average tree volume 4.32 m^3, and a total volume of 1112.57 m^3/ha.

VirtSilv automatically segmented 1339 with an overall accuracy of segmentation of 97.8% (Figure 3). In plot 442 the accuracy of segmentation was 99.1%, in plot 051A was 95.2%, and 98.1% in plot 050A.

Out of 30 trees which were unsegmented/missed by the segmentation algorithm of VirtSilv, the distribution per species was represented by Spruce 37%, Beech 37%, Sycamore 17%, and Fir 10%. Overall, an approximate balance between coniferous and deciduous was maintained. In the case of beech trees, the average DBH was 16.6 lower than the average 27.6. For spruce, the DBH was 42.4, closer to the average 40.9 especially due to the sample size, as spruce is the most represented in all three plots.

As a result of the digital twinning process, VirtSilv reconstructed 1339 trees, a population of data described by average DBH and H almost similar to the ones recorded in traditional measurements (Table 1). Regarding the volume, the average values were slightly different mostly to the fact that the volume calculated with VirtSilv was based on the unique 3D shape of each tree, unlike the traditional volume which was calculated using the specific equation based on species, DBH, and height. In terms of all descriptive statistics, both traditional (DBH_t, H_t, Vol_t) and virtual measurements (DBH_v, H_v, Vol_v) presented similar results (Table 1).

Statistical populations differ in the way they are measured based on different techniques used when measuring either traditional or virtual. As for DBH and H characteristics, both the traditional and virtual measurements are applying the same mathematical approach. This similar approach is seen in the standard error, median, mode, standard deviation, and curve distribution characteristics, as they are close to each other on both traditional and virtual measurements (Table 1). In the case of volumes, the characteristics of the populations are slightly different as the mathematical approach is different between traditional and virtual measurements. The traditional approach did not involve measuring the volume of the trees, as the formula for calculating the volume derives from trees' species, DBH, and height. These different approach changes are observed in the different range values, with a difference of 10%, or in a very high (75%) difference between the minimum values of volume populations (Table 1).

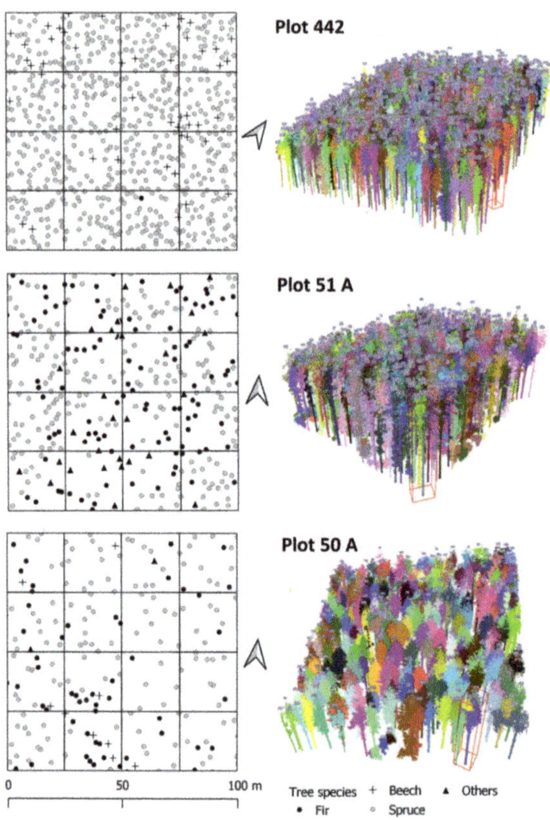

Figure 3. Results of segmentation for three plots, maps produced based on traditional measurements (left column), and 3D maps with digital twins produced with VirtSilv AI platform (right column).

Table 1. Results for parameters extracted (DBH, height, volume).

	DBH_t	DBH_v	H_t	H_v	Vol_t	Vol_v
Mean	42.57	43.24	33.50	33.66	2.50	2.76
Standard Error	0.43	0.39	0.19	0.18	0.05	0.06
Median	39.40	40.60	33.80	33.79	1.78	2.04
Mode	34.80	33.00	30.80	28.02	0.65	1.44
Standard Deviation	15.65	14.10	6.87	6.41	2.00	2.14
Sample Variance	244.82	198.83	47.24	41.11	4.00	4.56
Kurtosis	−0.21	0.08	2.01	2.79	1.84	1.43
Skewness	0.58	0.67	−1.00	−0.97	1.33	1.31
Range	94.30	79.80	42.30	42.86	13.25	11.95
Minimum	10.00	13.30	5.00	5.82	0.02	0.08
Maximum	104.30	93.10	47.30	48.68	13.27	12.03
Sum	57,005.30	57,901.20	44,849.93	45,072.73	3353.35	3696.87
Count	1339.00	1339.00	1339.00	1339.00	1339.00	1339.00
Confidence Level (95.0%)	0.84	0.76	0.37	0.34	0.11	0.11

According to the Breusch-Pagan test used on all measurements (DBH, H, and volume), heteroskedasticity is not present. The variance function and the $\chi2$-test were used to test the null hypothesis that heteroskedasticity is not present, and they show that the variability of the random disturbance is not different across elements of the vector (Table A1).

The Pearson correlation coefficient used to examine the strength and direction of the linear relationship between classical and virtual measurement continuous variables indicates values over 0.9 (very close to the absolute value of 1), which indicates a perfect linear relationship. The relationship with the highest correlation coefficient was identified at DBH measurements of 0.96 and the lowest was identified at height measurements of 0.9 (Figure 4).

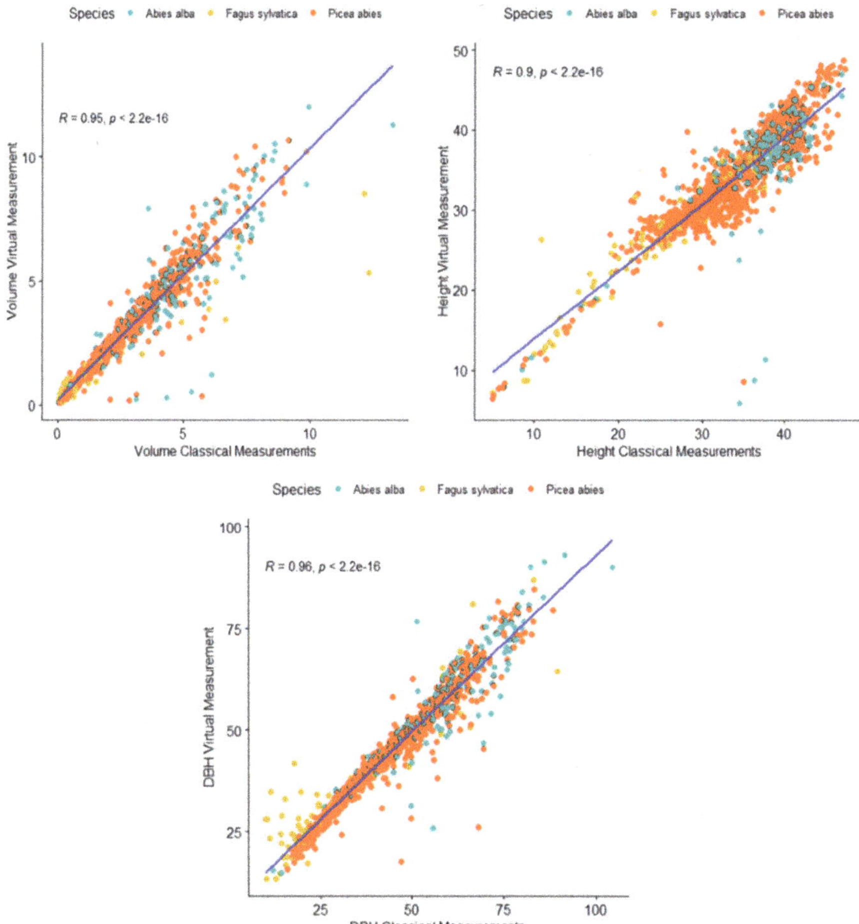

Figure 4. Correlations between classical measurements and virtual measurements.

The lower R value in height measurements is given by an overestimation in the virtual environment at dominated trees, as in some cases additional branches from neighboring trees interfered with the measurement process. In case of DBH, it can be observed that a cluster of Fagus sylvatica trees with lower diameters than 25 were overestimated due to high density of branches at 1.3 m on the stem (Figure 3).

The *p*-values for the correlation between classical and virtual measurements are less than the significance level of 0.05, which indicates that the correlation coefficients are significant (Figure 4).

Bland and Altman's limits of agreement plot (LoA) described how far apart measurements by two methods were more likely to be for most individuals. Inside LoA there were calculated the mean difference, the estimated bias, and the standard deviations of the differences to measure the random fluctuations around the mean. The mean value of the difference does not differ significantly from 0 on all observed characteristics. The differences within a mean of ± 1.96 SD appeared as not important, most of them remaining in the 95% limits of agreement for each comparison, concluding that the two methods may be used interchangeably and practically estimate the same results (Figure 5).

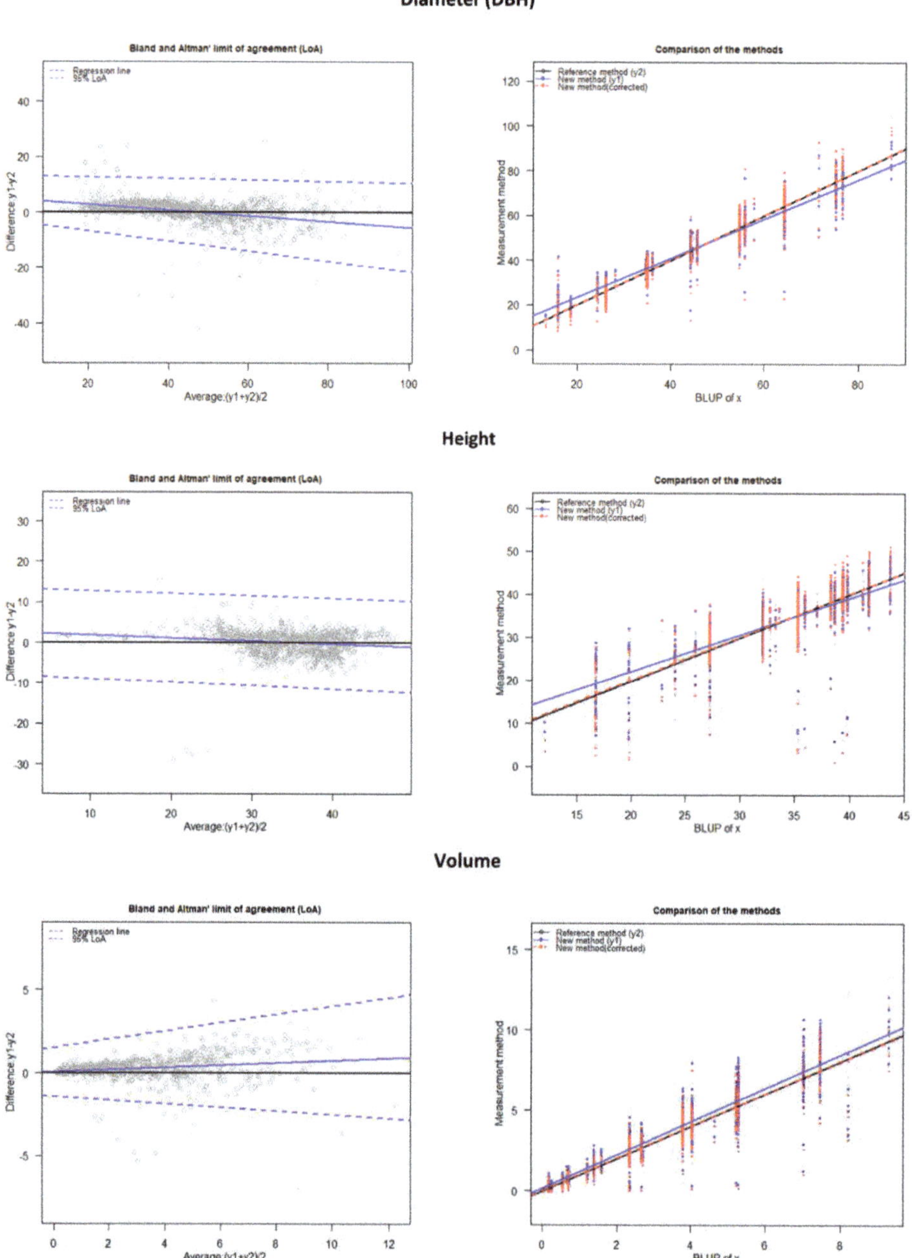

Figure 5. Bland and Altman plot for DBH, height, and volume.

To visually appraise the performance of the new method in the spirit of Bland and Altman's limits of the agreement, the bias plot and the precision plot were generated. These plots allow the visualization of the bias-corrected values (i.e., recalibrated values, variable y1_corr) of the new measurement method (Figure 5).

In the case of DBH, compared to the reference method, the new method has a differential bias of 6.098 and a proportional bias of 0.873. The variance of the new method is smaller than the one for the reference method. The scatter plot of the new method (virtual measurements) and reference method (traditional measurements) versus the best linear prediction (BLUP) with the two regression lines shows that the bias is decreasing with the increase of the DBH. The red bias line shows an inverse proportional trend as the bias of DBH is decreasing with the value from 4 cm in smaller diameters than 20 cm, to −4 cm in larger diameters over 80 cm. This shows a limitation of the virtual measurements as in younger trees the number of branches at 1.3 m is higher and affects the precision of measurement. The precision plot of the new measurement method shows that the standard deviation of measurement error is increasing with the increase of DBH (Figure 6).

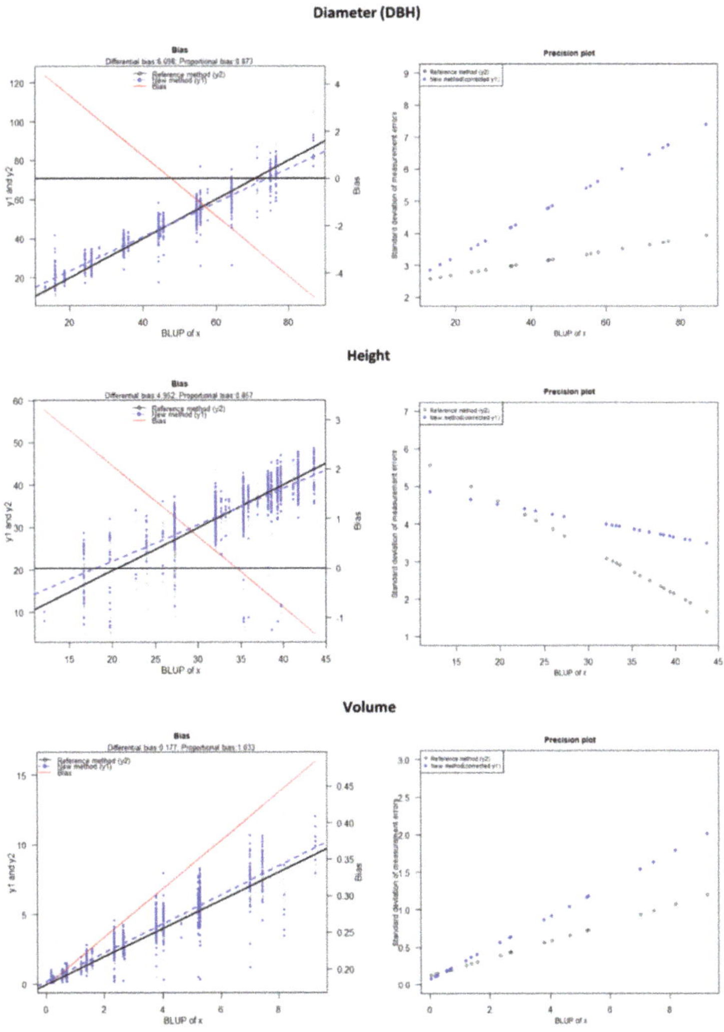

Figure 6. Bias and precision plot.

For height, compared to the reference method (traditional measurements), the new method (virtual measurement) has a differential bias of 4.952 and a proportional bias of 0.857. The red bias line shows an inverse proportional trend as the bias of H is decreasing with the value from 3 m in smaller heights than 15 m, to -1 m in higher trees over 40 m. The standard deviation of measurement error trend shows that it is decreasing with the increase of DBH (Figure 6).

For tree volume, compared with the traditional method, the virtual measurement has a differential bias of 0.177 and a proportional bias of 1.033. The red line bias is showing that it is increasing with the increase of volume and the standard deviation of measurement error is increasing with the increase of tree volume. The increase of bias with volume is explained by the different approach in calculating the volume, as for traditional measurements the generalized formula is not taking into consideration local sites conditions and stand density and mixture which in many cases leads to an underestimation of volume.

4. Discussion

The forestry sector is well-positioned to play a strong role in reaching the objectives of the European Green Deal, including the EU Biodiversity Strategy 2030, the new EU Strategy on Adaptation to Climate Change, and the upcoming new Forest Strategy (H1 2021). Climate-smart forestry principles include maintaining and enhancing environmental benefits, biodiversity, and ecosystem services, as well as specific actions for maintaining and enhancing forest characteristics, biodiversity, and ecosystem services. Moreover, the EU's Digital Strategy (launched in 2020) plans on transforming Europe into a digital single market by 2030. This strategy which covers the forestry sector should revolve around four key pillars: government, skills, infrastructure, and businesses. About 75% of EU businesses are expected to use cloud technology, artificial intelligence (AI), or big data by 2030, with more than 90% of SMEs expected to have at least a basic level of digital intensity by 2030 [47].

With two big challenges ahead (i.e., a green and digital transition) digital twinning and AI solutions in forestry are the next steps for more sustainable solutions that are resource-efficient and circular. At the same time, digital twinning will contribute to the European commitment to climate neutrality by 2050 [48]. Investing digital capabilities (including machine learning, artificial intelligence, and blockchain) may contribute to achieving EU Green Deals and digital transition objectives, including the forestry sector. Many global forestry operators and enterprises have already pioneered the progressive use of advanced technologies to enhance forest management results, particularly in plantation forestry, an approach that has become known as "precision forestry". However, it has not yet become an established part of business-as-usual sustainable forest management practices, especially due to the lack of key components such as mobile scanners and complete solutions for analysis.

In this context, the present paper validates a workflow in supporting the digitalization process of the forestry sector to better inform and enhance the implementation of climate-smart forestry sustainable management practices. Currently, it has been demonstrated that digital twinning in forestry can be carried out on large areas using terrestrial perspective, producing accurate and complete digital twins for each tree [49]. Despite the capacity of ALS (airborne lidar scanning) to cover large areas, TLS remains the complete solution for complex forests with multiple stories [17,29]. The terrestrial perspective is giving the optimal results in producing the complete 3D pointclouds of trees since the understory eye-of-sight is increasing the visibility of the scanned objects such as trunks and lower branches.

In this paper it was demonstrated that for digital twinning in forestry the entire workflow needs to take into consideration both the field measurements as they are the most important part of digital twinning. The efficiency of the segmentation technique developed in the VirtSilv platform for individual trees from an initial cloud of 3D points observed in the field proved to be very high. The capacity of VirtSilv AI algorithms to be customized into a user-friendly interface improved the results of segmentation. Therefore, the AI algorithms integrated into the system successfully identified unique tree shapes from complex forest environments such as natural mixed beech, spruce, and fir forest. Moreover, the quality of the digital twins in terms of comparing the traditional tree inventory values (DBH, height, volume) was very high and accurate. VirtSilv AI platform proved to be a reliable solution in setting up an automatic workflow of processing 3D pointclouds to produce digital twins for every tree in a specific large area forest. When combined with a fast mobile LIDAR scanner such as Zeb Horizon, the digital twinning process was reduced to several hours. This method, compared with traditional inventory, reduced the processing and analysis time by approximately four-fold.

For the segmentation technique developed in the platform to extract individual trees using 3D points observed in the field, an accuracy varying between 95 and 98% was recorded. This result was higher in accuracy than reported by other solutions such as 3D Forest (85%–89.9%) [50].

The 1339 digital twins produced by the platform were similar in terms of DBH, H, and volume derived from traditional measurements. Even though it has had a non-significant influence on the results, the bias of DBH and H was decreasing with the increase of the values. It was found that both the scanning device and segmentation procedure had some limitations. It has been documented in other papers that the higher bias on lower values in DBH is mainly due to the noise in the pointcloud generated by the mobile scanner [41,49,51,52]. This bias is explained by noise effects especially on lower DBH. In the case of height, the bias is explained mostly by the segmentation technique which is influenced by the quality of the pointclouds. At lower heights, the digital twin can be contaminated with points from a neighboring taller tree, and this explains the descending trend of the bias with the increase of the tree height.

The upper canopy (branches and leaves) obscures some of the visible parts of the trunk, resulting in incomplete records. This is a common problem caused by the lack of vertical visibility. In terms of forest management, the upper part of the trunk does not usually have an industrial value, this being too thin in diameter. Still, this limitation remains an issue when managing other forest ecosystem services which rely on finer information. This study demonstrated that the visibility problem can be solved with a mobile laser and leaf-off scanning season, which is effective even in mixed coniferous-deciduous stands (as revealed in other studies as well) [49,53].

The European Union policy framework on forests aims to preserve and restore biodiversity with sustainable forest management. The sustainability principles will cover the entire forest cycle, seeking further knowledge on the optimum integration of all forest services. Digital twinning offers a strong informational background to achieve this principle, extending the knowledge on forests to decision makers and managers, and AI platforms will be the digital backbone for implementing this strategy.

5. Conclusions

The future of digitalization relies on the high capacity and adaptability of mobile scanners to produce complete and accurate pointclouds over large areas of forests and the speed and accuracy of A.I. platforms to translate the raw data into products for decision-makers. The workflow based on this technology is now validated using GeoSLAM scanner and VirtSilv platform to produce results comparable with methodologies which were previously only for research due to their difficulty and high production costs. This new approach brings the forest sector one step closer to the big data needed for climate-smart sustainable forest management.

Funding: This work was supported by a grant of the Romanian Ministry of Education and Research, CNCS–UEFISCDI, project number PN-III-P4-ID-PCE-2020-0401, within PNCDI III.

Acknowledgments: The author would like to thank for access to an educational account in VirtSilv platform and field support in the collection of traditional measurements to Forest Design, supported by Action No.: 2018-1-91, FPCUP project.

Conflicts of Interest: The author declares no conflict of interest.

Appendix A

Table A1. Breusch-Pagan test.

Variable	Breusch-Pagan Test
DBH	Residuals: Min 1Q Median 3Q Max -22.87 -12.82 -10.46 -6.96 1535.81 Coefficients: Estimate Std. Error t value Pr($>$|t|) (Intercept) 2.1133 5.9225 0.357 0.7213 dbh_v 0.2751 0.1302 2.112 0.0348 * Residual standard error: 67.16 on 1337 degrees of freedom Multiple R-squared: 0.003326, Adjusted R-squared: 0.002581 F-statistic: 4.462 on 1 and 1337 DF, *p*-value: 0.03484
Height	Residuals: Min 1Q Median 3Q Max -35.73 -9.78 -3.73 3.11 783.40 Coefficients: Estimate Std. Error t value Pr($>$|t|) (Intercept) 51.7973 5.9211 8.748 $< 2 \times 10^{-16}$ *** h_v -1.3065 0.1728 -7.561 7.41×10^{-14} *** Residual standard error: 40.53 on 1337 degrees of freedom Multiple R-squared: 0.041, Adjusted R-squared: 0.04029 F-statistic: 57.16 on 1 and 1337 DF, *p*-value: 7.412×10^{-14}
Volume	Residuals: Min 1Q Median 3Q Max -1.766 -0.382 -0.219 -0.093 54.220 Coefficients: Estimate Std. Error t value Pr($>$|t|) (Intercept) -0.05168 0.10338 -0.500 0.617 vol_v 0.17775 0.02962 6.002 2.51×10^{-9} *** Residual standard error: 2.315 on 1337 degrees of freedom Multiple R-squared: 0.02624, Adjusted R-squared: 0.02551 F-statistic: 36.02 on 1 and 1337 DF, *p*-value: 2.509×10^{-9}

* $p < 0.1$, *** $p < 0.01$.

References

1. Zou, W.; Jing, W.; Chen, G.; Lu, Y.; Song, H. A Survey of Big Data Analytics for Smart Forestry. *IEEE Access* **2019**, *7*, 46621–46636. [CrossRef]
2. Wagner, N.S.; Son, L.H.; Joo, M. Complex evolutionary artificial intelligence in cognitive digital twinning. *J. Intell. Fuzzy Syst.* **2021**, *40*, 2013–2016. [CrossRef]
3. Dietz, M.; Pernul, G. Digital Twin: Empowering Enterprises Towards a System-of-Systems Approach. *Bus. Inf. Syst. Eng.* **2020**, *62*, 179–184. [CrossRef]
4. Raj, P. Empowering digital twins with blockchain. *Adv. Computers* **2021**, *121*, 267–283. [CrossRef]
5. Pache, R.-G.; Abrudan, I.V.; Niță, M.-D. Economic Valuation of Carbon Storage and Sequestration in Retezat National Park, Romania. *Forests* **2021**, *12*, 43. [CrossRef]
6. Verkerk, P.J.; Costanza, R.; Hetemäki, L.; Kubiszewski, I.; Leskinen, P.; Nabuurs, G.J.; Potočnik, J.; Palahí, M. Climate-Smart Forestry: The missing link. *For. Policy Econ.* **2020**, *115*, 102164. [CrossRef]
7. Nabuurs, G.-J.; Delacote, P.; Ellison, D.; Hanewinkel, M.; Hetemäki, L.; Lindner, M. By 2050 the Mitigation Effects of EU Forests Could Nearly Double through Climate Smart Forestry. *Forests* **2017**, *8*, 484. [CrossRef]
8. Bowditch, E.; Santopuoli, G.; Binder, F.; del Río, M.; La Porta, N.; Kluvankova, T.; Lesinski, J.; Motta, R.; Pach, M.; Panzacchi, P.; et al. What is Climate-Smart Forestry? A definition from a multinational collaborative process focused on mountain regions of Europe. *Ecosyst. Serv.* **2020**, *43*, 101113. [CrossRef]
9. Scholz, R.W.; Bartelsman, E.J.; Diefenbach, S.; Franke, L.; Grunwald, A.; Helbing, D.; Hill, R.; Hilty, L.; Höjer, M.; Klauser, S.; et al. Unintended Side Effects of the Digital Transition: European Scientists' Messages from a Proposition-Based Expert Round Table. *Sustainability* **2018**, *10*, 2001. [CrossRef]
10. Müller, F.; Jaeger, D.; Hanewinkel, M. Digitization in wood supply—A review on how Industry 4.0 will change the forest value chain. *Comput. Electron. Agric.* **2019**, *162*, 206–218. [CrossRef]

11. Manakos, I.; Tomaszewska, M.; Gkinis, I.; Brovkina, O.; Filchev, L.; Genc, L.; Gitas, I.Z.; Halabuk, A.; Inalpulat, M.; Irimescu, A.; et al. Comparison of Global and Continental Land Cover Products for Selected Study Areas in South Central and Eastern European Region. *Remote Sens.* **2018**, *10*, 1967. [CrossRef]

12. Moreno, A.; Neumann, M.; Hasenauer, H. Forest structures across Europe. *Geosci. Data J.* **2017**, *4*, 17–28. [CrossRef]

13. Drake, J.B.; Dubayah, R.O.; Knox, R.G.; Clark, D.B.; Blair, J.B. Sensitivity of large-footprint lidar to canopy structure and biomass in a neotropical rainforest. *Remote Sens. Environ.* **2002**, *81*, 378–392. [CrossRef]

14. Akay, A.E.; Oğuz, H.; Karas, I.R.; Aruga, K. Using LiDAR technology in forestry activities. *Environ. Monit. Assess.* **2009**, *151*, 117–125. [CrossRef] [PubMed]

15. McRoberts, R.E.; Tomppo, E.O. Remote sensing support for national forest inventories. *Remote Sens. Environ.* **2007**, *110*, 412–419. [CrossRef]

16. Tang, J.; Chen, Y.; Kukko, A.; Kaartinen, H.; Jaakkola, A.; Khoramshahi, E.; Hakala, T.; Hyyppä, J.; Holopainen, M.; Hyyppä, H. SLAM-Aided Stem Mapping for Forest Inventory with Small-Footprint Mobile LiDAR. *Forests* **2015**, *6*, 4588–4606. [CrossRef]

17. Paris, C.; Kelbe, D.; van Aardt, J.; Bruzzone, L. A Novel Automatic Method for the Fusion of ALS and TLS LiDAR Data for Robust Assessment of Tree Crown Structure. *IEEE Trans. Geosci. Remote Sens.* **2017**, *55*, 3679–3693. [CrossRef]

18. Lovell, J.L.; Jupp, D.L.B.; Culvenor, D.S.; Coops, N.C. Using airborne and ground-based ranging lidar to measure canopy structure in Australian forests. *Can. J. Remote Sens.* **2003**, *29*, 607–622. [CrossRef]

19. Parker, G.G.; Harding, D.J.; Berger, M.L. A portable LIDAR system for rapid determination of forest canopy structure. *J. Appl. Ecol.* **2004**, *41*, 755–767. [CrossRef]

20. Thies, M.; Pfeifer, N.; Winterhalder, D.; Gorte, B.G.H. Three-dimensional reconstruction of stems for assessment of taper, sweep and lean based on laser scanning of standing trees. *Scand. J. For. Res.* **2004**, *19*, 571–581. [CrossRef]

21. Henning, J.G.; Radtke, P.J. Ground-based laser imaging for assessing three-dimensional forest canopy structure. In Proceedings of the Photogrammetric Engineering and Remote Sensing. *Am. Soc. Photogramm. Remote Sens.* **2006**, *72*, 1349–1358. [CrossRef]

22. Hopkinson, C.; Chasmer, L.; Young-Pow, C.; Treitz, P. Assessing forest metrics with a ground-based scanning lidar. *Can. J. For. Res.* **2004**, *34*, 573–583. [CrossRef]

23. Maas, H.G.; Bienert, A.; Scheller, S.; Keane, E. Automatic forest inventory parameter determination from terrestrial laser scanner data. *Int. J. Remote. Sens.* **2008**, *29*, 1579–1593. [CrossRef]

24. Murphy, G.E.; Acuna, M.A.; Dumbrell, I. Tree value and log product yield determination in radiata pine (pinus radiata) plantations in australia: Comparisons of terrestrial laser scanning with a forest inventory system and manual measurements. *Can. J. For. Res.* **2010**, *40*, 2223–2233. [CrossRef]

25. Islam, M.N.; Kurttila, M.; Mehtätalo, L.; Haara, A. Analyzing the effects of inventory errors on holding-level forest plans: The Case of measurement error in the basal area of the dominated tree species. *Silva Fenn.* **2009**, *43*, 71–85. [CrossRef]

26. Li, R.; Weiskittel, A.R. Comparison of model forms for estimating stem taper and volume in the primary conifer species of the North American Acadian Region. *Ann. For. Sci.* **2010**, *67*, 302. [CrossRef]

27. Berger, A.; Gschwantner, T.; McRoberts, R.E.; Schadauer, K. Effects of measurement errors on individual tree stem volume estimates for the Austrian national forest inventory. *For. Sci.* **2014**, *60*, 14–24. [CrossRef]

28. Yu, X.; Liang, X.; Hyyppä, J.; Kankare, V.; Vastaranta, M.; Holopainen, M. Stem biomass estimation based on stem reconstruction from terrestrial laser scanning point clouds. *Remote Sens. Lett.* **2013**, *4*, 344–353. [CrossRef]

29. Kankare, V.; Liang, X.; Vastaranta, M.; Yu, X.; Holopainen, M.; Hyyppä, J. Diameter distribution estimation with laser scanning based multisource single tree inventory. *ISPRS J. Photogramm. Remote Sens.* **2015**, *108*, 161–171. [CrossRef]

30. Liang, X.; Kankare, V.; Yu, X.; Hyyppä, J.; Holopainen, M. Automated stem curve measurement using terrestrial laser scanning. *IEEE Trans. Geosci. Remote Sens.* **2014**, *52*, 1739–1748. [CrossRef]

31. Astrup, R.; Ducey, M.J.; Granhus, A.; Ritter, T.; von Lüpke, N. Approaches for estimating stand-level volume using terrestrial laser scanning in a single-scan mode. *Can. J. For. Res.* **2014**, *44*, 666–676. [CrossRef]

32. Newnham, G.J.; Armston, J.D.; Calders, K.; Disney, M.I.; Lovell, J.L.; Schaaf, C.B.; Strahler, A.H.; Mark Danson, F. Terrestrial laser scanning for plot-scale forest measurement. *Curr. For. Rep.* **2015**, *1*, 239–251. [CrossRef]

33. Tomșa, V.R.; Curtu, A.L.; Niță, M.D. Tree Shape Variability in a Mixed Oak Forest Using Terrestrial Laser Technology: Implications for Mating System Analysis. *Forests* **2021**, *12*, 253. [CrossRef]

34. Tansey, K.; Selmes, N.; Anstee, A.; Tate, N.J.; Denniss, A. Estimating tree and stand variables in a Corsican Pine woodland from terrestrial laser scanner data. *Int. J. Remote Sens.* **2009**, *30*, 5195–5209. [CrossRef]

35. Li, W.; Guo, Q.; Jakubowski, M.K.; Kelly, M. A new method for segmenting individual trees from the lidar point cloud. *Photogramm. Eng. Remote Sens.* **2012**, *78*, 75–84. [CrossRef]

36. Raumonen, P.; Kaasalainen, M.; Åkerblom, M.; Kaasalainen, S.; Kaartinen, H.; Vastaranta, M.; Holopainen, M.; Disney, M.; Lewis, P. Fast Automatic Precision Tree Models from Terrestrial Laser Scanner Data. *Remote Sens.* **2013**, *5*, 491–520. [CrossRef]

37. Wilkes, P.; Lau, A.; Disney, M.; Calders, K.; Burt, A.; Gonzalez de Tanago, J.; Bartholomeus, H.; Brede, B.; Herold, M. Data acquisition considerations for Terrestrial Laser Scanning of forest plots. *Remote Sens. Environ.* **2017**, *196*, 140–153. [CrossRef]

38. Disney, M.I.; Kalogirou, V.; Lewis, P.; Prieto-Blanco, A.; Hancock, S.; Pfeifer, M. Simulating the impact of discrete-return lidar system and survey characteristics over young conifer and broadleaf forests. *Remote Sens. Environ.* **2010**, *114*, 1546–1560. [CrossRef]

39. Henning, J.G.; Radtke, P.J. Detailed stem measurements of standing trees from ground-based scanning lidar. *For. Sci.* **2006**, *52*, 67–80. [CrossRef]

40. Disney, M.; Burt, A.; Calders, K.; Schaaf, C.; Stovall, A. Innovations in Ground and Airborne Technologies as Reference and for Training and Validation: Terrestrial Laser Scanning (TLS). *Surv. Geophys.* **2019**, *40*, 937–958. [CrossRef]
41. Zhong, L.; Cheng, L.; Xu, H.; Wu, Y.; Chen, Y.; Li, M. Segmentation of Individual Trees from TLS and MLS Data. *IEEE J. Sel. Top. Appl. Earth Obs. Remote Sens.* **2017**, *10*, 774–787. [CrossRef]
42. Forest Design VirtSilv. Available online: https://virtsilv.com/2020 (accessed on 14 November 2020).
43. Fick, S.E.; Hijmans, R.J. WorldClim 2: New 1-km spatial resolution climate surfaces for global land areas. *Int. J. Climatol.* **2017**, *37*, 4302–4315. [CrossRef]
44. Martin Bland, J.; Altman, D. Statistical methods for assessing agreement between two methods of clinical measurement. *Lancet* **1986**, *327*, 307–310. [CrossRef]
45. Carkeet, A. Exact parametric confidence intervals for bland-altman limits of agreement. *Optom. Vis. Sci.* **2015**, *92*, e71–e80. [CrossRef] [PubMed]
46. Taffé, P.; Peng, M.; Stagg, V.; Williamson, T. MethodCompare: An R package to assess bias and precision in method comparison studies. *Stat. Methods Med. Res.* **2019**, *28*, 2557–2565. [CrossRef]
47. European Commission. *The European Green Deal*; COM(2019) 640 final; European Commission: Brussels, Belgium, 2019.
48. European Union States. Declaration on Digital Day 2021. Available online: https://ec.europa.eu/newsroom/dae/items/705312 (accessed on 20 May 2021).
49. Gollob, C.; Ritter, T.; Nothdurft, A. Forest Inventory with Long Range and High-Speed Personal Laser Scanning (PLS) and Simultaneous Localization and Mapping (SLAM) Technology. *Remote Sens.* **2020**, *12*, 1509. [CrossRef]
50. Trochta, J.; Kruček, M.; Vrška, T.; Kraál, K. 3D Forest: An application for descriptions of three-dimensional forest structures using terrestrial LiDAR. *PLoS ONE* **2017**, *12*, e0176871. [CrossRef]
51. Koreň, M.; Hunčaga, M.; Chudá, J.; Mokroš, M.; Surový, P. The Influence of Cross-Section Thickness on Diameter at Breast Height Estimation from Point Cloud. *ISPRS Int. J. Geo-Inf.* **2020**, *9*, 495. [CrossRef]
52. Mokroš, M.; Mikita, T.; Singh, A.; Tomaštík, J.; Chudá, J.; Wężyk, P.; Kuželka, K.; Surový, P.; Klimánek, M.; Zięba-Kulawik, K.; et al. Novel low-cost mobile mapping systems for forest inventories as terrestrial laser scanning alternatives. *Int. J. Appl. Earth Obs. Geoinf.* **2021**, *104*, 102512. [CrossRef]
53. Gollob, C.; Ritter, T.; Nothdurft, A. Comparison of 3D Point Clouds Obtained by Terrestrial Laser Scanning and Personal Laser Scanning on Forest Inventory Sample Plots. *Data* **2020**, *5*, 103. [CrossRef]

Article

Monitoring of Carriageway Cross Section Profiles on Forest Roads: Assessment of an Ultrasound Data Based Road Scanner with TLS Data Reference

Michael Starke [1,*], Anton Kunneke [2] and Martin Ziesak [1]

[1] Division Forest Sciences, School for Agricultural, Forest and Food Sciences HAFL, Bern University of Applied Sciences, Länggasse 85, CH-3052 Zollikofen, Switzerland; martin.ziesak@bfh.ch
[2] Department of Forest and Wood Science, Stellenbosch University, Stellenbosch 7599, South Africa; ak3@sun.ac.za
* Correspondence: Michael.Starke@bfh.ch

Abstract: Forest roads are an important element in forest management as they provide infrastructure for different forest stakeholder groups. Over time, a variety of road assessment concepts for better planning were initiated. The monitoring of the surface cross-section profile of forest roads particularly offers the possibility to take early action in restoring a road segment and avoiding higher future costs. One vehicle-based monitoring system that relies on ultrasound sensors addresses this topic. With advantages in its dirt influence tolerance and high temporal resolution, but shortcomings in horizontal and vertical measuring accuracy, the system was tested against high resolution terrestrial laser scanner (TLS) data to find and assess working scenarios that fit the low- resolution measuring principle. In a related field test, we found low correct road geometry interpretation rates of 54.3% but rising to 91.2% under distinctive geometric properties. The further applied line- and segment-based method used to transform the TLS data to fit the road scanner measuring method allows the transfer of the road scanner evaluation principle to point-cloud or raster data of different origins.

Keywords: ultrasound sensors; road scanner; terrestrial laser scanning; TLS; forest road maintenance; forest road monitoring; crowned road surface

Citation: Starke, M.; Kunneke, A.; Ziesak, M. Monitoring of Carriageway Cross Section Profiles on Forest Roads: Assessment of an Ultrasound Data Based Road Scanner with TLS Data Reference. *Forests* **2021**, *12*, 1191. https://doi.org/10.3390/f12091191

Academic Editors: Robert Keefe, Andrea R. Proto, Mihai Nita and Stelian Alexandru Borz

Received: 23 July 2021
Accepted: 26 August 2021
Published: 2 September 2021

Publisher's Note: MDPI stays neutral with regard to jurisdictional claims in published maps and institutional affiliations.

1. Introduction

Information about forest road condition has become increasingly important. Not only basic road accessibility, but also an intensified use of forest roads by other forest stakeholder groups [1] can thereby influence the need for maintenance intensity and frequency [2]. Questions, from basic usability and stability, up to the assessment of high- quality road construction standards needed for, e.g., recreational aspects [3,4], can therefore be driving factors for collecting additional information about a road condition status to be able to take action within given financial constraints [2].

The therefore selected parameters describing the road condition vary with the quality standards and maintenance concepts. Thus, destruction-free monitoring concepts can include the road surface roughness in its different definitions and recording methods [3,5,6], direct wear expressions of the road surface [7–9], or the road geometry in comparison with a targeted road design. The road geometry, however, is an especially important part of road quality assessment. Its design determines the drainage of the road surface and is crucial for avoiding longitudinal water accumulation which can result in accelerated erosion effects [4,10–12]. Thus, it helps to identify potential construction problems even before severe damage in the form of wear expressions on a forest road surface appears.

One data source that is used to describe forest road geometries is originated from airborne laser scanning. Caused by the given spatial resolution, the area of application is found on larger-scale road geometries [13] or focuses on strong geometrical expressions such as the ditch as drainage system [7]. Higher potential and more possibilities in monitoring the road geometries can be achieved by changing the recording distance, and therefore the resolution of the data [14]. Specially equipped vehicles for road condition monitoring which are mostly based on LiDAR systems already exist, but are mostly designed for sealed road surfaces [15], creating data with low temporal resolution, as a separate, cost-intensive measuring vehicle is required for data collection. To utilise the advantage of the close-range recording with higher temporal resolution, alternative measuring principles have emerged in the forest sector with the aim to close the gap between temporal and spatial resolution to serve alternative monitoring concepts.

In this context, a different vehicle-based and low-cost road scanning concept, which is applied as an ultrasound sensor-based setup, was developed [16,17]. The aim of this system is to be used in the day-to-day business when mounted on the back of a forester's car to collect frequent information about the road condition status by describing the road surface geometry in combination with its surface roughness. In comparison with other near-range sensor setups suitable for forest use, most of which are based on LiDAR or photogrammetric systems [3,9], this system alternatively operates with ultrasound distance sensors to detect the cross-section profile of a forest road's carriageway surface, and so addresses annual maintenance concepts to detect and further restore a functional lateral water drainage of the carriageway [18]. The lens-less ultrasound measurement principle allows the user to continue measurements under muddy or dusty measuring conditions, but again limits the resolution of the measurement. This drawback was already noted in earlier tests conducted under laboratory conditions [19].

The present study focuses on the recording of the carriageway cross-section profiles on forest roads, particularly of a single-laned design. In a field testing, vehicle-based data from the ultrasound sensor setup of the road scanner are compared with high resolution data collected with a terrestrial laser scanning (TLS) system, to find areas of application to substitute high resolution data with low-cost and -quality data of high temporal resolution. For the data comparison, the recording method of the road scanner was adopted to TLS data to describe cross-sections within defined road segments in order to assess lateral water flow over parallel lane sections. The specific objectives of the study were:

(a) to adopt the lane-based measuring principle of the ultrasound sensor-based road scanner to high resolution LiDAR data, to assess the quality of the lower resolution data of the road scanner, and to transfer the lane-based recording method to other data sources; and

(b) to use data filtering to identify data application scenarios for using the road scanner setup for forest road surface monitoring purposes.

2. Materials and Methods

For the study, a gravel road surface was recorded with two measurement principles: the ultrasound-sensor-based, low resolution road scanner, installed at the back of a car and measuring in movement, and the terrestrial laser scanner in a static measuring setup. To accommodate the different data resolutions, the road was split into equally sized segments in longitudinal direction to obtain comparable road segments for further evaluation. The road characteristic per segment was then described through the height differences between sensor position related lanes on the road surface that are used to describe an inclined or crowned road surface profile. Subsequently, the algebraic sign of the lateral road inclination was compared between the measurement principles to describe the direction of potential water flow. The percentage of equal classification of both principles was then evaluated to find a comparable road description setup.

The study was carried out in Switzerland on a gravel road (46°59′27.6″ N 7°27′50.4″ E), separating two agricultural fields in an open area on flat terrain (Δz_{max} = 3.35 m for the whole road), to focus on the surface recording principle and to rule out forest canopy influences. The road had a total length of 440 m and a width of 2.2 m (outer edges of the visible lanes). It was straight, but with one 90-degree-exceeding corner at two thirds of the length (segments no. 29, 30). The cross-section profile did not follow a distinct crowned profile, but was characterized by existing or emerging vegetation in the middle of the road, combined with a beginning of rut expression. In the area of two local vertical drainage installations (segments no. 19, 29), road surface erosion expressions started appearing. There was no subsequent water drainage for lateral nor longitudinal direction.

The road was first measured with a TLS system followed by the low-resolution ultrasound sensors (US). With 5 US (MaxBotics Inc., Fort Mill, Brainerd, MN, USA: MaxSonar MB7040) that are built into the road scanner [19,20], the vertical distance between the scanner bar and the road surface was measured and the cross-section profile of the road recorded. The sensor distance was equally set up with a 0.45 m spacing (Figure 1) and the scanner was mounted in a height of 0.4 m on the car hitch for vertical measuring towards the road surface. Four of the sensors used were the MB7040 XLI2C-MaxSonar-WR type. At position 4, a MB7040 XLI2C-MaxSonar-WRC sensor equipped with a ceramic cone head was mounted and expected to provide a more focused measuring cone of the road surface. The sensors were triggered in a round-robin measurement principle and a 10 Hz trigger frequency of each sensor to minimize reflection interference between the sensors. All sensors provided a resolution of 1 cm in vertical direction [20] where the distance value is calculated by the sensor internally in a pre-processing step. Erratic values above 0.65 m were excluded in advance of the data evaluation step, as these values relate to technical errors and cannot be explained by specific, distance-related situations. With a u-blox NEO-M8N GNSS sensor, the spatial reference was added simultaneously to the measurements in a 1 Hz resolution, with location interpolation for in-between recordings. For higher accuracy, an external active magnetic antenna was used, which was mounted on the car roof. This setup reaches an accuracy of at least 1.5 m in driving direction, which was verified by a shock-inducing control point for the built-in acceleration sensor, that was placed on the test track. The road was then recorded in 14 overall passes (repetitions) at a strived constant driving speed of 20 km h^{-1}. In total, 13 passes contained valid GPS values for the further spatial evaluation.

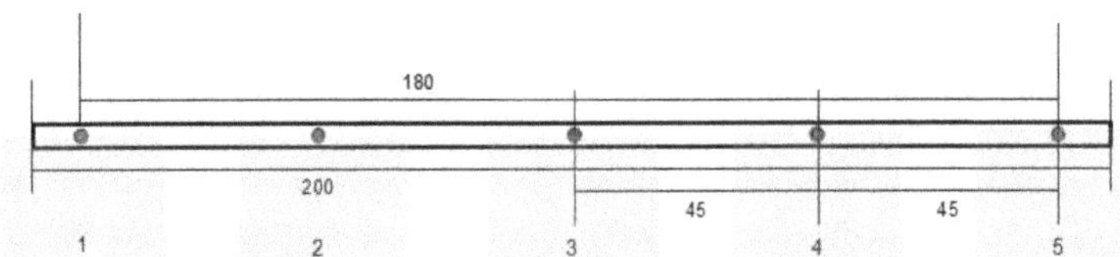

Figure 1. Ultrasonic sensor setup and dimensions of the mounted road scanner bar (cm), including the spacing of the ultrasound sensors (red), with a central mounting point at sensor 3, in combination with their maximal detection beam width (grey).

For reference measurements, the road was scanned with a TLS system FARO 3D X330 in a chained scanning setup of 21 scans, positioned on the carriageway in a scanning height of 1.5 m and a varying scanning distance of around 25 m. For the point-cloud registration and to improve the basic internal GPS referencing, the position of 13 scanning targets were additionally calculated from theodolite measurements, referenced to an official geographic survey point, located near the road entrance.

After the separate scans were combined to a single point cloud, road scanner related sensor lanes were constructed following the known sensor spacings built up after one manually defined sensor lane. After the lanes were located, subsets of the point cloud were extracted as 0.05 m wide strips (Figure 2).

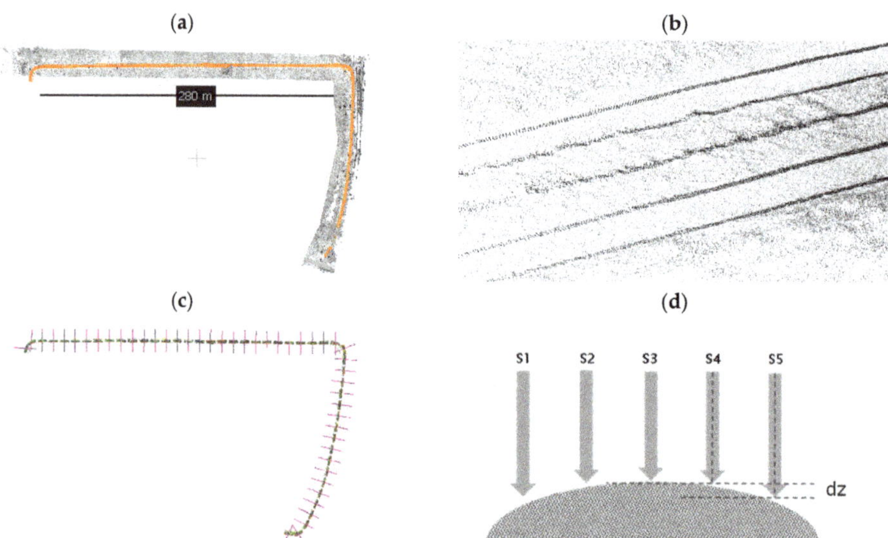

Figure 2. Data processing and evaluation steps: (**a**) combination of terrestrial laser scanning (TLS, grey background area) and road scanner measurements (orange); (**b**) reference sensor lane extraction (black) from the TLS data background (grey); (**c**) longitudinal segmentation of both data sources with 2 m spacing; and (**d**) calculating the mean height difference (dz) to represent the lateral water-flow between two sensor lanes, visualized between sensor 4 (S4) and sensor 5 (S5) of one segment.

Next, the recordings of both measurement types were similarly segmented into 44 sections of 8 m with a spacing of 2 m between the segments. The spacing was used to minimize GPS inaccuracies that may influence adjacent sections and could not be excluded with the open field setup. The minimum segment length of 8 m is limited by the number of sample points that are collected by the road scanner and are expected to be counted within one road segment. Despite the multiple repetition of the measurements of the road scanner, the created input data are differently characterised by number of data points, referenced to all 44 road segments (Table 1).

Table 1. Overview of the collected data, including the number of valid data points per sensor lane within the defined segments.

	TLS Measurements	**Road Scanner Measurements**
Recording date	4 July 2019	5 July 2019
Number of scans	21 single scans	13 repetitions
Number of segments	44	44
Number of data points	769'537 (for extr. sensor lanes)	25'814
Sum of valid data points	Sensor 1 → 174'711	Sensor 1 → 5'473
	Sensor 2 → 145'448	Sensor 2 → 4'996
	Sensor 3 → 137'993	Sensor 3 → 5'039
	Sensor 4 → 140'674	Sensor 4 → 5'324
	Sensor 5 → 170'711	Sensor 5 → 4'982

As the lateral geometrical expression of the road is the focus of the study, the relative height differences in z direction (dz) between the mean sensor values of two sensors each were compared (Figure 2). For each segment separately, the result could then be interpreted as direction for a potential lateral waterflow between the related two sensor lanes with an information gap reflecting uncertainty in between the sensor lanes and the data averaging for the segment length. For the evaluation, the algebraic sign of the dz value from the different data sources was used. This characterises the inclination direction with the minimum required resolution for water flow interpretation. When comparing the different data sources, a "positive match" was noted in case of a match of the sign of the relative heights. This means, for the US sensor data, respectively the TLS sensor lanes n, k = 1,2,3,4,5:

$$(S_{TLS}\,(n) - S_{TLS}\,(k)) \cdot (S_{US}\,(n) - S_{US}\,(k)) > 0 \Rightarrow \text{pos. match (1), with n} < \text{k,} \qquad (1)$$

$$(S_{TLS}\,(n) - S_{TLS}\,(k)) \cdot (S_{US}\,(n) - S_{US}\,(k)) < 0 \Rightarrow \text{no match (0), with n} < \text{k,} \qquad (2)$$

with S(n) as relative mean z value per segment of the road surface of the first, and S(k) the mean elevation of the second sensor (lane) considered.

As a last evaluation step, and to overcome low mean detection percentage, the introduction of an evaluation threshold as data filter was further tested. For this, all sensor combinations and segments were grouped for a combined dataset to separate thresholds that influence the height difference recognition.

All statistical analyses were carried out with the statistic software R (R Development Core Team 2020). To check the repetition accuracy of the recordings, the Dunnett's Test was used for multi-group comparison. Further, the effects of different scenarios that influence the matching rate of the data were tested with the Wilcoxon Rank Sum Test, which suited the testing preconditions.

3. Results

3.1. Data Quality and Mean Detection Rate

On average, 9.75 values (SD = 13.2, min = 1, max = 273, 1.22 points m^{-1}) per segment, repetition, and sensor were recorded with the road scanner. This point density equals 24.4 points m^{-2}, upscaled from the sensor lane area and ignoring spaces between the sensor lanes. For the TLS data, 3497.9 (SD = 3997.6, min = 45, max = 18549) values per segment and sensor lane (8745 points m^{-2}) were taken. The difference between the mean values of the segments were characterized with a SD = 2.17 cm for the road scanner and SD = 3.06 cm for the TLS data.

The repetition accuracy of the road scanner shows constant results. For all sensors, the repetition measurements do not differ significantly ($p < 0.05$) regarding the mean values per segment (Dunnett's Test, with first recording as control data). In a confidence interval with $\alpha = 0.1$, sensor 4 with the ceramic cone shows significant differences in the repetition in two cases. It has the lowest SD = 1.89 cm of vertical values compared to the other sensors (total sensor SD = 2.06 cm).

The average recorded height differences between the sensors and recording types for the entire road are shown in Figure 3 and Table 2. The crown profile (S1_3 and S3_5) is expressed with, on average, 3.54 cm between these sensor lanes, which equals a lateral inclination of 3.93% of the carriageway from the highest point in the middle to the lowest point on the outer lane. The related ultrasound sensors recognized a lateral height difference of 1.23 cm on average, with its highest peak on sensor 2 showing a different picture (Figure 3). The TLS-derived mean profile shows higher inter-quantile ranges for all sensors compared to the US-derived road profile.

For all possible sensor combinations, a matching of the sensor pairs of the different data origins averaged 54.3% (SD = 22.3) for the mean segment values. A possible connection between the detection percentage and the mean dz value is visible for the higher correlated sensors referring to a maximum expected dz value (S1_3, S3_5, Table 2, Figure 4). These sensor combinations reach a detection percentage of 72.1%.

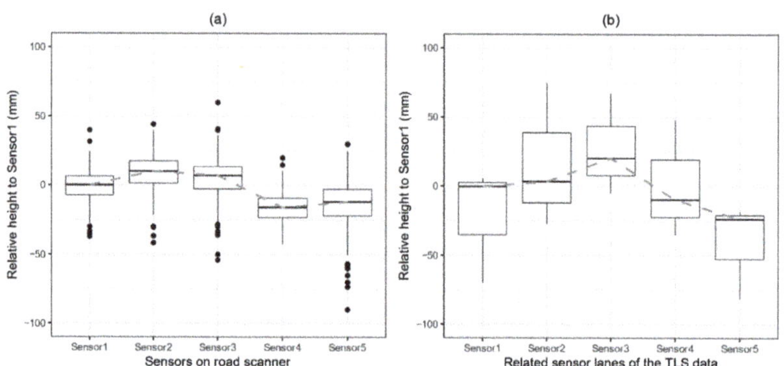

Figure 3. Visualization of the mean road cross-section profiles for the whole road of the different recording types: (**a**) low-resolution, ultrasound measurement and (**b**) TLS measurement.

Table 2. Mean sensor (S) height differences ($Sn\_k = S_n - S_k$) of the recording type road scanner (US), and terrestrial laser scanner (TLS), and the compartment on positive matches of sensor pairs considering a positive or negative height difference.

Sensor Combination	S1_2	S1_3	S1_4	S1_5	S2_3	S2_4	S2_5	S3_4	S3_5	S4_5
US data (cm)	−0.998	−0.637	1.510	1.191	0.361	2.508	2.189	2.146	1.828	−0.319
TLS data (cm)	−1.545	−3.497	−2.003	0.081	−1.952	−0.459	1.626	1.493	3.578	2.084
Correlation between US and TLS datasets	0.24	0.17	0.02	−0.06	0.01	−0.01	0.08	0.21	0.21	0.06
Positive detections (%)	67.3	71.3	31.2	43.6	38.2	44.7	55.1	72.3	73.0	46.3

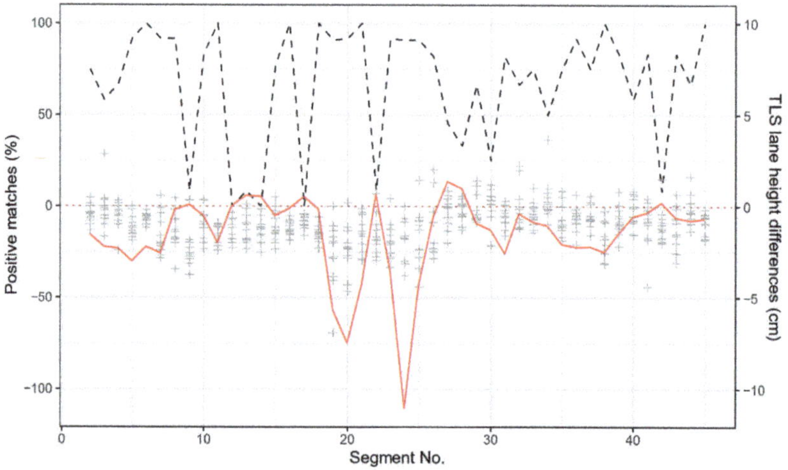

Figure 4. Positive inclination matches of sensor height differences between road scanner and TLS data, per repetition and segment (dashed line), including the absolute height difference (cm) derived from the TLS data (red) and from the road scanner data (grey crosses) for sensor combination S1_2 (mean dz_{TLS} = −1.54 cm, dz_{US} = −1.00 cm with 67.3% detection rate).

When plotting the dz value and the detection percentage of only one sensor pair, the correlation behaviour between the TLS and the US data becomes visible (Figure 4). When a certain dz value is exceeded, the detection rate tends to sharply increase (segments 17–24, Figure 4). A moderate positive correlation between the absolute, mean TLS–dz values per sensor pair and the matching percentage (0.48) supports this connection.

3.2. Threshold Value Filtering Effects

For a minimum evaluated mean difference of +1 cm, the detection rate rises significantly from 54.3% to 78.8% ($p = 0.005$, Wilcox Test) (Figure 5a). When applying an absolute value as a filter to consider negative and positive deviations, or setting the filter on the US data, no further sudden increments of the rate of detection matches are observed. The difference of setting filters on positive or absolute values on the lower resolution road scanner data has a minor effect compared on the TLS data, and is still characterized by a high standard deviation of values.

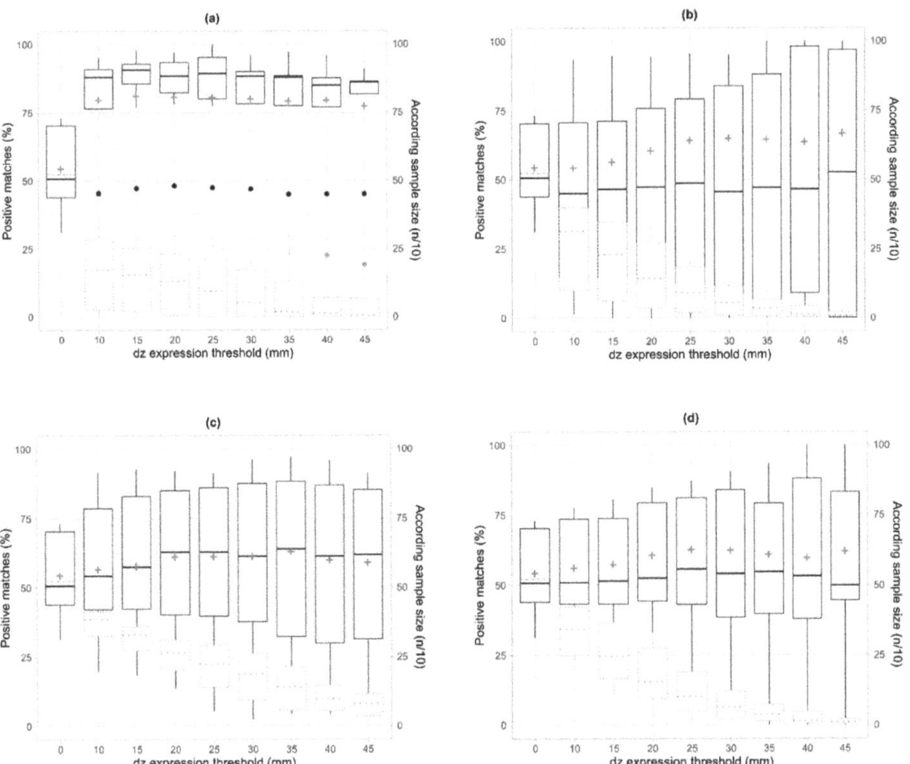

Figure 5. Development of positive matches (solid line boxes, cross marks the mean of all observations) and remaining sample segments (dash-lined boxes) between the ultrasound sensor data and the TLS data, when setting filters on a dz value for (**a**) only positive dz values of the TLS data, (**b**) only positive dz values on the ultrasound data, (**c**) absolute values of the TLS data, and (**d**) absolute values of the ultrasound data.

As for application purposes only internal data filter are usable; a maximum detection percentage with 62.5% can be reached by applying a 2.5 cm threshold for evaluation.

3.3. Sensor Lane Filtering Effect

As some sensor combinations are expected to show no height differences due to the road profile expression, these combinations can be excluded before applying the system when the basic road geometry is known. In a single-laned, crowned road profile, these combinations are the same height levelled sensor-pair 1 and 5 and, respectively, sensor-pair 2 and 4. The best results can thus be reached with 71.6% in combination with a 3 cm height difference filter (Figure 6).

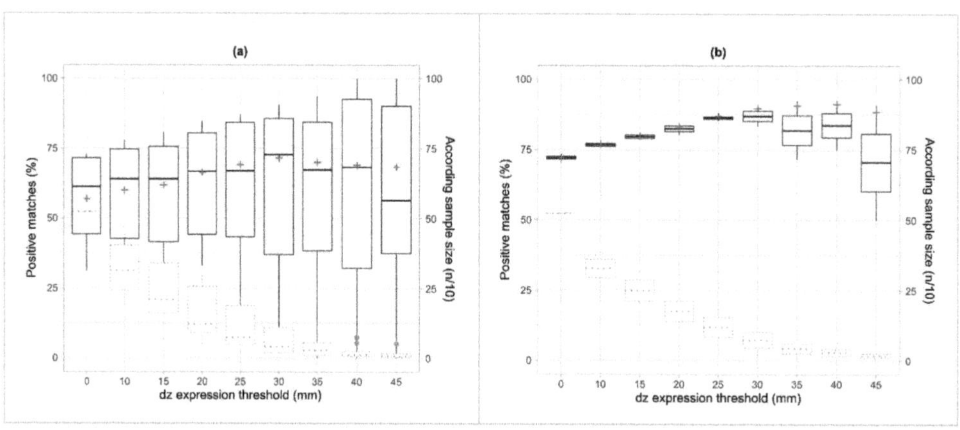

Figure 6. Development of positive matches (solid line box) and remaining sample size between the ultrasound sensor data and the TLS data, when setting filters on a dz value and (**a**) excluding horizontally aligned sensor pairs (S1_5 and S2_4) or (**b**) only considering max dz value sensor pairs (S1_3 and S3_5).

When differently applied, the maximum expected height difference of a crowned road profile is observed with the system over the highest and the lowest sensor point locations (S1_3 and S3_5), the detection rate rises from 54.3% without filter to 72.2%, and reaches a maximum of 91.2% with an additional 4 cm dz filter applied. For higher filter rates, the remaining sample size ($n = 35$) cannot be considered as high enough to keep up the trend. Evaluating a filter up to 4 cm, the according regression model shows a 5.2% higher matching rate per 1 cm dz filtering (adj. R-squared = 0.97, $p < 0.000$).

4. Discussion

With the segment- and lane-based method applied, we presented a way to simplify LiDAR data and make it comparable with different data sources as the road scanner measurements. The concept of directly describing the lateral inclination thereby substitutes the method of data comparison over quality parameters as used in further studies [21] due to the earlier integration of false positive and false negative values considered as incorrect interpretation.

Longitudinal geometry parameters that could override lateral geometry expressions in the evaluation process are minimized in advance, as the road inclination in driving direction is considered equal for all lanes. A balanced distribution of datapoints in longitudinal direction is thereby important for a successful data preparation. With the point density of 24.4 points per square meter, upscaled from an assumed 5 cm sample-stripe that represents the minimal detection width of the sensor, the road scanner data point density was relatively high in comparison with airborne laser scanning (ALS) data with up to 16 points per m^2 [22]. This sample drawing method could therefore also be a possible enhancement of ALS-based evaluation concepts in steeper terrain [8].

Between the different data sources, the mean road profiles showed basic similarities in their geometric expression, but with lower average values in z direction at the middle sensor position of the ultrasound sensor data. As this sensor lane was partly influenced by emerging vegetation, the relation between the ultrasound measuring principle and the vegetation could have caused that effect. This observation is supported by the results of earlier tests under laboratory conditions, where the detection of vegetation with the ultrasound sensors was also not possible [19]. This difference in the measuring principles between ultrasound and LiDAR can lead to an advantage of this sensor selection, as the direct road surface can potentially be separated with the ultrasound measurements.

The limited vertical resolution of the US sensors, caused by the sensor-internal distance calculation, made the comparison of the measurement principles challenging. When a filter from -1 to $+1$ cm as factory resolution of the sensor was set separately on each dataset, only small improvements of the matching data percentage were observed. Only with the filter set exclusively on positive height difference expressions did a significant rise of the matching percentage appear. This can be caused by multiple influence factors, such as a missing homogenous distribution of the sample data within the segments, in combination with a longitudinal road inclination, a broken sensor, or a missing horizontal alignment of the scanner bar. As no one-sided expressed road inclination was noted, and the sensor data showed no inconsistencies, the alignment of the scanner bar seems to cause the trend of the measured data.

To discuss a suitable application of the system with the identified measuring peculiarities, the consideration of typical forest influences and deviating road construction parameters seem necessary. As the study was conducted in open terrain, the aimed advantage of high accuracies achieved within the spatial referencing exceeded accuracies reachable under forest canopy conditions [23]. As the lane-based information of the road scanner is relative information dependent on the sensor spacing, this issue only affects the allocation accuracy in driving direction, or the basic spatial join of the data with the road assessed. Accuracy limitations are, thus, not crucial for an implementation of the system, as segment lengths can be adjusted independently of statistical evaluation intervals. Furthermore, to transfer the road scanner results to forest conditions, common lateral road inclinations of forest roads must also be known, as road construction variants that can influence the earlier noted data quality are common, related to the expressed dz values observed. Former studies mention that, in road construction, a carriageway inclination in lateral direction should be expressed with a slope of 5–8% for crowned road surfaces [24,25]. For the given sensor spacing, this would imply an expected dz value of 2.25 cm for neighboured sensor pairs, or 4.5 cm when only every second sensor is considered. In the present study, the mean dz value of the road was below 1.77 cm. The reference road that was selected for this study can thus be rated as ambitious, regarding its overall profile characteristics and the given measuring behaviour of the system. When applying the system in these situations, higher detection rates related to the raw data can therefore be assumed.

As a further measure to raise data reliability, filters can alternatively be set on dz values for specific sensor combinations to focus on the road's maximum dz expressions. When the road is designed as a single driving lane, and the vehicle used for the measurement can pass the road in the middle of the crowned surface profile, the highest dz values are given between the middle sensor and the outer sensors. Applied in this manner, the previously mentioned skipping of one sensor lane raised the detection percentages of the true geometry up to over 91%, and with this showed good results for application. On the downside, geometric information in between the longer sensor distances is thereby lost, which needs to be considered in the overall monitoring purpose.

A fundamental absence of a crowned profile, however, makes the application of the system challenging. As the detection rates rise with dz filters used, the targeted geometry or road damage of the observation must at least exceed the thresholds of the filters applied. Additionally, the valuably recognized sensor-skipping approach to raise the dz value can no longer be used, which forces the system to be used with US data filtering only.

5. Conclusions

The method of a striped and segment-based analysis for assessing different cross-section monitoring principles was demonstrated to also be possible on a LiDAR-based road recording concept. This method can be especially helpful for comparing measurement principles with one dataset missing exact horizontal spatial reference, or for overcoming longitudinal inclination influences on existing methods.

The road scanner presented itself as diverse working system. With its characteristic to screen vegetation on the road surface, advantages in comparison with the reference TLS measurements occurred that need further attention. Satisfying data quality for application was found for a geometry expression threshold of 3.5 cm. This is in accordance with literature-based suggestions of a lateral road inclination for single laned crowned road profiles in the existing sensor setup, which makes the system best fitted for these situations to record a forest road carriageway geometry with high temporal resolution.

Author Contributions: Conceptualization, M.S., A.K. and M.Z.; methodology, M.S.; data collection, M.S. and A.K.; validation, M.S.; formal analysis, M.S.; resources, M.Z.; data curation, M.S.; writing—original draft preparation, M.S.; writing—review and editing, M.S., A.K. and M.Z.; visualization, M.S.; and project administration, M.Z. All authors have read and agreed to the published version of the manuscript.

Funding: The APC was funded by the Bern University of Applied Sciences Open Science Office.

Data Availability Statement: The data presented in this study are available on request from the corresponding author.

Conflicts of Interest: The authors declare no conflict of interest.

References

1. Toscani, P.; Sekot, W.; Holzleitner, F. Forest Roads from the Perspective of Managerial Accounting—Empirical Evidence from Austria. *Forests* **2020**, *11*, 378. [CrossRef]
2. Dodson, E.M. Challenges in Forest Road Maintenance in North America. *Croat. J. For. Eng.* **2021**, *42*, 107–116. [CrossRef]
3. Marinello, F.; Proto, A.R.; Zimbalatti, G.; Pezzuolo, A.; Cavalli, R.; Grigolato, S. Determination of forest road surface roughness by Kinect depth imaging. *Ann. For. Res.* **2017**, *60*, 217–226. [CrossRef]
4. Dobias, J. Forest road erosion. *J. For. Sci.* **2012**, *51*, 37–46. [CrossRef]
5. Brown, M.; Mercier, S.; Provencher, Y. Road Maintenance with Opti-Grade®: Maintaining Road Networks to Achieve the Best Value. *Transp. Res. Rec. J. Transp. Res. Board* **2003**, *1819*, 282–286. [CrossRef]
6. Svenson, G.A.; Fjeld, D. The impact of road geometry and surface roughness on fuel consumption of logging trucks. *Scand. J. For. Res.* **2015**, *31*, 526–536. [CrossRef]
7. Kiss, K.; Malinen, J.; Tokola, T. Forest road quality control using ALS data. *Can. J. For. Res.* **2015**, *45*, 1636–1642. [CrossRef]
8. Kiss, K.; Malinen, J.; Tokola, T. Comparison of high and low density airborne lidar data for forest road quality assessment. *ISPRS Ann. Photogramm. Remote Sens. Spat. Inf. Sci.* **2016**, *III-8*, 167–172. [CrossRef]
9. Zhang, C. Monitoring the Condition of Unpaved Roads with Remote Sensing and Other Technology: Final Report for US DOT DTPH56-06-BAA-0002, Brookings. 2008. Available online: https://rosap.ntl.bts.gov/view/dot/36464/dot_36464_DS1.pdf (accessed on 22 January 2021).
10. Heinimann, H.R. Aggregate-surfaced Forest Roads—Analysis of Vulnerability Due To Surface Erosion. In Proceedings of the IUFRO/FAO Seminar on Forest Operations in Himalayan Forests with Special Consideration of Ergonomic and Socio-Economic Problems, Thimphu, Bhutan, 20–23 October 1997; pp. 30–37. Available online: http://www.uni-kassel.de/upress/online/frei/978-3-933146-12-0.volltext.frei.pdf#page=39 (accessed on 1 September 2021).
11. Napper, C. Soil and Water Road-Condition Index—Field Guide. 2008. Available online: https://www.fs.fed.us/t-d////pubs/pdf/08771806.pdf (accessed on 21 January 2021).
12. Moll, J.; Copstead, R.; Johansen, D.K. Traveled Way SurfaceShape. 1997. Available online: https://www.fs.fed.us/eng/pubs/html/wr_p/97771808/97771808.htm (accessed on 21 January 2021).
13. White, R.A.; Dietterick, B.C.; Mastin, T.; Strohman, R. Forest Roads Mapped Using LiDAR in Steep Forested Terrain. *Remote Sens.* **2010**, *2*, 1120–1141. [CrossRef]
14. Talbot, B.; Pierzchała, M.; Astrup, R. Applications of Remote and Proximal Sensing for Improved Precision in Forest Operations. *Croat. J. For. Eng.* **2017**, *38*, 327–336. Available online: http://www.crojfe.com/site/assets/files/4090/talbot.pdf (accessed on 21 July 2021).
15. Ahlin, K.; Granlund, J.; Lindstrom, F. Comparing road profiles with vehicle perceived roughness. *Int. J. Veh. Des.* **2004**, *36*, 270–286. [CrossRef]
16. Rommel, D.; Ziesak, M. Automatisierte Zustandserfassung von Güterwegen—Bedienbarkeitsaspekte des Instrumentariums [Automated condition assessment of freight roads—Usability aspects of the instrumentation]. In Proceedings of the Informatik in der Land-, Forst- und Ernährungswirtschaft: Fokus: Komplexität versus Bedienbarkeit, Mensch-Maschine-Schnittstellen; Referate der 35. GIL-Jahrestagung, Geisenheim, Germany, 23–24 February 2014; Ruckelshausen, A., Ed.; Gesellschaft für Informatik: Bonn, Germany, 2015; pp. 153–156, ISBN 978-3-88579-632-9. (In German)

17. Ziesak, M. System zur Ermittlung des Zustandes von Insbesondere Unbefestigten Fahrtrassen, Wie z. B. Forststraßen Oder Güterwegen [System for Determining the Condition of, in Particular, Unpaved Roadways, Such as Forest Roads or Freight Roads], 10 July 2014. Available online: https://register.dpma.de/DPMAregister/pat/register?AKZ=1020142134242 (accessed on 21 July 2021). (In German)

18. Begus, J.; Pertlik, E. Guide for Planning, Construction and Maintenance of Forest Roads. 2017. Available online: http://www.fao.org/3/i7051e/i7051e.pdf (accessed on 21 July 2021).

19. Starke, M.; Ziesak, M.; Rommel, D.; Hug, P. An automated detection system for (forest) road conditions. In Proceedings of the FORMEC 2016—From Theory to Practice: Challenges for Forest Engineering, Warsaw, Poland, 4–7 September 2016; pp. 261–266. Available online: https://www.formec.org/images/formec2016/proceedings-formec-poland-2016.pdf (accessed on 21 July 2021).

20. MaxBotix. Datasheet for the I2CXL-MaxSonar-WR Sensor Line. Available online: https://www.maxbotix.com/documents/I2CXL-MaxSonar-WR_Datasheet.pdf (accessed on 10 February 2015).

21. Azizi, Z.; Najafi, A.; Sadeghian, S. Forest Road Detection Using LiDAR Data. *J. For. Res.* **2014**, *25*, 975–980. [CrossRef]

22. Hrůza, P.; Mikita, T.; Janata, P. Monitoring of Forest Hauling Roads Wearing Course Damage Using Unmanned Aerial Systems. *Acta Univ. Agric. Silvic. Mendel. Brun.* **2016**, *64*, 1537–1546. [CrossRef]

23. Naesset, E.; Jonmeister, T. Assessing Point Accuracy of DGPS Under Forest Canopy Before Data Acquisition, in the Field and After Postprocessing. *Scand. J. For. Res.* **2002**, *17*, 351–358. [CrossRef]

24. Kuonen, V. *Wald- und Güterstrassen: Planung—Projektierung—Bau: Planung—Projektierung—Bau. [Forest and Freight Roads: Planning—Projecting—Construction]*; ETH Zurich: Pfaffhausen, Switzerland, 1983. [CrossRef]

25. Hölldorfer, B. Einfach, aber wirkungsvoll-das R2005-Gerät [Simple but effective-the R2005 device]. *LWF Aktuell* **2007**, *59*, 30–31. Available online: https://www.lwf.bayern.de/mam/cms04/service/dateien/a59_einfach_aber_wirkungsvoll_das_r2_geraet.pdf (accessed on 21 July 2021). (In German)

forests

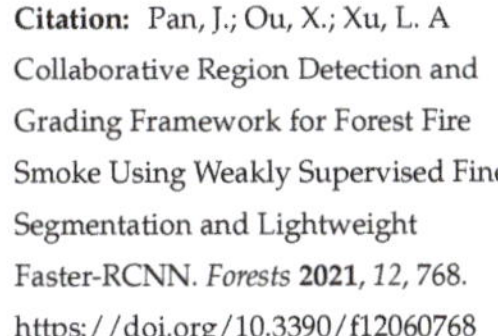

Article

A Collaborative Region Detection and Grading Framework for Forest Fire Smoke Using Weakly Supervised Fine Segmentation and Lightweight Faster-RCNN

Jin Pan [1,2], Xiaoming Ou [1] and Liang Xu [3,*]

1   College of Economics and Management, South China Agricultural University, Guangzhou 510642, China; panj@gzpyp.edu.cn (J.P.); ouxm0758@126.com (X.O.)
2   College of Finance and Economics, Guangzhou Panyu Polytechnic, Guangzhou 511483, China
3   School of Automation, Guangdong University of Technology, Guangzhou 510006, China
*   Correspondence: celiangxu@gdut.edu.cn

Abstract: Forest fires are serious disasters that affect countries all over the world. With the progress of image processing, numerous image-based surveillance systems for fires have been installed in forests. The rapid and accurate detection and grading of fire smoke can provide useful information, which helps humans to quickly control and reduce forest losses. Currently, convolutional neural networks (CNN) have yielded excellent performance in image recognition. Previous studies mostly paid attention to CNN-based image classification for fire detection. However, the research of CNN-based region detection and grading of fire is extremely scarce due to a challenging task which locates and segments fire regions using image-level annotations instead of inaccessible pixel-level labels. This paper presents a novel collaborative region detection and grading framework for fire smoke using a weakly supervised fine segmentation and a lightweight Faster R-CNN. The multi-task framework can simultaneously implement the early-stage alarm, region detection, classification, and grading of fire smoke. To provide an accurate segmentation on image-level, we propose the weakly supervised fine segmentation method, which consists of a segmentation network and a decision network. We aggregate image-level information, instead of expensive pixel-level labels, from all training images into the segmentation network, which simultaneously locates and segments fire smoke regions. To train the segmentation network using only image-level annotations, we propose a two-stage weakly supervised learning strategy, in which a novel weakly supervised loss is proposed to roughly detect the region of fire smoke, and a new region-refining segmentation algorithm is further used to accurately identify this region. The decision network incorporating a residual spatial attention module is utilized to predict the category of forest fire smoke. To reduce the complexity of the Faster R-CNN, we first introduced a knowledge distillation technique to compress the structure of this model. To grade forest fire smoke, we used a 3-input/1-output fuzzy system to evaluate the severity level. We evaluated the proposed approach using a developed fire smoke dataset, which included five different scenes varying by the fire smoke level. The proposed method exhibited competitive performance compared to state-of-the-art methods.

Keywords: region detection of forest fire; grading of forest fire; weakly supervised loss; fine segmentation; region-refining segmentation; lightweight Faster R-CNN

Citation: Pan, J.; Ou, X.; Xu, L. A Collaborative Region Detection and Grading Framework for Forest Fire Smoke Using Weakly Supervised Fine Segmentation and Lightweight Faster-RCNN. *Forests* **2021**, *12*, 768. https://doi.org/10.3390/f12060768

Academic Editors: Olga Viedma, Stelian Alexandru Borz, Andrea R. Proto, Robert Keefe and Mihai Nita

Received: 8 April 2021
Accepted: 9 June 2021
Published: 10 June 2021

Publisher's Note: MDPI stays neutral with regard to jurisdictional claims in published maps and institutional affiliations.

1. Introduction

Forest fires have become one of the major disasters causing serious ecological, social, and economic damage, as well as personal casualty loss [1–3]. In 2013, a forest fire burned a land area of approximately 1042 km^2 in California, causing USD 127.35 million of damage. In China, 214 forest fire events occurred alone in the Huichang County of the JiangXi province from 1986 to 2009, with an area of more than 460 km^2 being affected [4]. The statistical data provided in [5] show that fire disasters alone caused an overall damage of

USD 3.1 billion in 2015. To monitor the fire smoke, numerous image-based surveillance systems have been installed in forests. Therefore, rapid and accurate detection and grading of fire smoke is crucial and helpful for preventing and reducing the forest losses.

Forest fires commonly spread quickly and are difficult to rapidly control. Accurately identifying the fire region and evaluating the fire smoke severity, which helps firefighters to take proper measures and quickly control a fire's spreading, is a very challenging task. Moreover, most firefighters will need to decide how many resources to allocate to a particular forest fire according to useful extinguishing information, which contains the region, the location, and the severity (i.e., the grading of risk) of fire smoke. Therefore, a technique for a fire surveillance system which can give an early region detection of fire or smoke and evaluate the severity of fire or smoke is indispensable.

The traditional technologies for detecting fire and smoke use various sensors. A point sensor [6–8] works best in indoor spaces, as it only covers a small area, and is insensitive to fire in the outdoors and over a large range. An Unmanned Aerial Vehicle (UAV) technique can also be used to monitor forest fires [9]. Additionally, such techniques cannot provide important vision information to help firefighters quickly evaluate the severity of the fire and make appropriate decisions. Satellite sensors [10] only detect a large fire in a wide range, and they are not useful for the early detection of fire and smoke. Currently, with a large amount of image surveillance systems installed in forests, there is an appropriate alternative to the traditional techniques, and vison-based inspecting technologies for fires and smoke have been widely adopted due to their easy deployment and lower cost, insusceptibility to the weather, and long and short availability.

Vision-based technologies make full use of color and motion features for fire detection. Due to the conspicuous color of fire, Chen et al. [11] proposed RGB- and HIS-based color models are used to examine the dynamic behavior of fires, which can be applied to detect the irregular properties of fire. Additionally, the YUV color [12], RGB [13], and YCbCr color space [14], have been explored to classify the pixels in fire and non-fire regions. However, such methods have many limitations in various situations, e.g., due to the complexity of wild scenes, the diversity of fire and smoke in forests, irregular lighting, and low-contrast flame and smoke.

In recent years, convolutional neural networks (CNNs) have attracted attention due to their outstanding performance in image recognition. Some scholars have introduced CNN models—for example, AlexNet [15], GoogleNet [16], ZF-Net [17], VGG [18], and ResNet [19], into the field of the vision detection of fires. Regarding the use of these models, some scholars have also proposed improved CNN-based methods for fire or smoke detection, such as, smoke detection in a video based on a deep belief network using energy and intensity features [20], a video-based detection system using an object segmentation and efficient symmetrical features [21], a two-stream CNN model with the adaptive adjustment of the receptive field [22], and an object detection model incorporating environmental information [23]. Additionally, Liu et al. [24] also proposed a forest fire detection system based on ensemble learning to reduce false alarms. Nevertheless, the aforementioned methods are only applicable for the recognition of whether fire or smoke exists in an image, but such methods cannot provide more detailed information about fires, such as their location, shape, size, etc., which can be used to grade the level of fires or smoke. Sometimes, we need to focus on the fire/smoke spreading or the emerging regions; thus, the region detection of fire or smoke is a better solution to this issue.

Recently, rapid progress in this technology has been made using powerful models, such as DeepLab [25], U-Net [26], and fully convolutional networks (FCNs) [27]. However, the performance of these deep models heavily depends on a large amount of training data with expensive pixel labels. Due to the uncertain, complex, and changeable shape of fires and smoke, annotating such training data has become a bottleneck in applying these models to forest fire detection, as it is a time-consuming and arduous task to label each pixel on a large amount of fire images. One objective of our work was to loosen the supervision, i.e., by performing weakly supervised segmentation for forest fire detection using only image-level supervision.

It is well known that multi-model cooperation can improve the performance of any machine-learning algorithm. Therefore, model selection and collaborative strategies need to be considered carefully. To grade forest fire smoke, we need to collect and consider various factors, such as the fire region size, fire shape, location, etc., which may influence fire smoke and cause fires to change and spread.

Our work mainly concentrated on a region detection of forest fires or smoke at an early-stage and evaluation of the fires' smoke severity, thereby proposing a collaborative region detection and grading framework for forest fire smoke using a weakly supervised fine segmentation and a lightweight Faster R-CNN.

Our other contributions can also be summarized as follows.

(1) We proposed the weakly supervised fine-segmentation method, which consists of a segmentation network (LS-Net), used to simultaneously and accurately detect the region of forest fire smoke, and a novel two-stage weakly supervised leaning strategy, which includes a weakly supervised loss (WSL), a region-refining segmentation (RRS) algorithm, and an attention-based decision network (AD-Net) for fire smoke classification.

(2) To reduce the complexity of the Faster R-CNN, we introduce a knowledge distillation process to compress this model into a simple model. We also proposed a distillation strategy for this model. Moreover, we used a 3-input/1-output fuzzy system based on fuzzy logic to grade forest fire smoke.

(3) We developed a forest fire dataset from public resources to evaluate our method, which is composed of various situations, including large fires, small fires, dense smoke, light smoke, and other scenes.

The remainder of this paper is organized as follows. Section 2 introduces related work. Section 3 describes the proposed framework, including weakly supervised fine segmentation, lightweight Faster R-CNN, a collaborative learning strategy, and the grading method. Section 4 presents the experimental results and a discussion of them. Section 5 concludes the paper.

2. Related Work

The accurate and timely recognition of the region of forest fire smoke is an important task in preventing forest disasters and protecting the environment. To address this issue, many researchers have developed various techniques, such as wireless network sensors and satellite systems [8,10,28], robotic systems [29], intelligent techniques [30], and image processing techniques [11–14]. Due to the factors of deployment, utilization convenience of use, and a high detection rate, the image techniques have been used widely and have attracted the attention of many researchers, as they are more suitable for forest fire detection. Nevertheless, there are some limitations of the traditional image technologies that are used in real-world applications [11–14]. Recently, DL-based methods have become a mainstream technology for intelligent fire detection based on vision [23,31–35].

To improve the performance of fire detection, a deep normalization and convolutional neural network was proposed for automatic feature extraction and classification, avoiding hard-crafted features [31]. Different classic CNN models, Namozov et al. [32], proposed the use of an adaptive piecewise linear unit, instead of using traditional rectified linear units in the hidden layers of the network. In previous work [33], the authors tackled the overfitting and accuracy of CNN-based fire detection using a limited dataset and a deep convolutional generative adversarial network, which achieved a high accuracy in visual fire detection. To improve its implementation in a real-world surveillance network, Baik et al. [34] presented an energy-friendly and computationally efficient CNN architecture for the detection, localization, and scene understanding of fires. This model reduces the computational requirements to a minimum and obtains a better accuracy due to its increased depth. Furthermore, the authors proposed that an efficient CNN model via edge intelligence could be used to detect fires in uncertain surveillance scenarios [35]. This model utilized a lightweight deep network, instead of dense fully connected layers, which require expensive computation. Nevertheless, the CNN models have only been applied to classification tasks to predict whether fires exist in an image or not. Zhang et al. [23] proposed a forest fire smoke recognition method based on an anchor box adaptive generation.

The most important objective in predicting forest fire risk is to obtain more information regarding a fire in an image, such as its shape, size and location, and the CNN-based image segmentation techniques are a better solution for addressing this issue. Recently, several powerful baseline systems, such as the Fast/Faster/Mask R-CNN [36–38], have been proposed to drive the rapid progress in instance segmentation. Semantic segmentation methods highly related to instance segmentation, such as DeepLab [25] and U-Net [26], only recognize the category of each pixel, without distinguishing different detection object instances. While fully supervised methods can achieve an outstanding performance, they require a large amount of training data with expensive annotations, which causes inconvenience in practical applications. Thus, weakly supervised image segmentation using relaxed supervision—i.e., image-level annotations—is a better solution to the issues.

Weak supervision is an inexact pattern [39]. Every pixel in an image should ideally be annotated in image segmentation. However, we usually have only coarse-grained labels, instead of pixel-wise labels. Weakly supervised image segmentation refers to the training of models with coarse-grained labels to obtain pixel-wise segmentation results. This has been explored in image segmentation primarily to reduce the effort of establishing training data. Many multiple instance learning (MIL) techniques have been investigated for weakly supervised image segmentation. Pathak et al. [40] proposed an MIL formulation of multi-class semantic segmentation learning using a fully convolutional network. Pinheiro et al. [41] investigated a CNN-based model with MIL, which was constrained during training to put more weight on the pixels that were important for classifying an image.

To further accurately segment objects, graphical energy minimization techniques have been extensively used to regularize image segmentation due to their inherent optimality guarantees. In previous studies, many researchers have proposed diverse solutions for image segmentation using this technique. Boykov et al. [42] used graph cuts to find the globally optimal segmentation position. A GrubCut method was proposed to segment images with bounding box annotations by iteratively updating the parameters of a Gaussian mixture model (GMM) [43]. Fully connected CRF models were implemented by an efficient inference algorithm, defining pairwise edge potentials by a linear combination of Gaussian kernels [44]. To overcome the poor localization property of deep networks, Chen et al. [45] proposed a new final layer to be combined with DenseCRF.

3. Methods

In this paper, we focus on the region detection of forest fire smoke and the grading of fire smoke severity in a surveillance image. Considering a better solution which integrates several simple models into a framework to obtain a better performance, we propose a novel collaborative region detection and grading framework for forest fire smoke using a weakly supervised fine segmentation and a lightweight Faster R-CNN (called FireDGWF), as shown in Figure 1. This framework consists of a detection process and a grading process.

In the detection process, we use 3 different models— classification (for example, ShuffleNet [46]), region detection (weakly supervised fine segmentation, also known as WSFS), and region-proposal (lightweight Faster R-CNN) methods, which can locate, segment, and predict the location, region, and category of forest fire or smoke in an input image.

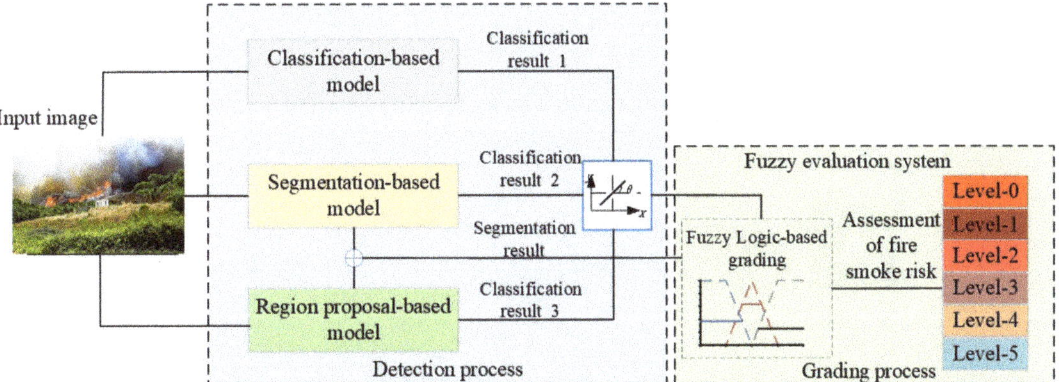

Figure 1. Our proposed framework.

Moreover, compared with the approach that some scholars proposed of using a bounding box to achieve relaxed supervision [47,48], we propose the weakly supervised fine segmentation to obtain a greater segmentation accuracy, which is achieved by applying "pixel-level labels" in our method.

The Faster R-CNN has better detection accuracy than does the one-stage object detection, such as YOLOv3 [49], SSD [50], DSSD [51], and RefineNet [52]. However, the greater number of parameters and time-consuming training has become a bottleneck in applying this method to fire smoke detection in forests. In this paper, we first introduce a knowledge distillation approach, which was proposed by Hinton [53], to reduce the complexity of this model, i.e., the lightweight Faster R-CNN.

For the grading process, we utilize a fuzzy evaluation system, which can synthetically evaluate the fire smoke level based on 3 inputs, including the classification prediction, the region detection (segmentation result), and the location.

In the following section, we separately introduce the region detection using the WSFS, lightweight Faster R-CNN, collaborative learning strategy, and grading method with fuzzy logic.

3.1. Region Detection Using Weakly Supervised Fine Segmentation

The time-consuming effort involved in pixel-wise annotations makes the identification of forest fire regions an inconvenient and challenging task in real-world applications. To address this issue, we propose the novel weakly supervised fine segmentation approach for the detection of the region of forest fire smoke, including the segmentation network (LS-Net) and the decision network (AD-Net), as shown in Figure 2. LS-Net obtains pixel-level segmentation results for fire or smoke regions that are difficult to identify in complex forest scenarios and provides good semantic features for AD-Net. AD-Net uses the deepest feature and the segmentation results of LS-Net and predicts the probability of fire smoke existing in an image. To improve the classification performance, AD-Net focuses on the fire or smoke pixels, the weights of which are determined by the segmentation results provided by LS-Net.

To address the problem that pixel-wise annotations are expensive, we introduce the two-stage weakly supervised learning strategy, including the weakly supervised loss (WSL) and the region-refining segmentation (RRS). LS-Net can only roughly locate fire or smoke regions using the weakly supervised loss only with image-level labels. However, using weak annotation solely at the image level is insufficient for training a high-quality segmentation model. After training the model with the weakly supervised loss, we propose the RRS to fine-tune LS-Net for accurate region segmentation.

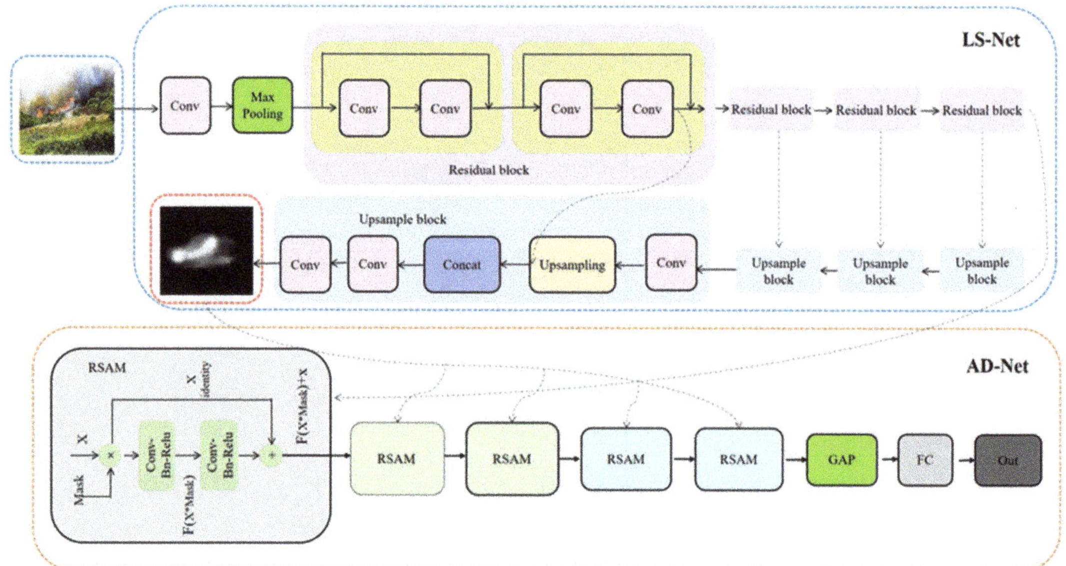

Figure 2. Proposed framework for weakly supervised fine segmentation.

3.1.1. Structure of LS-Net

Table 1 lists the structural parameters of LS-Net. The segmentation network (LS-Net), based on the U-Net structure, is composed of an encoder and decoder, and the encoder is improved using a modified ResNet18 [54]. The encoder uses a hierarchical structure and residual connections to extract multi-level features from an input image, and these have different spatial sizes and semantic information. The first convolutional layer in the original ResNet18 used a 7 × 7 kernel, with a step size of 2. We use 3 3 × 3 convolutional layers, with a step size of 1, to reduce computational costs and maintain the spatial size. The other layers are the same as in the original ResNet18. In the forward process, features of different sizes are outputted from the encoder. The decoder makes full use of the multi-level features of the encoder and outputs a single-channel image to segment fires at the pixel level. The decoder has 4 upsampling layers and multiple convolutional layers. Upsampling is implemented with bilinear interpolation. To capture small fires, each convolutional layer uses a 3 × 3 kernel, with a step size of 1. Batch normalization and nonlinear ReLU layers are included after each convolutional layer.

Table 1. LS-Net architecture.

Encoder			Decoder		
Name	**Layers**	**Output Size**	**Name**	**Layers**	**Output Size**
Layer1	Conv3_3 × 3	[h, w]	Layer1	Upsampling, Conv3_3 × 2	[h//8, w//8]
Layer2	Residual block × 2	[h//2, w//2]	Layer2	Upsampling, Conv3_3 × 2	[h//4, w//4]
Layer3	Residual block × 2	[h//4, w//4]	Layer3	Upsampling, Conv3_3 × 2	[h//2, w//2]
Layer4	Residual block × 2	[h//8, w//8]	Layer4	Upsampling, Conv3_3 × 2	[h, w]
Layer5	Residual block × 2	[h//16, w//16]	Out-layer	Conv1_1, Sigmoid	[h, w]

3.1.2. Structure of AD-Net

Table 2 shows the structure of AD-Net. In this study, an attention mask is introduced into the residual branch of a residual spatial attention model (RSAM) to focus more attention on heavily weighted regions, as shown in Figure 2. The RSAM modules utilize the results of LS-Net as an attention mask to cause AD-Net to focus more on small object regions from the segmentation results. The number of RSAM modules is increased from 1 to 5 to enhance the contrast between fire and non-fire regions. In Figure 2, AD-Net includes 5 RSAM modules, 1 global average pooling layer (RRS) and 1 fully connected layer. As in LS-Net, batch normalization and ReLU are used. This pooling layer squeezes the spatial dimensions and extracts abstract semantic information. Because the output features only depend on the channel of the input feature for this pooling layer, a fixed input image size is unnecessary. The fully connected layer with a sigmoid activation function takes the output of this pooling layer as the input and predicts the probability that an object exists in an image.

Table 2. AD-Net architecture.

Name	Layers	Output Size
Layer1	RSAM	$[h//16, h//16]$
Layer2	RSAM	$[h//32, h//32]$
Layer3	RSAM	$[h//32, h//32]$
Layer4	RSAM	$[h//64, h//64]$
Layer5	RSAM	$[h//64, h//64]$
Layer6	GAP	-
Layer7	Denser Layer	-

3.1.3. Weakly Supervised Loss

In this section, we introduce the novel weakly supervised loss. To simplify the description, Table 3 firstly presents the variables and descriptions used in this loss.

Table 3. Variables and descriptions.

Variable	Description
X	Input image
$Y(X)$	Image label for X
Φ	Model parameters
$p(Y|(X, \Phi))$	Predicted model for X
w, h	Height and width of X
χ	Pixel set for X
x_i	A pixel in χ
y_i	Pixel label for x_i
$q(y_i|(x_i, \Phi))$	Predicted model for x_i

This weakly supervised loss is used to train LS-Net from scratch with only image-level annotations to simultaneously locate and segment the fire or smoke regions. This loss includes a positive loss, negative loss, image-level loss, and pixel-level loss. The positive and negative losses are used for positive (fire or smoke) and negative (non-fire or non-smoke) samples, respectively, during training. The positive and negative loss guides LS-Net to recognize object (fire or smoke) and non-object pixels. The image-level loss allows for an easier and faster convergence [55,56], and the pixel-level loss forces LS-Net to provide a clear prediction for each pixel. The weakly supervised loss can be written as:

$$Loss_{WSL} = Loss_{negative} * \beta + Loss_{positive} + Loss_{image} + Loss_{pixel} \tag{1}$$

where $Loss_{negative}$ is the target loss for negative samples. Since there are only non-object pixels in negative samples, $Loss_{negative}$ is proposed to guide LS-Net to classify the pixels in negative samples as non-objects. The negative samples satisfy $Y(X) = 0, y_i = Y(X) = 0, \forall(x_i, y_i) \in (\mathcal{X}, \mathcal{Y})$. The cross-entropy loss is used to optimize the negative samples, and the negative loss can be rewritten as:

$$Loss_{negative} = (1 - Y(X)) * E_{(x_i, y_i) \in (\mathcal{X}, \mathcal{Y})} - \log(1 - p(y_i|x_i, \Phi)) \tag{2}$$

where $Loss_{positive}$ is the target loss for positive samples. We know that both object and non-object pixels exist in the positive sample image. We sort the pixels in positive samples by their prediction values from LS-Net and assume that the largest α proportion pixels are fire or smoke. α is defined as the ratio of the area of fire or smoke pixels to the total area. We choose the largest double α proportion as the region of interest R. Let R denote the set of pixels in $\mathcal{X}_R$. $\mathcal{Y}_R$ is the set of labels corresponding to $\mathcal{X}_R$. When all the predictions in the positive sample are correct, the expectation entropy corresponding to $\mathcal{X}_R$ reaches its maximum. This can be expressed as:

$$Loss_{positive} = \left[\begin{array}{c} P(\mathcal{Y}_R|\mathcal{X}_R,\Phi)\log(P(\mathcal{Y}_R|\mathcal{X}_R,\Phi)) \\ +(1 - P(\mathcal{Y}_R|\mathcal{X}_R,\Phi))(\log(P(\mathcal{Y}_R|\mathcal{X}_R,\Phi))) \end{array} \right] * Y(X), \tag{3}$$

where the prediction expectation corresponding to $\mathcal{X}_R$ is:

$$P(\mathcal{Y}_R|\mathcal{X}_R,\Phi) = E_{(x_i,y_i)\in(\mathcal{X}_R,\mathcal{Y}_R)}p(y_i|x,\Phi). \tag{4}$$

where $Loss_{image}$ is the image-level loss. For positive samples, at least one pixel should be labeled as fire or smoke. No pixels in negative samples should be labeled as objects. Based on this idea, we minimize the KL divergence of image-level labels and the maximum of the image's segmentation result by cross-entropy loss. This can be expressed as:

$$Loss_{image} = Y * \log(p_M) + (1 - Y) * \log(1 - p_M), \tag{5}$$

where $p_M = \text{Max}_{(x_i,y_i)\in(\mathcal{X},\,y)}(p(y_i|x_i,\Phi))$ is the maximum of an image's pixel predictions.

Ideally, each pixel in an image is an object (fire or smoke) or non-object; i.e., each value segmentation result should be close to 1 or 0. To ensure that LS-Net provides a clear prediction for each pixel, we minimize the entropy values of pixel predictions during training. This can be expressed as:

$$Loss_{pixel} = E_{(x_i,y_i)\in(\mathcal{X},\mathcal{Y})} H[\,p(y_i|x_i,\Phi)\,] = \frac{1}{h * w} \sum_{i=1}^{w*h} H[\,p(y_i|x_i,\Phi)\,], \tag{6}$$

where the entropy of pixel (x_i, y_i) is:

$$\begin{aligned} H[\,p(y_i|x_i,\Phi)\,] = &-p(y_i|x_i,\Phi)\log(p(y_i|x_i,\Phi)) - \\ &(1 - p(y_i|x_i,\Phi))(\log(1 - p(y_i|x_i,\Phi))). \end{aligned} \tag{7}$$

3.1.4. Region-Refining Segmentation

The region detection (segmentation result) using LS-Net only identifies the rough regions of fire and smoke. Furthermore, the fine segmentation results are implemented on the RRS algorithm, which is proposed to fine-tune LS-Net. This can be regarded as an iterative energy minimization method like GrubCut [43]. We build a DenseCRF on LS-Net, and based on these methods, we formulate an energy minimization problem to estimate the latent pixel labels, which are called pseudo-pixel labels. We consider these as the supervision to fine-tune LS-Net with a small learning rate. RRS alternates between estimating pseudo-pixel labels and using them to optimize LS-Net. In other words, DenseCRF improves the segmentation results, and RRS training can be seen as a self-evolution process of LS-Net, as shown in Figure 3.

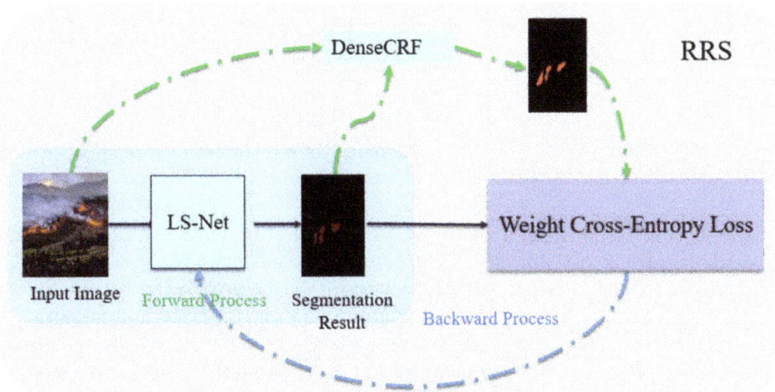

Figure 3. Proposed region-refining segmentation (RRS) algorithm.

The training stage has two key phases: label estimation and model updating.

In the label estimation stage, DenseCRF is built, and pseudo-pixel labels can be acquired by minimizing the energy function [44].

$$E(x) \; = \; \sum_i \psi_u(f_i) + \sum_{i<j} \psi_p(f_i, f_j) \tag{8}$$

The pairwise regularization term $\psi_p(f_i, f_j)$ penalizes label differences for pixels i and j, and typically has the form:

$$\psi_p(f_i, f_j) \; = \; \mu(x_i, x_j) \; * \; k(v_i, v_j) \tag{9}$$

where μ is a label compatibility function given by the Potts model, $\mu(x_i, x_j) \; = \; [x_i \neq x_j]$, and k represents the linear combinations of Gaussian kernels [44].

In the model updating phase, LS-Net is fine-tuned using the pseudo-pixel labels as supervision. We know that all pixels in negative samples are non-fire or non-smoke, so those pixel labels are set to zero, and pseudo-pixel labels are not used. The detailed process of this algorithm is described in Algorithm 1.

Algorithm 1. Region-Refining Segmentation algorithm (RRS)

Input: dataset $D = \{x_i\}_{i=1}^{N}$, $y(x_i) \in [0, 1]$
Initialization: model Φ trained by the weakly supervised loss; parameters of RRS ω_a, ω_s, θ_α, θ_β, and θ_γ; number of steps N; learning rate lr; batch size b;
For step = 1, 2, … , N do:
   $\leftarrow$ sample b images x from D
Predict labels $y(x)$ of samples with Φ
Get pseudo-pixel labels $\acute{y}(x)$ with DenseCRF
update Φ by minimizing loss of $y(x)$ and $\acute{y}(x)$
End for

3.1.5. Training Strategy

Our method has a phased training strategy for the WSFS. The LS-Net is trained with the weakly supervised loss and the region-refining segmentation algorithm successively, and it obtains a good initialization, so it can accurately locate and segment fire or smoke pixels. The next step is to train the AD-Net, during which the LS-Net outputs remain unchanged, and the cross-entropy loss is used.

3.2. Lightweight Faster R-CNN

In this section, we introduce a knowledge distillation technique to simplify the complexity of the Faster R-CNN.

3.2.1. Knowledge Distillation

Knowledge distillation [55] aims to compress a complex model into a simpler model that is much easier to deploy. The main goal of knowledge distillation is to train a small network model to imitate a pre-trained effective and complex network. Our proposed lightweight Faster R-CNN is implemented using a teacher stream and a student stream. For the sake of simplicity, we define the key components of the two streams as follows:

The teacher stream uses a complex CNN structure, with a set of parameters pre-trained as a feature extractor. Here, we assume that this model has absorbed the rich knowledge encoded in generous high-resolution forest fire smoke images with labels. Generally, the training dataset is very large and may be invisible to the student stream. The student stream is a much simpler CNN network that does not need too many parameters for recognizing low-resolution forest fire smoke.

In contrast to the classical Faster R-CNN, we use the ResNet 50 as the teacher stream and the ResNet 18 as the student stream to substitute for VGG-16 for proposal and detection. The features extracted by the teacher stream are used to distill the knowledge. The feature loss—i.e., L2 loss—is based on both the eigenvectors obtained from the student stream and the eigenvectors distilled from the teacher stream. In the process of forward propagation, the loss of the whole network includes the feature loss, RPN loss, and RCNN loss.

3.2.2. Loss Function

To train the lightweight Faster R-CNN, we propose a novel loss $Loss_{all}$, which includes $Loss_{RPN}$, $Loss_{RCN}$, and $Loss_{backbone}$. $Loss_{RPN}$ and $Loss_{RCN}$ represent the loss of the RPN module and the loss of the RCNN module, respectively. $Loss_{backbone}$ is the loss in extracting features, which is expressed as

$$Loss_{all} = Loss_{RPN} + Loss_{RCN} + Loss_{backbone} \tag{10}$$

Specially, $Loss_{RPN}$ and $Loss_{RCN}$ are the classification loss and object regression loss, respectively, which are defined as

$$Loss_{RPN} = \frac{1}{N_{cls}} \sum Loss_{cls}^{RPN} + \lambda \frac{1}{N_{reg}} \sum p^* Loss_{reg}^{RPN}$$
$$Loss_{RCN} = \frac{1}{M_{cls}} \sum Loss_{cls}^{RCN} + \lambda \frac{1}{M_{reg}} \sum p^* Loss_{reg}^{RCN} \tag{11}$$

where $p^* = 1$ while an anchor is positive; and $p^*=0$ while an anchor is negative. N_{cls} and N_{reg} are the batch size of the RPN and anchor location, respectively; M_{cls} and M_{reg} are the batch size of the RCNN and anchor location, respectively.

The classification loss of RPN and RCNN can use a cross-entropy loss. The regression loss of RPN and RCNN is calculated using a smooth L1 loss. A parameter is added to control the smooth area, which is expressed as:

$$smooth_{L1}(x) = \begin{cases} 0.5x^2 \; if |x| < 1\sigma^2 \\ |x| - \frac{0.5}{\sigma^2} \; otherwise \end{cases} \tag{12}$$

$Loss_{backbone}$ is based on a calculation of the KL divergence between the teacher network and the student network. However, before calculating the KL divergence, it is necessary to ensure that the dimensions of the feature map between the teacher network and the student network are consistent:

$$Loss_{backbone}(y_T, y_S) = \sum_{i=1}^{n} y_{S_i} \times \log\left(\frac{y_{S_i}}{y_{T_i}}\right) \tag{13}$$

where y_t is the output feature of the teacher network, and y_s is the output feature of the student network.

The distilling process of the lightweight Faster R-CNN is shown in Algorithm 2.

Algorithm 2. Distilling Process of Faster R-CNN

Input: Parameters of the features of the teacher network: T_Parameter, Dataset with labels: S_Input$\{P, T\}$
Procedure Iteration process
1. Get map from Teacher-Net: T_Feature = Teacher_backbone(T_Parameter)
2. Transform dimension: T_Map_trans = Trans (T_ Feature)
3. Generate map from Student-Net: S_Map = Student_backbone(S_Input)
4. Caculate distill_loss with $L2_{loss}$: $Loss_{backbone} = L^{T \rightarrow S}$(T_Map, S_Map)
5. Compute the loss of detection with RPN and RCN: $Loss_{rpn}$ = RPN (P_i, S_Map, T_i), $Loss_{RCN}$ = RCN (S_i, S_Map, T_i)
6. Reverse the loss $L_{all} = L^{RPN} + L^{RCN} + L^{backbone}$ for the student network
7. Update the parameters and continue

3.3. Collaborative Learning Strategy

In the detection process, we adopt 3 different methods, which are the classification-based model (ShuffleNet [46]), segmentation-based model (WSFS), and the region-proposal-based model (lightweight Faster R-CNN), to predict the probability of fire smoke existing in an image. The final result is determined by the estimation of 3 model reasoning results, which are the classification result 1, the classification result 2, and the classification result 3, as shown in Figure 1.

Thus, the final prediction results can be calculated using a logistic regression, which is expressed as

$$P = \alpha + \sum_{i=1}^{n} \beta_i \times x_i \tag{14}$$

where P is the final prediction; x_i ($i = 1, 2, \ldots, n$) is the deterministic variables related to the probability of a model; α is a constant; β_i ($i = 1, 2, \ldots, n$) is a coefficient in Equation (14); and i represents the i-th model. In this paper, the 3 models are combined to predict the results. Therefore, we need to determine the 3 coefficients, β_1, β_2, and β_3, corresponding to the 3 inputs of models x_1, x_2, and x_3. Thus, we proposed a learning strategy, which is based on a stacking method [57], to fit Equation (14) in order to compute these coefficients.

We can perform a 5-fold cross validation on the training dataset, assuming that the training dataset includes 500 images which are divided into 5 subsets represented as T_{data1}, T_{data2}, T_{data3}, T_{data4}, and T_{data5}, and that each subset includes 100 images. The testing dataset also includes 100 images. The models to be combined are represented as C, S, and R corresponding to the classification-based model, the segmentation-based model, and the region proposal-based model, respectively.

Eventually, datasets TC, TS, and TR are considered as the input value. Then, taking a real label as a guide, the models C, S, and R can learn the importance of the different models, and the models are assigned the weight of every algorithm (β_1, β_2, and β_3). The datasets DC, DS, and DR are further used to verify the models to achieve the best results. The detailed process of this strategy is described in Algorithm 3.

Algorithm 3. Collaborative Learning Strategy

Input: Five training subsets and one test dataset
Procedure Learning process for model x = C//x indicates C, S, and R.
1: Utilize T_{data2}-T_{data5} to train model x. T_{data1} is used as the test data, the obtained result is saved as TC1, and the verified result of the test dataset is saved as D_{S1};
2: Utilize Tdata1 and T_{data3}-T_{data5} to train model x. T_{data2} is used as the test data, the obtained result is saved as TC2, and the verified result of the test dataset is saved as D_{S2};
3: Continue to train and verify model x through the above process, and datasets TC3, TC4, and TC5, as well as D_{S3}, D_{S4}, and D_{S5}, are obtained;
4: Compute the average value of TC1 to TC5 stored as $T_{average}$;
5. The results of 5-fold cross validation is saved as TC = [TC1, TC2, TC3, TC4, TC5];
6. Calculate the average value of D_{S1} to D_{S5} and obtain DC;
7: Repeat this process for model S and R and obtain TS, TR, DS, and DR.

3.4. Grading Method Using Fuzzy Logic

The influencing factors of the level of fire or smoke include many aspects, such as the category (fire or smoke), region size, location, temperature, and wind, among others. In this paper, we only study the fire or smoke category (CT), region size (RS), and location (LT). The former category is predicted by Equation (14). The value of the region size is computed by Equation (16). The value of the location is divided into two types: dry forest or wet forest, which is determined by the color in the background of the input image, where yellow and green correspond to dry or wet forest, respectively.

The value of region size is expressed as:

$$S_{seg} = Area_{segmentaion} \cap Area_{region-proposal} \tag{15}$$

$$Region_{fire\ or\ smoke} = \begin{cases} FV, & S_{seg} > \theta \\ 0, & other \end{cases} \quad j = 1, 2, \ldots, m \tag{16}$$

where S_{seg} represents the size of segmentation. $Area_{fire\ or\ smoke}$ is calculated by the segmentation results using the WSFS. $Area_{region-proposal}$ is taken from the segmentation results using Lightweight Faster R-CNN; $Region_{fire\ or\ smoke}$ represents the region size of fire or smoke; FV is the corresponding value of this region; and FV $\in$ [big, middle, small]. θ is an empiric value.

To assess the fire or smoke level, a fuzzy strategy is designed to weigh the variables CT, RS, and LT. This strategy is similar to that employed in our previous work [58,59]. The ambiguity $Level = f(i) \in [0, 1, 2, 3, 4, 5]$ guides the evaluation of the possibility of fire or smoke level. Here, the numbers correspond to the fire or smoke level as follows: 0 = very high, 1 = high, 2 = middle, 3 = low, 4 = very low, and 5 = no fire/no smoke. The CT is developed by defining extreme alarm (i.e., fire, H_fire), alarm (i.e., smoke, M_smoke), and normal situation (Norm) fuzzy sets. The RS is developed by defining large, middle, and small fuzzy sets. The LT is developed by defining high (yellow) and low (green) fuzzy sets. A conventional trapezoid or triangle is selected as the membership function, since they have few parameters and are easily optimized. The ambiguity term $Level$ determines the final fire smoke level. The Mamdani model is applied as a reasoning engine [58,59], because it is suitable for capturing and coding expert-based knowledge.

4. Experiments

4.1. Dataset Description

We evaluated our approach on a developed forest fire smoke dataset, named FS-data, which was set up using some image search engines, such as Google and Baidu. The entire fire dataset contains 4856 samples distributed into 5 categories: large fire, small fire, dense smoke, small smoke, and other scenes, e.g., scenes of forests in different seasons. Some examples are shown in Figure 4.

Figure 4. Samples of the forest fire dataset (the first row shows normal images (non-fire), and the other images show fire and smoke, which are, from top left to bottom right, large fire, dense smoke1, dense smoke2, light smoke, small fire, and other scenes).

The first and second rows in Figure 4 contain non-fire negative samples and positive samples for fires and smoke. Each image was labeled at the pixel level. These images had a resolution of 256 × 256 pixels, and 1323 images clearly exhibited visible fire or smoke and served as positive samples. The remaining 3533 images were non-fire negative samples. We divided the dataset into training and testing examples, as shown in Table 4.

Table 4. Distribution of training and testing examples in the datasets.

Dataset	Training Examples		Testing Examples	
	Positive	Negative	Positive	Negative
Subset A (large fire)	150	309	41	138
Subset B (small fire)	176	454	48	197
Subset C (dense smoke)	245	546	65	207
Subset D (light smoke)	280	755	78	267
Subset E (other scenes)	201	455	39	205

4.2. Performance Metrics

In this section, to obtain a better evaluation of our proposed method, 2 indicators, which are intersection over union (IOU) and average precision (AP), are used for the model performance evolution.

The intersection over union (IOU) [60,61] is a commonly used performance metric for semantic segmentation tasks. The decision prediction is based on the construction of a confusion matrix. True positives (TP) and true negatives (TN) belong to correct predictions. False positives (FP) are negative samples misreported as positive (fire), and false negatives (FN) are positive samples misreported as negative (non-fire). Our segmentation metrics are $IOU = TP/(TP + FN + FP)$.

The average precision (AP) is used to assess the classification accuracy [23]. AP is determined using the calculated area under the precision-recall curve to obtain a precise representation of the comprehensive model performance at different thresholds, particularly when the dataset contains large numbers of negative (non-fire) samples.

4.3. Experiment Environment

We conducted the experiments with the Python language under Pycharm, and the network model was implemented in Pytorch. We used PyDenseCRF (https://github.com/lucasb-eyer/pydensecrf, accessed on 12 September 2020) to construct a DenseCRF model. The simulations were conducted on a PC with an Intel Core i7-7820X CPU running Windows 10, with two GTX1080Ti GPUs (total 22 GHz) and 32 GB of RAM.

4.4. Evaluation of WSFS

In this section, we evaluate the performance of our WSFS approach on FS-data. The WSFS consists of LS-Net and AD-Net. Therefore, we conducted several experiments to validate the performance of LS-Net with WSL and RRS, as well as AD-Net.

4.4.1. The Region Detection Result Using LS-Net with WSL and RRS

We conducted experiments using FS-data to validate the performance of the weakly supervised loss (WSL) by visualizing the segmentation results (region detection) and comparing them with the evaluation index mentioned above. After training with the weakly supervised loss, LS-Net could roughly locate fire areas.

Figure 5 shows the visualization results of FS-data using our method to segment fire or smoke from an original image. In Figure 5, the red region represents the fire region (segmentation result), and the gray region represents the smoke area segmented by LS-Net. The first and second columns show the original images and pixel labels annotated manually, respectively. Columns 3–8 show that the segmentation regions change at different α values of 0.01, 0.02, 0.04, 0.06, 0.08, and 0.1, respectively. The results shown in Figure 5 indicate that LS-Net can predict fire regions under the guidance of the weakly supervised loss, and that the areas of fire regions and α values are positively correlated.

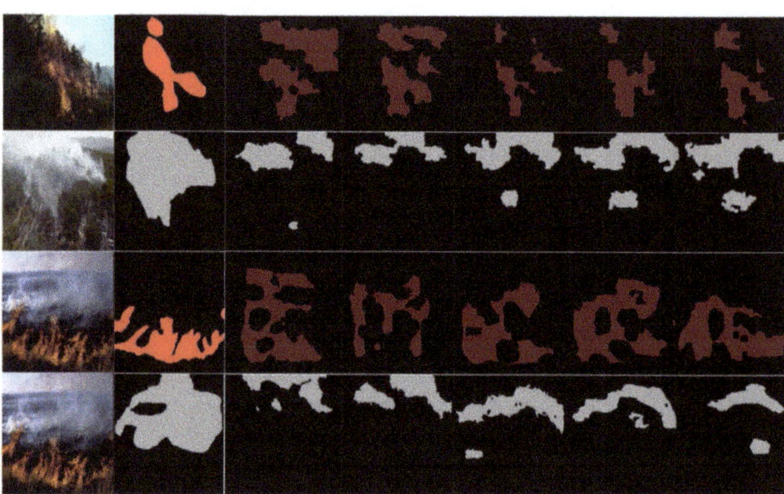

Figure 5. Segmentation results of the weakly supervised loss on FS-data.

We also used the IOU to evaluate the performance (segmentation accuracy). In Figure 6, the graph shows variations in IOU value at different values of α. The IOU value increases initially and then decreases with an increase in α. The IOU value is at a maximum at $\alpha = 0.004$, which means that the segmentation accuracy is the best. The IOU then decreases with an increase in α, because an increasing number of pixels around the fire regions are wrongly identified as fires. This illustrates that the WSL can effectively identify fire or smoke pixels.

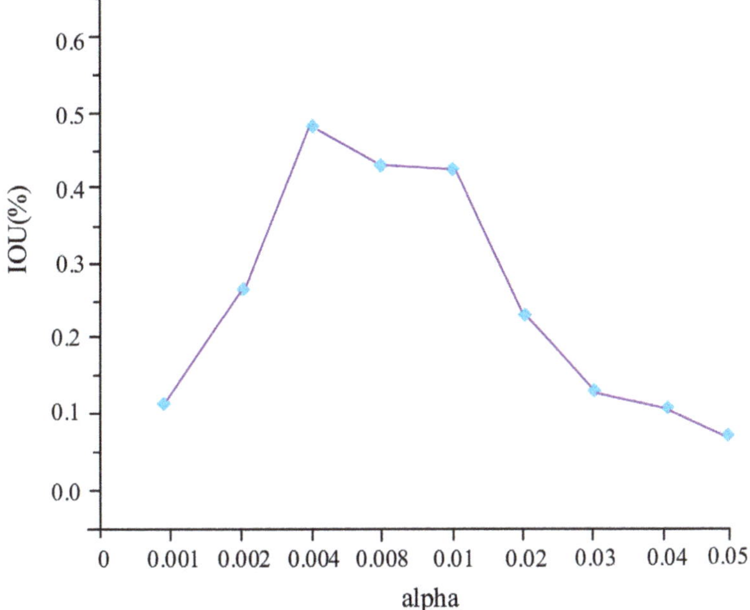

Figure 6. Changes in IOU at different α value.

However, the rough regions were only implemented using LS-Net. To further refine the detection region, we also implemented region-refining segmentation (RRS) on the FS-data. To evaluate the performance of RRS, the related algorithms were tested for comparison. These algorithms included 2 supervised learning segmentation methods, DeepLabV3+ [25] and U-Net [26]. The third method was the use of the weakly supervised loss alone. The fourth method adopted RRS to post-process the results of the weakly supervised loss. The same configurations were used in all experiments.

The detection results and metrics are shown in Figure 7. The first and second columns show the original images and pixel labels, respectively. The third and fourth columns show the results of a segmentation network trained in a supervised mode (U-Net) and a new supervised segmentation model, e.g., DeepLab V 3+. The segmentation results of LS-Net trained only using WSL are shown in column 5. The last column shows the segmentation results of WSFS trained successively using the 2-stage training strategy (WSL and RRS). The results show that our method, WSFS, has obtained a competitive performance in the segmentation result of fire or smoke.

The performance of these methods, evaluated using Boxplot descriptions, is shown in Figure 8. According to Figure 8, the U-Net method using pixel-wise labels obtains the maximum IOU for FS-data. The use of WSL alone returns the worst results, and RRS greatly improves the segmentation results of training with WSL; its results are closest to those of supervised training.

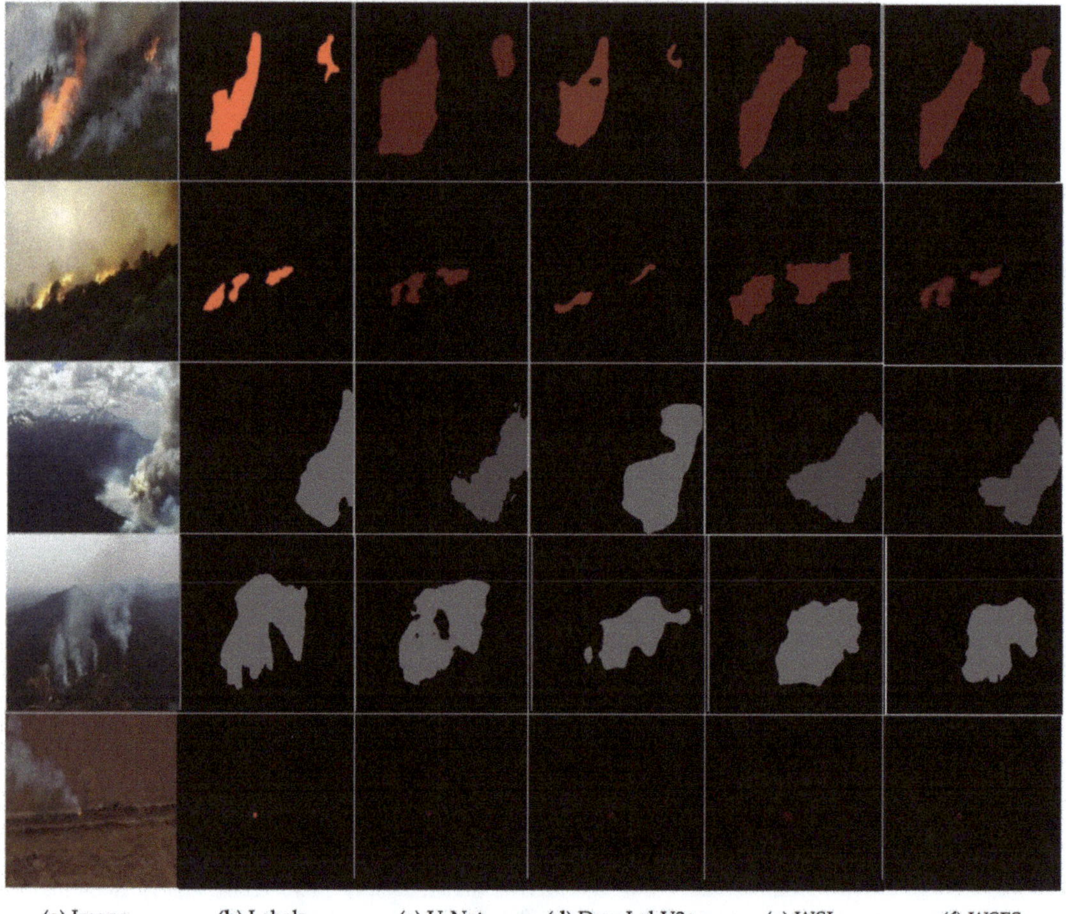

(a) Image   (b) Labels   (c) U-Net   (d) DeepLabV3+   (e) WSL   (f) WSFS

Figure 7. Visualization of the segmentation results of different methods. The (**a**,**b**) show the original images and pixel labels, respectively. (**c**–**f**) represent the experimental results using various methods which are U-Net, DeepLabV3+, WSL, and WSFS from the third column to the last column, respectively. (The red region represents the fire region, and the gray region represents the smoke area, which regions are segmented using the methods).

4.4.2. AD-Net

In this section, we performed controlled trials on FS-data to evaluate the classification performance of the AD-Net, while the ResNet18 and the ShuffleNet were selected as comparison methods. AP was used as an evaluation indicator, with values from the FS-data, as shown in Figure 9.

In Figure 9, the AP values of ResNet18, ShuffleNet, and our proposed method show better results for Subset A, Subset B and Subset C, in which the AP values are 99.9%,98.8%, and 98.9%, respectively. However, the AP value of these methods show worse results for Subset E and Subset D. The reason is that the image from Subset D and Subset E has a complex background, with low-contrast between objects and the background compared with other subsets. The AD-Net with the RSAM module built according to the result of LS-Net have shown the best performance for the five subsets.

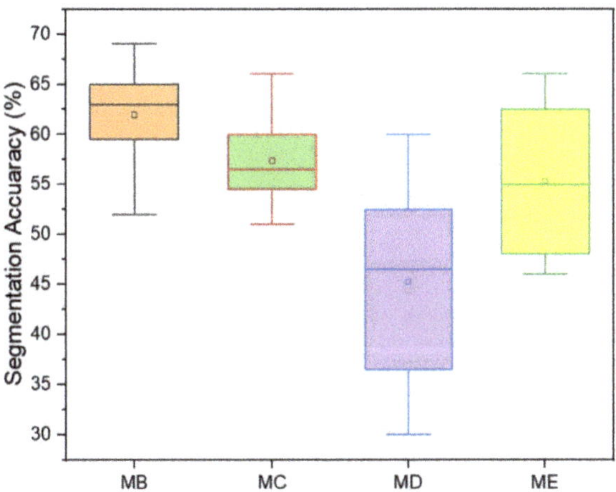

Figure 8. Boxplots of the segmentation results of different training methods: MB-U-Net, MC-DeepLabV3+, MD WSL (Ours), and ME- WSFS (Ours).

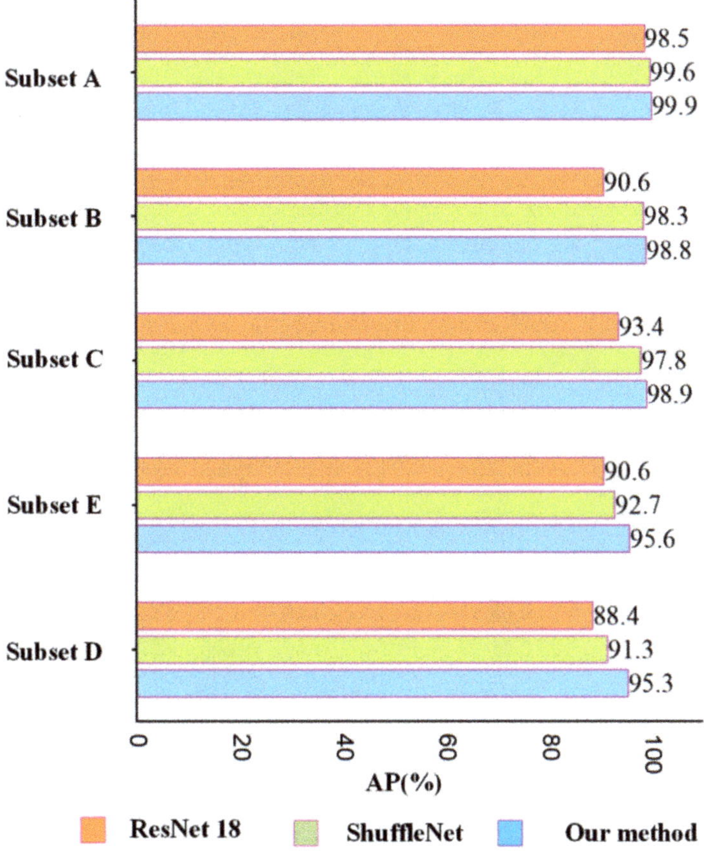

Figure 9. Comparison of the classification results of the different models on the FS-data.

4.5. Comparison with State-of-the Art Methods

We compared several state-of-the-art approaches with our method. For a fair comparison, the data and training setups were kept the same. The numerical experimental results for the fire dataset are summarized in Table 5, which involved the same experimental configurations.

Table 5. Comparison of our method with other methods using the FS-data.

Method	mIOU(%)	AP (%)
U-Net [26]	69.6	96.1
DeepLab v3+ [25]	67.4	98.4
Segmentation decision approach [62]		98.2
WSL (Ours)	60.5	97.9
WSFS (Ours)	68.8	98.6
FireDGWF (Ours)	70.2	99.6

Table 5 shows that the performance of region detection for DeepLabe V3+ is the worst of the tested methods and 67.4% in mIOU. The WSL methods achieve at 60.5% in mIOU. A small improvement of 8.3% is obtained by WSFS, due to RRS adopted by this method. However, the best performance of region detection, which reaches 70.2% in mIOU, is achieved by FireDGWF because the intersection between the detection regions obtained with the two methods (WSFS and lightweight Faster R-CNN) corresponds to the results.

The U-Net-based segmentation decision network produces an AP of 96.1%. The segmentation decision method based on DeepLab v3 + produces an average accuracy of 98.4%. Tabernik et al. [62] proposed a supervised 2-stage segmentation-based network that was trained with pixel labels to improve its classification performance. The WSL method obtains 97.9% in AP. Compared to these methods, the proposed method (WSFS) achieves 98.6% in AP. Moreover, FireGDWF shows a better ability to discriminate between fire-objects and non-fire objects, with an accuracy of 99.6% for 763 images from the FS-data. These results show that multi models can improve the performance of the algorithms.

4.6. Result for Grading of Forest Fire and Smoke

To evaluate forest fire severity, we designed a 3-input/1-output fuzzy evaluation system. These 3-inputs included the classification results, the segmentation region size, and the location. The single output was the forest fire smoke level: level 0, level 1, … , level 5. In our fuzzy evaluation system, the membership function must be decided first, which is similar to our previous work [58,59]. Once the membership degree of every input has been determined, a set of rules, which are defined in the system, as shown in Table 6, can be used to explain the evaluation results. Then, the detailed procedure for the example using the fuzzy evaluation system to grade the forest fire smoke level can be provided.

Table 6. Examples of the knowledge rules.

Rule No	Knowledge Rule
Rule1	*If* CT is H_fire and RS is Big and LT is Low, *then Level* = 0
Rule2	*If* CT is H_fire and RS is Big and LT is High, *then Level* = 1
Rule3	*If* CT is H_fire and RS is Middle and LT is Low, *then Level* = 1
Rule4	*If* CT is H_fire and RS is Middle and LT is Hight, *then Level* = 2
Rule5	*If* CT is H_fire and RS is Small and LT is Low, *then Level* = 2
Rule6	*If* CT is H_fire and RS is Small and LT is High, *then Level* = 1
Rule7	*If* CT is M_smoke and RS is Big and LT is Low, *then Level* = 2
Rule8	*If* CT is M_smoke and RS is Middle and LT is Low, *then Level* = 3
Rule9	*If* CT is M_smoke and RS is Small and LT is High, *then Level* = 3
Rule10	*If* CT is M_smoke and RS is Small and LT is Low, *then Level* = 4
Rule11	*If* CT is Norm, *then Level* = 5

When the system was run at a point of CT = fire, RS = 21,845 (pixel number), and LT = dry forest, and Rule 1 was activated (see Table 6), then, the result was put into the implication process and used to determine the output fuzzy set. The result given by the defuzzication process indicated that the output was *Level* = 0, as shown in Figure 10. Thus, we used the fuzzy evaluation system to determine different scenes of forest fire smoke, which could assess these fire smoke levels from level 1 to level 4. Level 5 indicates non-fire/non-smoke. These processes are shown in Figure 10.

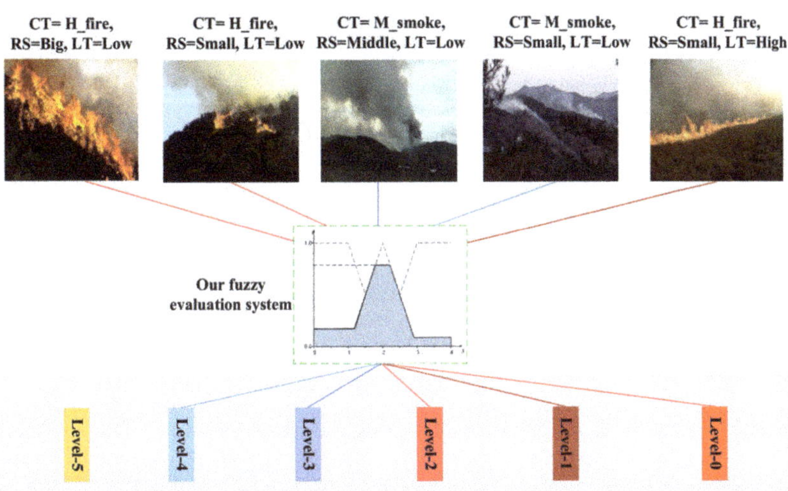

Figure 10. An example of grading forest fire smoke using the fuzzy evolution system.

Additionally, the response time to region detection of forest fire using our FireDGWF was evaluated on FS-data. The experimental results are shown in Table 7. Our method, FireDGWF, included 2 processes: detection and grading. The detecting time of FireDGWF for an image is 0.128 s, which is the maximum time among the 3 methods, and the grading time for an image is 0.023 s. Therefore, a total response time to region detection and grading of forest fire is 0.151 s.

Table 7. Response time to detection and grading of forest fire using our method.

Method	Detecting Time	Grading Time
FireDGWF	0.128 s	0.023 s

4.7. Analysis and Discussion

Several experiments were conducted using our proposed approach, and its performance was evaluated, indicating that our approach performed competitively for the region detection and grading of forest fire and smoke. In the detection process, we proposed a weakly supervised fine segmentation method to effectively detect fire regions in a coarse-to-fine way, similar to a human-like recognition process. A two-stage weakly supervised learning strategy consisting of a weakly supersized loss and a region-refining segmentation algorithm was proposed to train the segmentation network. The negative and positive loss allowed the segmentation network to identify fire areas that differed from the background (forest) in training. Our method (WSFS) has achieved a better segmentation result, with a 68.8% mIOU. Experiments on the FS-data demonstrated that the region-refining segmentation algorithm obviously improved the performance of the segmentation network, without increasing the inference time. To further improve segmentation performance, we combined the Lightweight Faster R-CNN with the model to obtain a small improvement of 8.3%, for the mIOU values.

We also evaluated the decision network on the FS-data and compared it to state-of-the-art CNN-based models. Our method has obtained a competitive performance in the prediction of the category of an input image. This is because the RSAM module utilizes the segmentation result of LS-Net as an attention mask to cause AD-Net to focus more on object regions.

In the grading process, a 3-input and 1-out fuzzy system was developed to assess the level of forest fire, with the classification result, the region detection results, and the location as input values of the system.

The experimental results showed that multi models can improve the performance of our algorithm in terms of the detection accuracy and segmentation accuracy. We synthetically utilized three values: the classification result, the segmentation region, and the location. These were fed into our designed fuzzy evaluation system to obtain the forest fire smoke level, which could help humans to take proper precautionary measures in a timely manner.

5. Conclusions

In this paper, we introduced a collaborative region detection and grading framework for forest fire and smoke using a weakly supervised fine segmentation and a lightweight Faster R-CNN. This framework can detect the region and grade of fire and smoke in forests. To obtain the accurate region of fire and smoke, we propose a weakly supervised fire-segmentation model, which is trained using only image-level labels. A distillation strategy is used to reduce the complexity of the Faster R-CNN. Our proposed method has achieved an excellent performance and outperformed state-of-the-art CNN-based models in terms of detection accuracy (99.6%) and segmentation accuracy (70.2%). The final latency of our proposed method is only 151ms, which shows an excellent balance between detection performance and efficiency. Moreover, our fuzzy evaluation system can be used to assess the forest fire smoke level in a timely manner.

In future works, we plan to study an attention mechanism to improve the weakly supervised fine-segmentation method in the detection performance. To overcome the insufficient training data in a real-world application, a data augmentation technique, which is based on a Generative Adversarial Networks [63], will be introduced into our model. Furthermore, a possible improvement to our method is the incorporation of Multi-scale Adversarial Erase to substantially improve the detection rate. Additionally, we will work on developing a forest fire and smoke assessment system for risk level, which can identify different types, locations, sizes, and levels of fires or smoke. This system can track the evolution, spread, and grade of forest fires and smoke.

Author Contributions: J.P.: Data curation, Investigation, Resources, Method, Visualization, and Writing. X.O.: Supervision, Project Administration, and Funding Acquisition. L.X.: Conceptualization, Software, and Funding Acquisition. All authors have read and agreed to the published version of the manuscript.

Funding: This research was funded by Key project of National Social Science Foundation of China (grand number 17AGL018) and National Natural Science Foundation of China (grant number 71773031 and 21376091).

Data Availability Statement: The data presented in this study are available on request from the corresponding author.

Acknowledgments: We thank MDPI for its linguistic assistance during the preparation of this manuscript.

Conflicts of Interest: The authors declare no conflict of interest.

References

1. Eugenio, F.C.; Santos, A.D.; Fiedler, N.C.; Ribeiro, G.A.; Silva, A.; Santos, Á.D.; Paneto, G.G.; Schettino, V.R. Applying GIS to develop a model for forest fire risk: A case study in Espírito Santo, Brazil. *J. Environ. Manag.* **2016**, *173*, 65–71. [CrossRef]
2. de Azevedo, A.R.; Alexandre, J.; Pessanha, L.S.P.; da S.T. Manhães, R.; de Brito, J.; Marvila, M.T. Characterizing the paper industry sludge for environmentally-safe disposal. *Waste Manag.* **2019**, *95*, 43–52. [CrossRef]
3. Çolak, E.; Sunar, F. Evaluation of forest fire risk in the Mediterranean Turkish forests: A case study of Menderes region, Izmir. *Int. J. Disaster Risk Reduct.* **2020**, *45*, 101479. [CrossRef]
4. Tang, X.; Machimura, T.; Li, J.; Liu, W.; Hong, H. A novel optimized repeatedly random undersampling for selecting nega-tive sam-ples: A case study in an SVM-based forest fire susceptibility assessment. *J. Environ. Manag.* **2020**, *271*, 1–13. [CrossRef] [PubMed]
5. Guha-Sapir, D.; Philippe, H. Estimating Populations Affected by Disasters: A Review of Methodological Issues and Research Gaps. In *Global Sustainable Report*; School of Public Health Université Catholique de Louvain: Brussels, Belgium, 2015; pp. 1–15.
6. Marshall, S.C.A.D. Fire detection using smoke and gas sensors. *Fire Saf. J.* **2007**, *42*, 507–515.
7. Qiu, X.; Wei, Y.; Li, N.; Guo, A.; Zhang, E.; Li, C.; Peng, Y.; Wei, J.; Zang, Z. Development of an early warning fire detection system based on a laser spectroscopic carbon monoxide sensor using a 32-bit system-on-chip. *Infrared Phys. Techn.* **2019**, *96*, 44–51. [CrossRef]
8. Varela, N.; Díaz-Martinez, J.L.; Ospino, A.; Zelaya, N.A.L. Wireless sensor network for forest fire detection. *Procedia Comput. Sci.* **2020**, *175*, 435–440. [CrossRef]
9. Sudhakar, S.; Vijayakumar, V.; Kumar, C.S.; Priya, V.; Ravi, L.; Subramaniyaswamy, V. Unmanned Aerial Vehicle (UAV) based Forest Fire Detection and monitoring for reducing false alarms in forest-fires. *Comput. Commun.* **2020**, *149*, 1–16. [CrossRef]
10. Fernandes, A.M.; Utkin, A.B.; Lavrov, A.V.; Vilar, R.M. Development of neural network committee machines for automatic forest fire detection using lidar. *Pattern Recognit.* **2004**, *37*, 2039–2047. [CrossRef]
11. Chen, T.-H.; Wu, P.-H.; Chiou, Y.-C. An early fire-detection method based on image processing. In Proceedings of the 2004 International Conference on Image Processing, Singapore, 24–27 October 2004; Volume 3, pp. 1707–1710.
12. Marbach, G.; Loepfe, M.; Brupbacher, T. An image processing technique for fire detection in video images. *Fire Saf. J.* **2006**, *41*, 285–289. [CrossRef]
13. Habiboğlu, Y.H.; Günay, O.; Çetin, A.E. Covariance matrix-based fire and flame detection method in video. *Mach. Vis. Appl.* **2012**, *23*, 1103–1113. [CrossRef]
14. Çelik, T.; Demirel, H. Fire detection in video sequences using a generic color model. *Fire Saf. J.* **2009**, *44*, 147–158. [CrossRef]
15. Tao, C.; Jian, Z.; Pan, W. Smoke Detection Based on Deep Convolutional Neural Networks. In Proceedings of the 2016 International Conference on Industrial Informatics—Computing Technology, Intelligent Technology, Industrial Information Integration (ICIICII), Wuhan, China, 3–4 December 2016; pp. 150–153.
16. Muhammad, K.; Ahmad, J.; Mehmood, I.; Rho, S.; Baik, S.W. Convolutional Neural Networks Based Fire Detection in Sur-veillance Videos. *IEEE Access* **2018**, *6*, 18174–18183. [CrossRef]
17. Yin, Z.; Wan, B.; Yuan, F.; Xia, X.; Shi, J. A Deep Normalization and Convolutional Neural Network for Image Smoke Detec-tion. *IEEE Access* **2017**, *5*, 18429–18438. [CrossRef]
18. Sharma, J.; Granmo, O.-C.; Goodwin, M.; Fidje, J.T. Deep Convolutional Neural Networks for Fire Detection in Images. *Commun. Comput. Inf. Sci.* **2017**, *744*, 183–193. [CrossRef]
19. Hu, Y.; Lu, X. Real-time video fire smoke detection by utilizing spatial-temporal ConvNet features. *Multimed. Tools Appl.* **2018**, *77*, 29283–29301. [CrossRef]
20. Kaabi, R.; Bouchouicha, M.; Mouelhi, A.; Sayadi, M.; Moreau, E. An Efficient Smoke Detection Algorithm Based on Deep Belief Network Classifier Using Energy and Intensity Features. *Electronics* **2020**, *9*, 1390. [CrossRef]
21. Islam, R.; Amiruzzaman; Nasim, S.; Shin, J. Smoke Object Segmentation and the Dynamic Growth Feature Model for Video-Based Smoke Detection Systems. *Symmetry* **2020**, *12*, 1075. [CrossRef]
22. Lu, P.; Zhao, Y.; Xu, Y. A Two-Stream CNN Model with Adaptive Adjustment of Receptive Field Dedicated to Flame Region Detection. *Symmetry* **2021**, *13*, 397. [CrossRef]
23. Zhao, E.; Liu, Y.; Zhang, J.; Tian, Y. Forest Fire Smoke Recognition Based on Anchor Box Adaptive Generation Method. *Electronics* **2021**, *10*, 566. [CrossRef]
24. Xu, R.; Lin, H.; Lu, K.; Cao, L.; Liu, Y. A Forest Fire Detection System Based on Ensemble Learning. *Forests* **2021**, *12*, 217. [CrossRef]
25. Chen, L.C.; Papandreou, G.; Kokkinos, I.; Murphy, K.; Yuille, A.L. DeepLab: Semantic Image Segmentation with Deep Con-volutional Nets, Atrous Convolution, and Fully Connected CRFs. *IEEE Trans. Pattern Anal. Mach. Intell.* **2017**, *40*, 834–848. [CrossRef] [PubMed]
26. Ronneberger, O.; Fischer, P.; Brox, T. U-net: Convolutional networks for biomedical image segmentation. In Proceedings of the International Conference on Medical Image Computing and Computer-Assisted Intervention, Cham, Switzerland, 5–9 October 2015; pp. 234–241.
27. Long, J.; Shelhamer, E.; Darrell, T. Fully convolutional networks for semantic segmentation. *IEEE Trans. Pattern Anal. Mach. Intell.* **2017**, *39*, 640–651.
28. Alkhatib, A.A.A. A Review on Forest Fire Detection Techniques. *Int. J. Distrib. Sens. Networks* **2014**, *10*, 1–14. [CrossRef]
29. Yuan, C.; Liu, Z.; Zhang, Y. Vision-based forest fire detection in aerial images for firefighting using UAVs. In Proceedings of the 2016 International Conference on Unmanned Aircraft Systems (ICUAS), Arlington, VA, USA, 7–10 June 2016; pp. 1200–1205.

30. Ren, X.; Li, C.; Ma, X.; Chen, F.; Wang, H.; Sharma, A.; Gaba, G.; Masud, M. Design of Multi-Information Fusion Based Intelligent Electrical Fire Detection System for Green Buildings. *Sustainability* **2021**, *13*, 3405. [CrossRef]

31. Zhao, Y.; Ma, J.; Li, X.; Zhang, J. Saliency Detection and Deep Learning-Based Wildfire Identification in UAV Imagery. *Sensors* **2018**, *18*, 712. [CrossRef]

32. Namozov, A.; Cho, Y.I. An Efficient Deep Learning Algorithm for Fire and Smoke Detection with Limited Data. *Adv. Electr. Comput. Eng.* **2018**, *18*, 121–128. [CrossRef]

33. Xu, Z.; Guo, Y.; Saleh, J.H. Tackling Small Data Challenges in Visual Fire Detection: A Deep Convolutional Generative Ad-versarial Network Approach. *IEEE Access* **2020**, *9*, 3936–3946. [CrossRef]

34. Muhammad, K.; Ahmad, J.; Lv, Z.; Bellavista, P.; Yang, P.; Baik, S.W. Efficient Deep CNN-Based Fire Detection and Locali-zation in Video Surveillance Applications. *IEEE Trans. Syst. Man Cybern. Syst.* **2019**, *49*, 1419–1434. [CrossRef]

35. Muhammad, K.; Khan, S.; Elhoseny, M.; Ahmed, S.H.; Baik, S.W. Efficient Fire Detection for Uncertain Surveillance Environment. *IEEE Trans. Ind. Inform.* **2019**, *15*, 3113–3122. [CrossRef]

36. Ren, S.; He, K.; Girshick, R.; Sun, J. Faster R-CNN: Towards Real-Time Object Detection with Region Proposal Networks. *IEEE Trans. Pattern Anal. Mach. Intell.* **2017**, *39*, 1137–1149. [CrossRef]

37. Girshick, R. Fast R-CNN. In Proceedings of the IEEE International Conference on Computer Vision (ICCV), Santiago, Chile, 7–13 December 2015; pp. 1440–1448.

38. He, K.; Gkioxari, G.; Dollár, P.; Girshick, R. Mask R-CNN. In Proceedings of the IEEE International Conference on Computer Vision (ICCV), Venice, Italy, 22–29 October 2017; pp. 2980–2988.

39. Zhou, Z.-H. A brief introduction to weakly supervised learning. *Natl. Sci. Rev.* **2018**, *5*, 44–53. [CrossRef]

40. Pathak, D.; Shelhamer, E.; Long, J.; Darrell, T. Fully Convolutional Multi-Class Multiple Instance Learning. *arXiv* **2014**, arXiv:1412.7144.

41. Pinheiro, P.O.; Collobert, R. From image-level to pixel-level labeling with Convolutional Networks. In Proceedings of the 2015 IEEE Conference on Computer Vision and Pattern Recognition (CVPR), Boston, MA, USA, 7–12 June 2015; pp. 1713–1721.

42. Boykov, Y.Y.; Jolly, M.-P. Interactive graph cuts for optimal boundary & region segmentation of objects in N-D images. In Proceedings of the Eighth IEEE International Conference on Computer Vision ICCV 2001, Vancouver, BC, Canada, 7–14 July 2001; pp. 105–112.

43. Rother, C.; Kolmogorov, V.; Blake, A. "GrabCut": Interactive foreground extraction using iterated graph cuts. *ACM Trans. Graph.* **2004**, *23*, 309–314. [CrossRef]

44. Krähenbühl, P.; Koltun, V. Efficient Inference in Fully Connected CRFs with Gaussian Edge Potentials. *Adv. Neural Inf. Process. Syst.* **2012**, *24*, 109–117.

45. Chen, L.C.; Papandreou, G.; Kokkinos, I.; Murphy, K.; Yuille, A.L. Semantic Image Segmentation with Deep Convolutional Nets and Fully Connected CRFs. *arXiv* **2014**, arXiv:1412.7062.

46. Zhang, X.; Zhou, X.; Lin, M.; Sun, J. ShuffleNet: An Extremely Efficient Convolutional Neural Network for Mobile Devices. In Proceedings of the 2018 IEEE/CVF Conference on Computer Vision and Pattern Recognition, Salt Lake City, UT, USA, 18–23 June 2018; pp. 6848–6856.

47. Khoreva, A.; Benenson, R.; Hosang, J.; Hein, M.; Schiele, B. Simple Does It: Weakly Supervised Instance and Semantic Seg-mentation. In Proceedings of the IEEE Conference on Computer Vision and Pattern Recognition 2017, Honolulu, HI, USA, 21–26 July 2017; pp. 876–885.

48. Hsu, C.C.; Hsu, K.J.; Tsai, C.C.; Lin, Y.Y.; Sinica, A. Weakly Supervised Instance Segmentation using the Bounding Box Tightness Prior. *Adv. Neural Inf. Process. Syst.* **2019**, *32*, 6582–6593.

49. Redmon, J.; Farhadi, A. YOLOv3: An Incremental Improvement. *arXiv* **2018**, arXiv:1804.02767.

50. Liu, W.; Anguelov, D.; Erhan, D.; Szegedy, C.; Reed, S.; Fu, C.; Berg, A.C. SSD: Single Shot MultiBox Detector. In *European Conference on Computer Vision*; Springer: Cham, Switzerland, 2016; pp. 21–37.

51. Fu, C.; Liu, W.; Ranga, A.; Tyagi, A.; Berg, A.C. DSSD: Deconvolutional Single Shot Detector. *arXiv* **2017**, arXiv:1701.06659.

52. Zhang, S.; Wen, L.; Lei, Z.; Li, S.Z. RefineDet++: Single-Shot Refinement Neural Network for Object Detection. *IEEE Trans. Circuits Syst. Video Technol.* **2021**, *31*, 674–687. [CrossRef]

53. Hinton, G.; Vinyals, O.; Dean, J. Distilling the knowledge in a neural network. *arXiv* **2015**, arXiv:1503.02531.

54. He, K.; Zhang, X.; Ren, S.; Sun, J. Deep residual learning for image recognition. In Proceedings of the IEEE Conference on Computer Vision and Pattern Recognition, Las Vegas, NV, USA, 27–30 June 2016; pp. 770–778.

55. Bearman, A.; Russakovsky, O.; Ferrari, V.; Fei-Fei, L. What's the Point: Semantic Segmentation with Point Supervision. In *European Conference on Computer Vision*; Springer: Amsterdam, The Netherlands, 2016; pp. 549–565.

56. Laradji, I.H.; Rostamzadeh, N.; Pinheiro, P.O.; Vazquez, D.; Schmidt, M. Where are the Blobs: Counting by Localization with Point Supervision. In *Computer Vision—ECCV 2018*; Springer: Munich, Germany, 2018; pp. 560–576.

57. Kim, J.; Kim, J.; Kwak, N. StackNet: Stacking feature maps for Continual learning. In Proceedings of the 2020 IEEE/CVF Conference on Computer Vision and Pattern Recognition Workshops (CVPRW), IEEE, Seattle, WA, USA, 14–19 June 2020; pp. 1–8. [CrossRef]

58. Xu, L.; He, X.-M.; Li, X.X.; Pan, M. A machine-vision inspection system for conveying attitudes of columnar objects in packing processes. *Measurement* **2016**, *87*, 255–273. [CrossRef]

59. Xu, L.; Wang, Y.; Li, R.; Yang, X.; Li, X. A Hybrid Character Recognition Approach Using Fuzzy Logic and Stroke Bayesian Program Learning with Naïve Bayes in Industrial Environments. *IEEE Access* **2020**, *8*, 124767–124782. [CrossRef]
60. Garcia-Garcia, A.; Orts-Escolano, S.; Oprea, S.; Villena-Martinez, V.; Garcia-Rodriguez, J. A review on deep learning tech-niques applied to semantic segmentation. *arXiv* **2017**, arXiv:1704.06857.
61. Branco, P.; Torgo, L.; Ribeiro, R. A Survey of Predictive Modeling on Imbalanced Domains. *ACM Comput. Surv.* **2016**, *49*, 1–50. [CrossRef]
62. Tabernik, D.; Ela, S.; Skvar, J.; Skoaj, D. Segmentation-based deep-learning approach for surface-defect detection. *J. Intell. Manuf.* **2020**, *31*, 759–776. [CrossRef]
63. Goodfellow, I.J.; Pouget-Abadie, J.; Mirza, M.; Xu, B.; Warde-Farley, D.; Ozair, S.; Courville, A.; Bengio, Y. Generative Adversarial Networks. *Adv. Neural Inf. Process. Syst.* **2014**, *3*, 2672–2680. [CrossRef]

forests

Article

Development of a Modality-Invariant Multi-Layer Perceptron to Predict Operational Events in Motor-Manual Willow Felling Operations

Stelian Alexandru Borz 

Department of Forest Engineering, Forest Management Planning and Terrestrial Measurements, Faculty of Silviculture and Forest Engineering, Transilvania University of Brasov, Şirul Beethoven 1, 500123 Brasov, Romania; stelian.borz@unitbv.ro; Tel.: +40-742-042-455

Abstract: Motor-manual operations are commonly implemented in the traditional and short rotation forestry. Deep knowledge of their performance is needed for various strategic, tactical and operational decisions that rely on large amounts of data. To overcome the limitations of traditional analytical methods, Artificial Intelligence (AI) has been lately used to deal with various types of signals and problems to be solved. However, the reliability of AI models depends largely on the quality of the signals and on the sensing modalities used. Multimodal sensing was found to be suitable in developing AI models able to learn time and location-related data dependencies. For many reasons, such as the uncertainty of preserving the sensing location and the inter- and intra-variability of operational conditions and work behavior, the approach is particularly useful for monitoring motor-manual operations. The main aim of this study was to check if the use of acceleration data sensed at two locations on a brush cutter could provide a robust AI model characterized by invariance to data sensing location. As such, a Multi-Layer Perceptron (MLP) with backpropagation was developed and used to learn and classify operational events from bimodally-collected acceleration data. The data needed for training and testing was collected in the central part of Romania. Data collection modalities were treated by fusion in the training dataset, then four single-modality testing datasets were used to check the performance of the model on a binary classification problem. Fine tuning of the regularization parameters (α term) has led to acceptable testing and generalization errors of the model measured as the binary cross-entropy (log loss). Irrespective of the hyperparameters' tunning strategy, the classification accuracy (CA) was found to be very high, in many cases approaching 100%. However, the best models were those characterized by α set at 0.0001 and 0.1, for which the CA in the test datasets ranged from 99.1% to 99.9% and from 99.5% to 99.9%, respectively. Hence, data fusion in the training set was found to be a good strategy to build a robust model, able to deal with data collected by single modalities. As such, the developed MLP model not only removes the problem of sensor placement in such applications, but also automatically classifies the events in the time domain, enabling the integration of data collection, handling and analysis in a simple less resource-demanding workflow, and making it a feasible alternative to the traditional approach to the problem.

Keywords: big data; automation; artificial intelligence; multi-modality; acceleration; classification; events; performance; motor-manual felling; willow; Romania

Citation: Borz, S.A. Development of a Modality-Invariant Multi-Layer Perceptron to Predict Operational Events in Motor-Manual Willow Felling Operations. *Forests* **2021**, *12*, 406. https://doi.org/10.3390/f12040406

Academic Editor: José Borges

Received: 22 February 2021
Accepted: 26 March 2021
Published: 29 March 2021

Publisher's Note: MDPI stays neutral with regard to jurisdictional claims in published maps and institutional affiliations.

1. Introduction

Short-rotation willow crops (SRWC) are seen nowadays as a valuable alternative to produce renewable energy, contributing also to the rural development, job market diversification, carbon sink, biodiversity and diversification of agricultural crops and bio-products. They are commonly established on agricultural lands and share many features with the traditional forestry, in particular the silvicultural practices [1]. Moreover, willow was found to be suitable for other engineering purposes, commonly exhibiting features such as a rapid growth rate, high biomass production, increased coppicing ability and tolerance to high planting densities [2].

When grown to produce biomass as a feedstock for the energy industry, one of the final SRWC production steps consists of harvesting operations. Several fully-mechanized operational systems were developed, tested and are currently in use for large-scale commercial willow harvesting purposes [3,4]. Still, the increased costs associated with harvesting operations [5] are seen as a limiting factor of profitability, which requires optimization [6]. While other operations which are common to SRWC cultivation can be done directly by their owners using equipment and machines of general agricultural purpose, it has been shown that, irrespective of the SRWC scale, the farmers cannot afford to own and operate expensive harvesting equipment [3,5]. In many cases, the lack or the limited availability of such equipment [7] has been tackled by the use of motor-manual means [8,9], which seem to be more adapted to harvesting operations carried out on small and dispersed plots [10], especially in those geographic regions in which the cost of the manual labor is still affordable, and may compensate for lower productivities [11].

Irrespective of the harvesting system used, its optimization requires at least data on production outputs and resources used (i.e., time, fuel, money) [12], and it is quite typical to implement time studies to evaluate its operational performance [13] as a common input for optimization. There are many examples, including those referenced in this paper, of using time-and-motion studies to evaluate the performance of operations. Unfortunately, most of the currently used methods to get time consumption data are resource intensive [14], requiring qualified personnel and specific logistics to collect, process, analyze and interpret it.

In relation to the use of motor-manual equipment to harvest SRWCs, some progress has been made by the use of dataloggers equipped with acceleration sensors to automate field data collection, which was then coupled with Global Positioning System's (GPS) data to infer the operational behavior in such operations [7,10]. In those studies, the analytical procedures of accounting and categorizing the time consumption on operational tasks were done by human intervention, requiring prior knowledge of the process mechanics and sensors' response and, more importantly, a hands-on approach to data processing, summarization and categorization. To eliminate much of that effort, Artificial Intelligence (AI) techniques have been lately used for a number of applications with excellent results in the recognition and classification of operational time, based on signals produced by various sensors, including accelerometers [15–18]. Nevertheless, the applicability of the developed models stays well inside their intended application, mainly due to the labelling outcomes, which are application-specific, and to the sensing modality which could produce contrasting responses in magnitude, calling for the location preservation when using unimodal sensing designs [17,18]. To this end, a sensing modality could be characterized by the type of physical parameter measured by a sensor and by the location of sensing it on a given study object. As such, multimodal sensing may involve the use of at least two sensors measuring the same physical parameter at different locations or the use of at least two sensors measuring different physical parameters at the same location.

Going back to the use of accelerometers to collect signals which may be useful for operational activity recognition in motor-manual felling of willow, unimodal sensing becomes important to save resources, and the use of a single sensor is desirable and affordable. However, there are many possible locations in which an accelerometer could be placed on a given tool, such as a brush cutter, making it likely to get signals characterized by a high variability in magnitude due to the sensing location. In addition, there is a high variability given by the operational conditions themselves, which may change even in the same harvested plot, and by the operational behavior of the workers. In such conditions, the models developed by training an AI algorithm need to have an acceptable generalization ability, in such a way that at least the location of signal collection would become irrelevant for a given classification algorithm. In our knowledge, a robust model able to reliably deal with unimodally sensed acceleration signals, i.e., acceleration sensed at a single location on the tool irrespective of the location, was not studied so far, while developing it from multimodally sensed data could help overcome the limitations stated above; if such a model would prove to be reliable irrespective of the sensing location, then it will contribute to resource saving while easing the data collection, handling and analysis workflow.

The goal of this study was to develop a modality-invariant operational prediction model with application in motor-manual felling of willow by brush cutters. The problem was approached by the means of training and testing a Multi-Layer Perceptron (MLP) with backpropagation on acceleration signals. Acknowledging that there could be many other approaches to the problem, the choice of the MLP as a technique to be used, as well as of the acceleration data as input signals, was mainly based on the author's experience with the MLP algorithms and the availability of acceleration signal data; in addition, the choice was guided by the findings of recent work, repeatedly showing excellent results in task recognition applications when using acceleration signals as inputs in the AI algorithms. By a modality-invariant model we are referring to a model able to acceptably generalize from any acceleration signal collected by the same sensor type, anywhere of an observed tool, while preserving the highest possible classification performance and the lowest possible generalization error.

2. Materials and Methods

2.1. Data Collection

2.1.1. Study Location and Crop Layout

The data used in this study were collected in the center of Romania, from three locations (Table 1), where motor-manual willow felling operations were observed in the Spring of 2017. The plots taken into study were located in an intra-mountainous depression at an altitude of ca. 600 m a.s.l. The climate in the area is characterized by a strong continentalism with warm summers and cold winters. All the plots (Poian 1, Poian 2 and Belani, Covasna County, Romania) were planted according to the European willow planting layout, in which the planting is done in twin rows distanced at 75 cm, and each twin row is distanced at 1.5 m from the next one; commonly, the distance between the cuttings used for planting is of 60 cm [19].

Table 1. Locations of data collection and the summarized description of the datasets.

Location	Dataset [a]	Size [b] [s]	Date of Collection	Coordinates
Poian 1	TRAIN_E	18,377	04/11/2017	46°04.373′ N–26°10.924′ E
	TRAIN_S	18,377	04/11/2017	46°04.373′ N–26°10.924′ E
Poian 2	TEST_E1	26,493	02/27/2017	46°04.354′ N–26°10.904′ E
	TEST_S1	22,885	02/27/2017	46°04.354′ N–26°10.904′ E
Belani	TEST_E2	11,859	03/02/2017	46°03.362′ N–26°11.208′ E
	TEST_S2	9285	03/02/2017	46°03.362′ N–26°11.208′ E
TOTAL	-	107,276	-	-

Note: [a] TRAIN means that the dataset was used in the training phase of the MLP, TEST means that the dataset was used in the testing phase of the MLP, E means that the data was collected by datalogger placement on the tool's engine, S means that the data was collected by the datalogger placement on the tool's transmission shaft, 1 means that the dataset was the first of the same modality class, 2 means that the dataset was the second of the same modality class; [b] size refers to the number of one-second sampled observations retained in the training and testing datasets.

Willow crops are becoming a common land use feature in the landscape of the study area, though they are typically established on small and dispersed plots whose previous use was agricultural [20]. In addition to the size and dispersion of the plots, the cultivation practice in the area is strongly influenced by the available technology for planting, cutback and harvesting operations, which are partly mechanized [7,9,10,19–21]; many of them, such as harvesting, are relying to a great extent on motor-manual operations which are usually done by the use of brush cutters [7,9,10]. This situation is often leading to practicing rotations of 2–3 years, and typical for the area is that motor-manual willow felling operations are done in the early spring. Most of the operations related to willow cultivation in the area are done by employing local people on a daily basis.

2.1.2. Tool Description, Work Organization and Relevant Process Mechanics

The tool used for felling was a brush cutter made by Husqvarna (Model 545 RX, Husqvarna AB, Stockholm, Sweden), featuring an engine output of 2.1 kW at 9000 rpm and a transmission shaft that enables the connection and power transfer between the engine and the cutting device (Figure 1); tools from this class are assumed to produce a noise level of 100 db(A) [22]. Harvesting work is typically done by motor-manual felling followed by manual bunching, transportation and chipping at a biomass terminal, or by bunching and chipping on site [23]. Brushcutters are commonly used in the study area to motor-manually fell the willow, being tools that can be adapted easily to a variety of jobs, simply by changing their active cutting devices [22,24]; when used for willow felling, they are usually equipped with steel saw blades (discs).

Figure 1. Work organization, tool description and instrumentation of data collection. Legend: (**a**)—the typical work organization (1—bunch of stems to be felled, 2—bunches of felled stems, 3—wooden stick used to direct the felling, 4—throttling control, 5—engine, 6—transmission shaft, 7—steel blade, 8—triaxial accelerometer, 9—helmet equipped with a sound pressure level datalogger), (**b**)—the general layout of the felling operations (example from Poian 1 location, close to finishing the felling work in a plot), (**c**)—detail of the acceleration datalogger (8) placed on the shaft, (**d**)—details on the instrumentation used to collect the data (8—triaxial accelerometers, 9—sound pressure level datalogger).

The felling work is commonly done by two workers (Figure 1) of which one is the brush cutter operator who is in charge of the mechanical felling tasks and tool maintenance, and the other one assists the felling by a wooden stick [7,9,10,19].

While being rather simple, the organization of felling work is influenced by the capability limits of the used tools, layout of the crops and weather conditions [7,19], and it needs to be done with much attention and caution to ensure the safety of both workers. As such, felling direction is commonly adopted toward the exterior of the crop, felling work is progressing on a single row (one of the twins), and in such cases in which the length of the crop is very long, transversal corridors are often practiced to shorten the distances covered per turn, to be able to refuel and maintain the tool [19]. The assistant needs to place himself at a considerable distance behind the feller and he interacts with the stems to be felled only by the wooden stick, while the feller needs to be able to coordinate and control his motions on very short trajectories. The relevant work elements which may occur in such operations are the effective on-row felling, moving at the headlands or on a transversal corridor to approach a new willow row at the opposite side of the crop, maintenance and refueling, rest and meal breaks, as well as other kind of delays [7,9]. The distinctive feature of these operations is that, excepting the felling, the rest of work elements are typically characterized by engine non-use. Therefore, monitoring the engine working time makes it possible to accurately monitor the main work time consumption, which stands for a category of time in which the direct transformation of the work object takes place [13], and which is also useful and important to account for the fuel intake as specific to motor-manual operations [24,25].

Looking at a finer scale, however, the felling consists of worker's advancement on the row with various engine running regimes, combined with movements of the active cutting device of the tool toward outside and inside the crop to make the cuts, which are likely to produce variability in the responses given by the use of various sensors. Another relevant issue is the placement location of the dataloggers. For instance, accelerometer dataloggers could be placed at different locations on the engine block, as well as at different locations of the transmission shaft, therefore making it possible to receive different magnitudes of the acceleration during engine use; however, when the engine is switched off, the responses collected by accelerometers placed at different locations of the tool could be similar.

In the study area, felling work is always done by workers having an extensive experience in SRWC felling operations, gained on already more than a decade of SRWC management in Romania. Field data collection, which was done in 2017, was based on the informed consent of the observed workers and of the SRWCs' owner. They were informed about the intended use of the data and agreed to be observed when performing their jobs.

2.1.3. Instrumentation

Two datalogger types were used to collect the field data used in this study. Acceleration response, as the main data stream used, was measured and recorded using two Extech® VB300 triaxial dataloggers (Extech Instruments, FLIR Commercial Systems Inc., Nashua, NH, USA). Irrespective of the study location, the acceleration dataloggers were set by the use of the dedicated software to collect data in the motion detection mode (threshold of 1 g), at a sampling rate of one second. One of them was placed on the engine's block and the other one was placed on the transmission shaft, at locations that were chosen carefully in such a way that they would not interfere with work safety. Both dataloggers were reinforced on the tool using highly-resistant plastic straps and were checked for holding before running the experiments and during the work breaks. An Extech® 407,760 sound pressure level datalogger (Extech Instruments, FLIR Commercial Systems Inc., Nashua, NH, USA) was used to collect additional data needed for labelling purposes. It was set to continuously collect the sound pressure level on the dB(A) scale, at a sampling rate of one second, and it was placed on the helmet worn by the brush cutter operator. The main technical features of the used dataloggers are available on the producing company's website [26,27]. Figure 1 shows the approach used to equip the tool and the worker with the used dataloggers.

2.1.4. Datasets

Six acceleration datasets were collected in the three locations taken into study (Table 1) and the intention was to get for each of them a time overlapping sound pressure level dataset. However, due to a battery malfunction, sound pressure level data was lost in the case of Belani location. By the construction and setup of the acceleration dataloggers, the data is collected and stored as discrete triaxial (X, Y and Z) time-labelled responses; they are further summarized in the form of vector magnitudes, also known as the Euclidian Norm (EN), which is some sort of data fusion [28], and which is enabled by the dedicated software. The EN, which is named by the dedicated software under the generic term of "vector sum", may be written as in Equation (1) and it allows for a first normalization of the data, making it invariant to the axis movement. In fact, raw acceleration signals contain movement, gravity and noise components [29], while the instruments used to collect them respond well to vibration, a property which was used in this study.

$$A_i = \sqrt{x_i^2 + y_i^2 + z_i^2} \tag{1}$$

where A_i is a discrete value, in the form of Euclidian Norm (vector magnitude, vector sum), computed for a given sampling rate (adopted to one second in this study), and x_i, y_i and z_i are the accelerometer's raw responses on the axis X, Y and Z, respectively, for the observation i.

Sound pressure level data was collected and outputted in a similar way, being time and date labelled, and showing the sound pressure level measured in dB(A) at a sampling rate set at one second. In addition, both dataloggers can output data in computer-friendly formats such as the Microsoft Excel® (Microsoft, Redmond, WA, USA). CSV files, and both of them provide data ID's and some summary statistics placed at the beginning of each file. Figures A1–A5 are showing the patterns of A_i in the datasets used for training and testing of the MLP, emphasizing the amplitude and magnitude differences due to the location of the datalogger and engine working regimes.

2.2. Data Preprocessing Workflow

2.2.1. Data Pairing, Segmentation and Labelling

To ease the effort of labelling the training data, as well as to compare the multimodal responses collected by the two acceleration dataloggers placed on the same tool, data pairing procedures were applied to the first two datasets (TRAIN_E and TRAIN_S) based on their time labels. This procedure was necessary to be able to label both datasets at once. Data pairing was done in Microsoft Excel®, and it accounted for those observations which were present in both datasets and shared the same time label, an issue which was computationally approached, assessed and solved using logical functions. For example, if an observation from a given training dataset did not have had a corresponding time-labeled observation in the other training dataset, then it was deleted. This process was also run vice versa, until reaching a double set of observations sharing their time labels.

Labelling of the training datasets was done by considering the responses recorded by the acceleration and sound pressure level dataloggers (Figure 2), based on known experience on their responses in terms of magnitude. Two states were documented by labelling and segmentation, namely the engine running (labelled in the database by the string code ON) and the engine turned off (labelled in the database by the string code OFF).

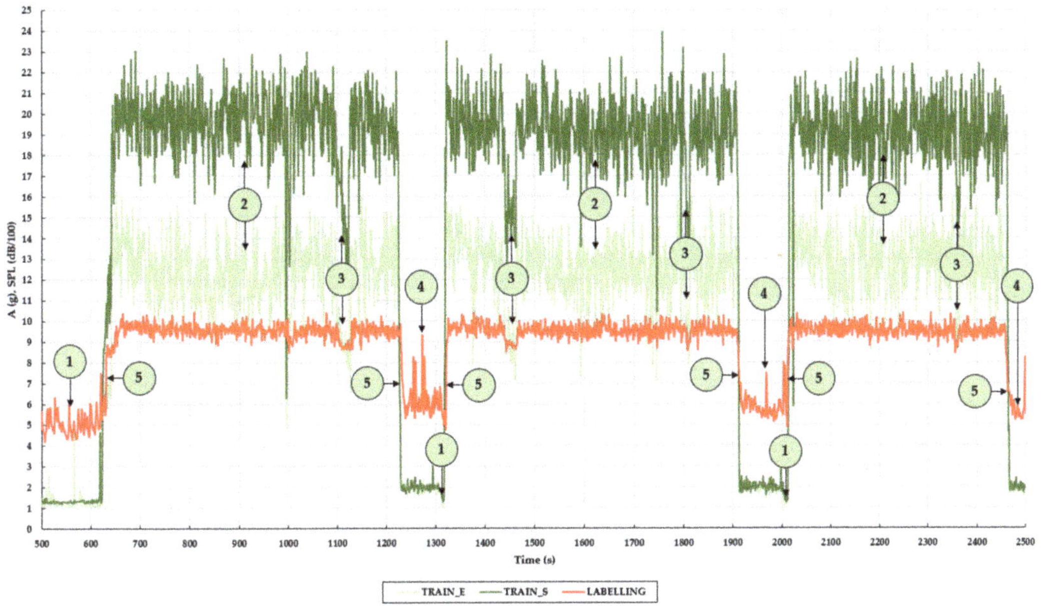

Figure 2. A sample from the acceleration and sound pressure level datasets used jointly to label the data. Note: for convenience, the sound pressure level data was downscaled by a factor of 100 to help in data comparison and labelling tasks. Legend: TRAIN_E stands for the training dataset collected by the placement of acceleration datalogger on the tool's engine, TRAIN_S stands for the training dataset collected by the placement of acceleration datalogger on the tool's transmission shaft, LABELLING stands for the sound pressure level data downscaled by a factor of 100; Events: 1—engine off and no movement of the worker (labelled as OFF), 2—engine on and felling (labelled as ON), 3—engine on and not felling (labelled as ON), 4—engine off and movement of the worker (labelled OFF), 5—data segments which were transient (inter-class variability) between the two engine states (labelled as ON).

For instance, sound pressure levels close to those described by the manufacturer for the operation of the tool (ca. 100 dB(A)) have pointed that the engine was on and throttled, therefore indicating that the worker was engaged in the effective felling operations. Drops in the magnitude of the sound pressure level (as shown in Figure 2 by the data labelled with 3), were considered to be the events in which the engine was on but no felling was done (idle running); these events were labelled as ON. Moreover, acceleration responses in the range of 1.1–3.0 g were compared to the data on sound pressure level, generally leading to their classification as engine OFF events. Transient events (Figure 2, data labelled by 5) were included in the engine working category as well. However, due to the acceleration data collection mode (motion detection) and pairing procedures used, which have led to some missing data, the sound pressure level dataset was paired by doing some adaptations in the time domain such as removing some data or moving some data segments to pair them with the acceleration data. This was done for approximately 10% of the joint dataset, then the patterns generated by the magnitude of acceleration data were used for further labelling.

Based on the experience gained during the labelling and segmentation tasks done on the training dataset, the distributions of data in specific patterns were used as a condition to label the data in the rest of the datasets, which were used for MLP testing (Figures A2–A5). Prior to the labelling and segmentation tasks, these datasets were preserved to their original number of observations, as they were outputted by the acceleration dataloggers. Therefore, the datasets shown in Figures A1–A5 contained the final number of one-second observations as described in Table 1, and each observation contained in them was the Euclidian Norm computed according to the Equation (1).

2.2.2. Fusion of the Training Datasets

All the datasets used in this study were subjected to early data fusion by the computation of Euclidian Norm. However, to simultaneously capture both the local dependencies over time and the spatial dependencies over modalities of collection, the approach was similar to that described in [30], and consisted of fusing the training datasets by a procedure referred as vertical stacking [28]. In particular, it was assumed that a more accurate data representation in the trained model, which could be achieved by the inclusion of spatial dependencies over the modalities of collection, could be important for the evaluation of datasets coming from other experiments using a single modality for field data collection, enhancing the trained model's recognition capacity. In addition, the procedure was assumed to improve the data representation in the trained model by actually doubling the size of the training dataset.

Procedurally, data fusion followed a simple procedure, by keeping the dataset collected on the engine as it was, and by merging the dataset collected on the transmission shaft at the end of the first dataset, resulting in the fused dataset (Figure A1). Following data merging, the ID's of the observations were updated, and the resulting data vector was used as input for data normalization.

2.2.3. Data Normalization

Data normalization is commonly done by transforming the original data, and it aims at giving all the attributes an equal weight; in MLP applications with backpropagation it also helps in speeding up the learning process [31]. A min-max normalization procedure was used in this study, according to the Equation (2), which performs a linear transformation of the data, outputting values in a new range (0, 1), while preserving the relationships among the original data values [31]. Although there are many other procedures that may be used to scale the data, the choice of this normalization procedure was based on its simplicity and ease of use.

$$An_{ij} = (A_{ij} - Amin_j)/(Amax_j - Amin_j) \tag{2}$$

where An_{ij} is the normalized value of observation i coming from the dataset j (An_{ij} can takes values between 0 and 1, inclusively), A_{ij} is the Euclidian Norm of the observation i coming from the dataset j, $Amin_j$ is the minimum value of the Euclidian Norm coming from the dataset j, $Amax_j$ is the maximum value of the Euclidian Norm coming from the dataset j.

The use of Equation (2) required the computation of the minimum and maximum values of A_i in each dataset j ($j = 5$), then it was applied to all observations from each dataset, using for this purpose the Microsoft Excel® software. The transformed data was saved as new datasets, then it was used for training and testing purposes of the MLP model.

2.3. Setup of the MLP

2.3.1. Software Used and General Architecture of the MLP

The software used for training and testing of the MLP was the freely-available open-source Orange Visual Programming Software (version 3.27.1) [32], which holds functionalities of implementing a multi-layer perceptron with backpropagation. All the training and testing tasks were run on a computer architecture that included the following features: system type—Alienware 17 R3, processor—Intel® Core™ i7-6700 HQ CPU, 2.60 GHz, 2592 MHz, 4 cores, 8 Logical Processors, installed physical memory (RAM)—16 GB, operating system—Microsoft Windows 10 Home.

The size of the MLP was set in advance of the training and testing tasks to the highest values of depth and width enabled by the software used, based on the author's experience, practical recommendations formulated by [33], and recent results showing the effect of MLP's architecture on the classification performance for similar equipment [16]. Three hidden layers (depth) of 100 units each (width, as the number of neurons) were set for the MLP's architecture, and the number of iterations was set at 1,000,000. Training and scoring were done by cross-validation using a stratified approach and a number of folds set at 20. The recommendations of [33], as well as the information available in the recent literature, were used to choose the activation function and the optimization algorithm. One of the most popular activation functions in the rectified linear unit function (ReLu), which is supposed to provide high performances in solving complex, nonlinear problems [34,35], and it was chosen for this study. In simple words, an activation function takes the weighted inputs of a node (neuron), adds a bias and based on its result decides whether or not that node should be activated (fired); typically, ReLu makes such decisions when the results are positive. The optimization algorithm chosen for the MLP architecture was the stochastic gradient descent-based optimizer (Adam), which is one of the recently developed and used solvers due to its low training costs [36].

2.3.2. Tunning and Error Metric Used to Evaluate the Generalization Ability

A manual tuning approach was taken to check the training and testing performance of the MLP, and it aimed at altering the α parameter of the regularization term (L2 penalty regularization), by a trial-and-error approach. By doing so, the intention was to check what regularization strategy would reduce the generalization error [33] in combination with the architecture of the MLP and hyperparameters already set as described in Section 2.3.1. In MPL applications, the regularization term helps in avoiding overfitting by penalizing weights with large magnitudes; α is a parameter of the regularization term, whose increased values may fix high variance while decreased values may fix high bias [33,37]. Values of the α parameter were set successively at 0.0001, 0.001, 0.01, 0.1, 1 and 10, then MLPs were trained and tested over all four testing datasets, accounting each time for the training and generalization error. The error metric chosen for the evaluation of generalization ability was the binary cross-entropy (Equation (3)), which is commonly used in binary classification problems. A detailed worked example can be found at [38]. Its calculation is enabled by the used software and it works based on predicted probabilities assigned to the observations.

$$H_p(q) = -\frac{1}{N} \sum_{i=1}^{N} l_i \times \log(p(l_i)) + (1 - l_i) \times \log(1 - p(l_i)) \tag{3}$$

where $H_p(q)$ is the binary cross-entropy (log loss) function, N is the number of observations in a given dataset, l_i is the label of a given observation i ($i = 0, 1$), and $p(l_i)$ is the predicted probability of an observation being ON for all the observations (N). Note: the label ON received the value of 1 and the label OFF received the value of 0.

For instance, if the label of an observation is ON, therefore $l_i = 1$, then Equation (3) will add $\log(p(l_i))$ to the loss, which is the probability of that instance of being ON; if the label of an observation is OFF, therefore $l_i = 0$, then it will add $\log(1 - p(l_i))$ to the loss, which is the probability of that instance of being OFF. Training and testing results of the binary cross-entropy function were used in conjunction to choose the best model in terms of training and testing generalization capacity. Since training and testing was run on a number of 5 models (1 for training and 4 for testing), the values of binary cross-entropy were plotted against those of the tuned α parameter. Then, minimum and maximum values of each repetition done for each α value were computed, and the range found at the minimum value was used as a criterion to keep the best performing models.

2.3.3. Classification Performance Metrics

In addition to the log loss function, the software used for training and testing enables the computation of the training and testing time, area under curve (AUC), classification accuracy (CA), F1 metric, precision (PREC), recall (REC) and specificity (SPEC). The meaning and the possibility of use for these metrics is comprehensibly described in papers such as [39,40], therefore their complete definitions and formulae are not given herein. While all of these metrics were computed at the class (ON, OFF) and overall (dataset) level, in both, training and testing phase, the focus was on the classification accuracy (CA) and recall (REC) metrics; in binary classification problems, the first one stands for the number of correctly classified true positives and negatives of the total number of observations in a dataset, and the second one stands for the number of true positives classified as such of the total number of positives in a given dataset [39,40].

2.4. Evaluation

The best performing models in terms of error rate minimization and generalization ability were retained as final and selected for an additional evaluation. The additional evaluation consisted of a more detailed description of the misclassifications in the training and testing datasets as well as of developing plots to depict the predicted probabilities of the data. Misclassification issues were addressed by exporting the outputs of the training and testing phases into Microsoft Excel® files, followed by the application of logical functions to extract the number of correctly classified datapoints (true positives—TP and true negatives—TN), false positives (FP) and false negatives (FN), based on a paired comparison of the ground truth against the predictions made on the training and testing datasets. This new data was summarized in the form of tables and plotted as graphs in the time domain, in the form of Euclidian Norm (Equation (1)) against misclassifications. Probability plots were developed by mapping the original data on Euclidian Norms (Equation (1)) against their predicted probability of falling in either the ON or OFF classes.

3. Results

3.1. Description of the Labelled Datasets

The datasets used in this study accounted for a cumulated size of 107,276 s (ca. 30 h, Table 2) of which the fused dataset used for training (TRAIN) represented ca. 34%. Datasets used for testing accounted (in their order shown in Table 2), for ca. 25%, 21%, 11% and 9%, respectively.

Table 2. Statistics of the used datasets.

Location	Dataset [a]	Size [b] [s]	Class Size [s]		Class Share [%]	
			ON	OFF	ON	OFF
	TRAIN_E	18,377	13,980	4487	76.07	23.93
Poian 1	TRAIN_S	18,377	13,980	4487	76.07	23.93
	TRAIN	36,754	27,960	8794	76.07	23.93
Poian 2	TEST_E1	26,493	20,579	5914	77.68	22.32
	TEST_S1	22,885	20,404	2481	89.16	10.84
Belani	TEST_E2	11,859	6806	5053	57.39	42.61
	TEST_S2	9285	5886	3399	63.39	36.61
TOTAL	-	107,276 [c]	81,635 [c]	25,641 [c]	-	-

Note: [a] meaning is similar to that from Table 1; [b] size refers to the number of one-second sampled observations retained in the training and testing datasets; [c] calculated on the basis of the fused training dataset (TRAIN) and testing (TEST_E1, TEST_S1, TEST_E2, TEST_S2) datasets.

Excepting the dataset TEST_E2, data distribution on classes was found to preserve different degrees of class imbalance. Irrespective of the dataset, more than 57% of the data was labelled as ON, a class that accounted for ca. 90% of the TEST_S1 dataset's size. While from the perspective of developing robust MLPs, this is a common issue to be solved [28], from an operational point of view this kind of data distributions emulates very well the practice of motor-manual willow felling, where the effective felling itself dominates.

3.2. Model Selection and Classification Performance

Values returned by the binary cross-entropy error as a function of the regularization parameter's tunning are shown in Figure 3. Irrespective of the tunning strategy used, or the dataset in question, up to a value of α set at 1, the training and generalization errors were found to be less than 0.074 (7.4%, TEST_E2), showing, in general, a good generalization ability of the trained model. For values of α set from 0.0001 to 0.1, both the training

(TRAIN) and generalization (TEST_E1, TEST_S1, TEST_E2, TEST_S2) errors were low, with the lowest ones found for $\alpha = 0.0001$ and $\alpha = 0.1$. Beyond this threshold ($\alpha = 0.1$) the error started to noticeably increase at least for one of the testing datasets (Figure 3, TEST_E2). The lowest differences in terms of errors were found in the case of $\alpha = 0.0001$ and $\alpha = 0.1$ irrespective of the values compared (training and testing data or just testing data). For instance, when setting α at 0.0001, the value of the log loss in the case of training data was of 0.005 (0.5%) and it corresponded to a maximum value of 0.036 (3.6%) found in the TEST_E2 dataset. The figures were similar for $\alpha = 0.1$, for which the error found for the training data was of 0.006 (0.6%), which corresponded to a maximum value of 0.037 (3.7%), found in the same testing dataset (TEST_E2). In term of errors, TRAIN and TEST_E1 datasets returned similar values for the range set for α between 0.0001 and 0.1. For the same range set for α, TEST_S1 and TEST_S2 datasets have returned a similar pattern in terms of errors.

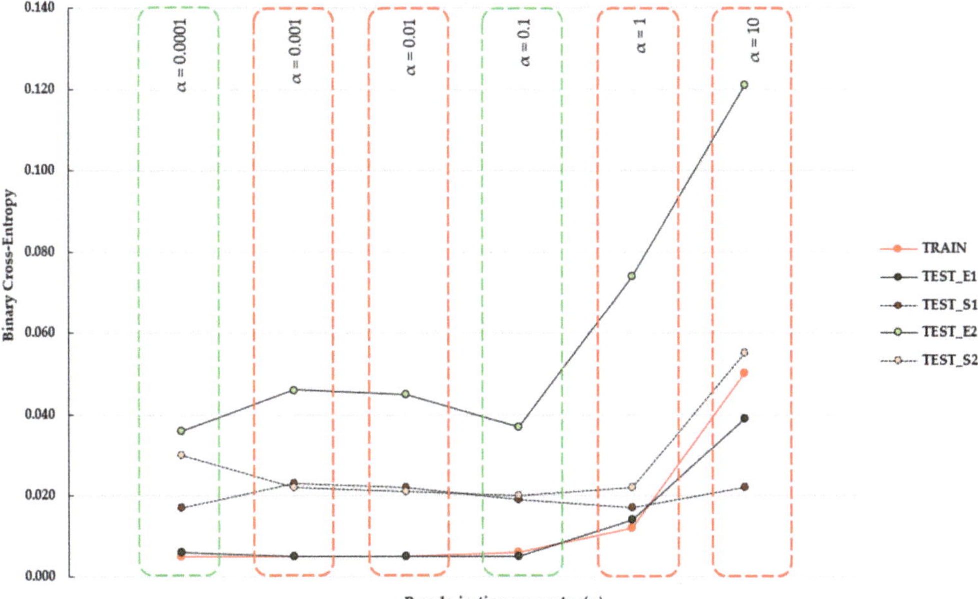

Figure 3. Log loss (binary cross-entropy) of the training and testing data as a function of the regularization parameter term (α). Legend: TRAIN—fused training dataset, TEST_E1 and TEST_E2—testing datasets collected on the tool's engine, TEST_S1 and TEST_S2—testing datasets collected on the tools' transmission shaft. Note: Values shown are computed by using Equation (3), based on the normalized data (Equation (2)); models retained for further analysis are bordered by green dashed lines.

Figure 4 is showing a comparison of the classification accuracy (CA) metric for the training and testing datasets, reflecting the effect that the value set for the α term had on this metric. In the training phase, all of the attempts to tune the regularization parameter term (α) returned very high classification accuracies. However, the classification accuracy of the training phase was preserved at the highest values (0.999, 99.9%) only in the range set for α between 0.0001 to 0.01, and it started to decrease as the regularization parameter approached the value set at 10. Moreover, the classification performance of the testing datasets was preserved to its highest values for α set in between 0.0001 and 0.01. However, the models selected for further assessment were those having this parameter set at 0.0001 and 0.1, based on the results returned by the log loss function, which are shown in Figure 3.

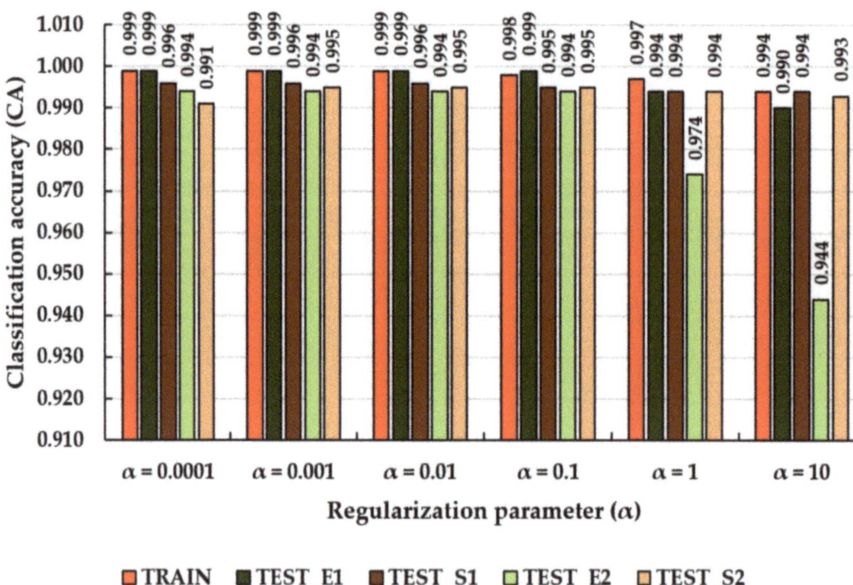

Figure 4. Classification accuracy scored at the training (TRAIN) and testing (TEST_E1, TEST_S1, TEST_E2, TEST_S2) phases, as a function of the value of the regularization parameter (α).

Tables A1–A3 are showing the detailed classification performance metrics at the overall (dataset) level, as well as on classes (ON, OFF). Irrespective of the class, the minimum values of classification accuracy (CA) metric were of 0.944 (94%), indicating a high share of correct predictions for the worst prediction case. The minimum values of the F1 metric, which stands for the harmonic mean of precision (PREC) and recall (REC), were of 0.944 (94%), 0.948 (95%) and 0.938 (94%) for the overall, ON and OFF data. In the same order, the minimum values of classification precision (PREC) and recall (REC) were of 0.950 (95%), 0.988 (99%), 0.884 (88%) and of 0.903 (90%), 0.903 (90%) and 0.988 (99%), respectively, where precision stands for the fraction of true positives from the total of positives (TP and FP) and recall stands for the fraction of correctly classified true positives from the total positives. Accordingly, these metrics returned high values for the worst prediction cases, with evident differences as a result of the regularization parameter tunning. Training time of the MLP varied in between ca. 261 and 482 s, and it was of ca. 261 and 443 s for the models trained for α = 0.0001 and 0.1, respectively. A more detailed comparison of the classification accuracy for the former models is given in Table 3, showing some of the highest values of the CA among the set of regularization terms used.

Table 3. Classification accuracy of the selected models.

Class	TRAIN		TEST_E1		TEST_S1		TEST_E2		TEST_S2	
	α = 0.0001	α = 0.1	α = 0.0001	α = 0.1	α = 0.0001	α = 0.1	α = 0.0001	α = 0.1	α = 0.0001	α = 0.1
ON	0.999	0.998	0.999	0.999	0.996	0.995	0.994	0.994	0.991	0.995
OFF	0.999	0.998	0.999	0.999	0.996	0.995	0.994	0.994	0.991	0.995
OVERALL	0.999	0.998	0.999	0.999	0.996	0.995	0.994	0.994	0.991	0.995

Excepting the TEST_S2 dataset, no significant differences were found in terms of classification accuracy as an effect of tuning the regularization parameter. In addition, classification performance was found to be very high in the case of most of the testing datasets, and in terms of classification accuracy (CA), its values ranged from 99.1% (TEST_S2, $\alpha = 0.0001$) to 99.9% (TEST_E1), proving a high generalization ability of the trained models.

3.3. Missclassification and Probability Plots

The correctly classified observations in the training dataset (TRAIN, Table 4) were close in terms of relative frequency. In absolute numbers, however, the model using a regularization term set at 0.1 misclassified more (25 observations) compared to that of α set at 0.0001. When checked for the testing datasets (TEST_E1, TEST_S1, TEST_E2, Table 4), the number of misclassifications was relatively tied in relation to the regularization parameter term used, excepting the last testing dataset (TEST_S2, Table 4) which returned a better performance for α set at 0.1. Figure 5 is giving a representation of misclassified data points in the training and testing datasets for a regularization parameter term set at 0.0001. Irrespective of the dataset, the misclassified datapoints shared a common feature, namely their location in terms of magnitude in the transient data segments characterizing inter-class variability. These segments were those mostly identified for operational events such as turning on or off the tool's engine, and which were formally included in the ON class. However, no attempts were taken to separate another class given the results obtained on classification performance and error metrics (Figures 3 and 4, Tables 3 and 4), which were considered to be acceptable. In addition, the number of observations which were found to be misclassified due to their belonging to these events is typically low in applications such as that studied herein (Table 4).

Table 4. Descriptive statistics of misclassifications.

Regularization Term	Dataset [a]	Size [b] [s]	Correctly Classified		Misclassified as			
					False Positives		False Negatives	
			N	Share (%)	N	Share (%)	N	Share (%)
$\alpha = 0.0001$	TRAIN	36,754	36,711	99.88	28	0.08	15	0.04
	TEST_E1	26,493	26,465	99.89	2	0.01	26	0.10
	TEST_S1	22,885	22,795	99.61	40	0.18	50	0.22
	TEST_E2	11,859	11,788	99.40	15	0.13	56	0.47
	TEST_S2	9285	9202	99.11	73	0.78	10	0.11
	TOTAL	107,276	-	-	-	-	-	-
$\alpha = 0.1$	TRAIN	36,754	36,686	99.82	23	0.06	45	0.12
	TEST_E1	26,493	26,469	99.90	12	0.05	12	0.05
	TEST_S1	22,885	22,787	99.57	21	0.09	77	0.34
	TEST_E2	11,859	11,784	99.37	8	0.07	67	0.56
	TEST_S2	9285	9237	99.48	26	0.28	22	0.24
	TOTAL	107,276	-	-	-	-	-	-

Note: [a] meaning is similar to that from Table 1; [b] size refers to the number of one-second sampled observations retained in the training and testing datasets.

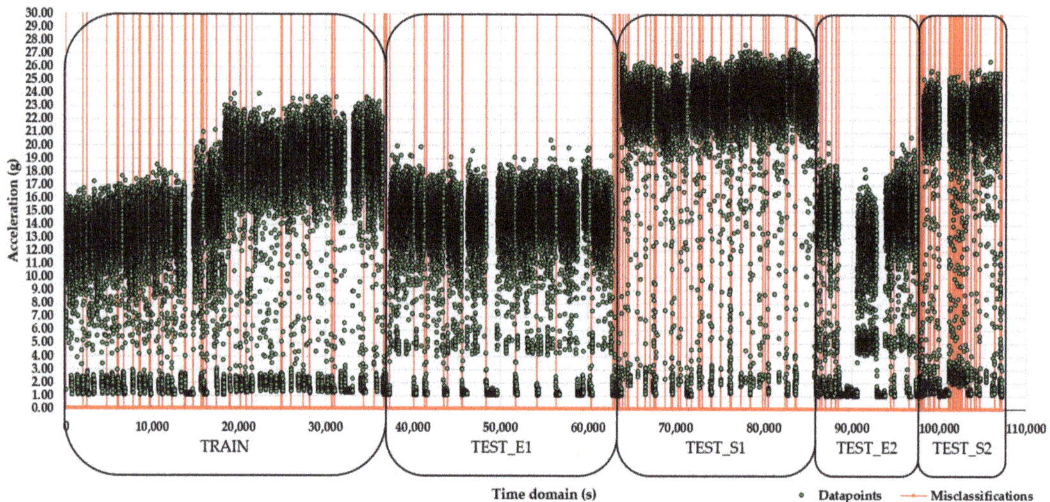

Figure 5. Distribution of misclassifications in the training (TRAIN) and testing (TEST_E1, TEST_S1, TEST_E2, TEST_S2) datasets for a regularization term (α) set at 0.0001. Legend: green points stand for the Euclidian Norm (EN); red lines stand for correctly classified datapoints when drawn horizontally at an EN of 0, and for misclassified datapoints when drawn vertically. Note: for convenience the datasets were merged in their order of analysis.

Figure 6 is showing a selection of predicted probability plots in a comparative approach. The data shown stands for the dataset used for training (TRAIN), as well as for datasets TEST_E2 and TEST_S1 used for testing. It compares the predicted probabilities of the datapoints from the abovementioned datasets of belonging to the classes ON and OFF, respectively, against the values of those datapoints computed according to the Equation (1).

For a value of the regularization parameter term set at 0.0001, the minimum values of the Euclidian Norm found to be predicted as ON were close to 3 g in all the datasets (detailed statistics are not shown herein, and Figure 6 shows only a selection of predicted probability plots). Accordingly, the maximum values of the Euclidian Norm found to be predicted as OFF were close to 3 g in most of the datasets. In comparison, for a value of the regularization parameter term set at 0.1, the minimum and maximum threshold values (as described above) of the Euclidian Norm were close to 4 g in most of the datasets. These statistics can be followed quite easy in Figure 6, where in the left panels (α = 0.0001) the predicted probability data is split for a probability set at 0.5 by a value close to 3. Accordingly, the left panels of the figure (α = 0. 1) split the predicted probability data, at the same probability threshold (0.5), by a value of the Euclidian Norm close to 4 g.

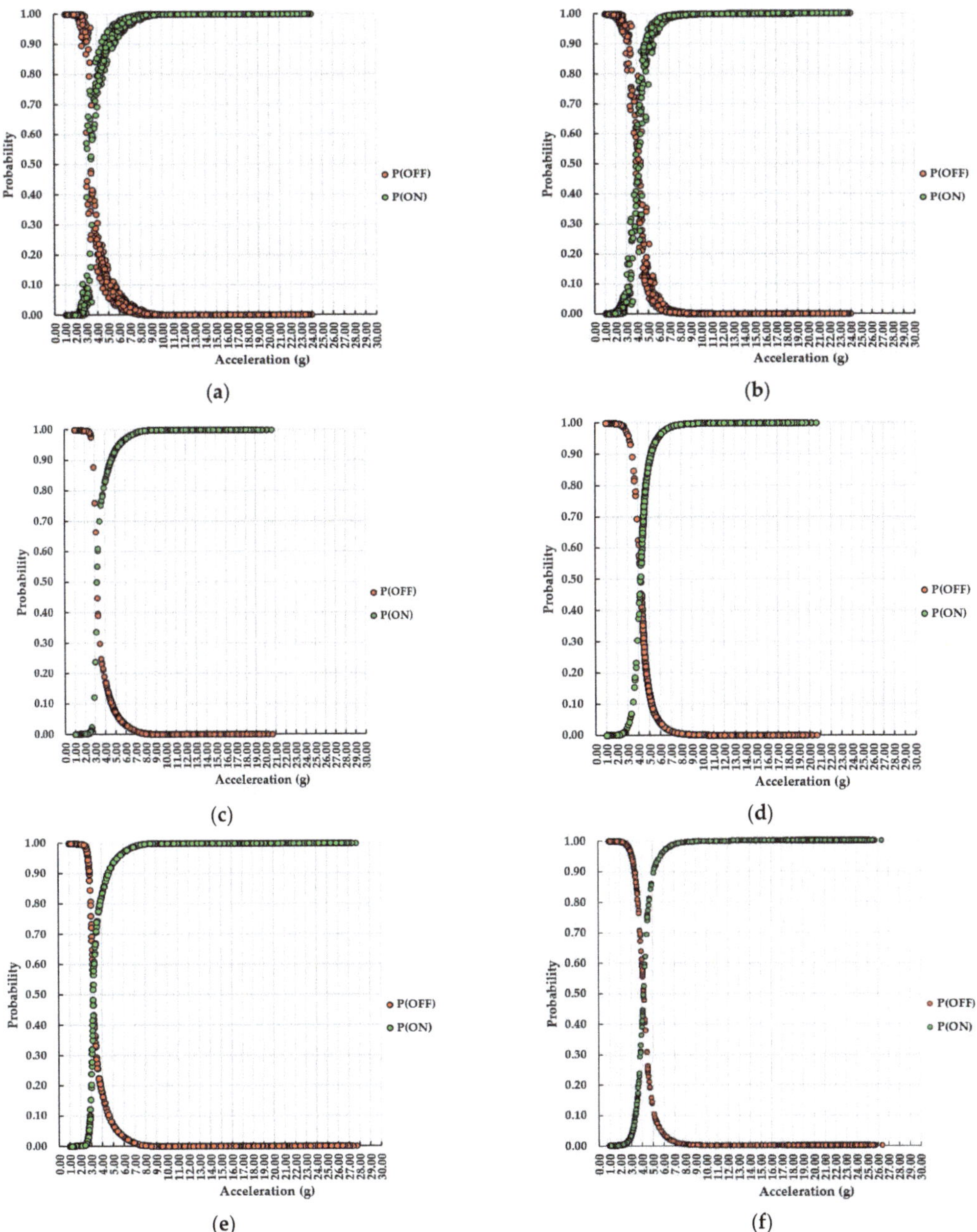

Figure 6. Selected plots showing the predicted classification probability. Legend: P(OFF), shown in red, stands for the predicted probabilities of the datapoints being OFF and P(ON), shown in green, stands for the predicted probabilities of the datapoints being ON. Note: (**a**,**b**)—predicted classification probability for the TRAIN dataset; (**c**,**d**)—predicted classification probability for the TEST_E2 dataset; (**e**,**f**)—predicted classification probability for the TEST_S1 dataset; left figure panels show the data for $\alpha = 0.0001$ and the right figure panels show the data for $\alpha = 0.1$.

4. Discussion

Monitoring the operational performance is one of the common ways to get the data needed for sound decisions on running and improving the way that various businesses work. It is already a fact that many manufacturing industries are currently collecting sensor-based data to improve their operations and to respond by informed decisions to various production anomalies and problems [41], enabling them to be more competitive, responsive and resilient. In forest and SRWC operations, getting monitoring data was traditionally based on observing workers, tools and machines by time-and-motion studies [12–14], which have evolved from pen-and-paper to various sensing-based techniques; the latter are often implementing an external rather than a built-in sensor system e.g., [7,10,15,17,18,20,23,42–46] mainly due to their purpose for collecting such data, which was often purely scientific. Although the modern machines may incorporate production monitoring systems that may work in real time, there are still few options to collect and handle such data for hand-operated tools. Recent studies have shown that the acceleration sensors may be successfully used to collect long term operational monitoring data e.g., [7,10,16,17,23,43,45,47] including by the use of platforms such as the smartphones [15]. In many cases, however, such data comes as modality-variant, unannotated sets, requiring significant resources to process and analyze it [7,10,47]. In this regard, the merit of this study is that it developed data collection invariant models able to automatically and accurately classify, analyze and interpret signals collected by triaxial accelerometers, enabling the possibility to extend their applicability to new coming datasets. As such, the implementation of MLP can serve to automatically classify new data recorded by triaxial accelerometers, irrespective of the datalogger placement on the tool.

One of the relevant issues for discussion is the sensing modality itself. Dealing with sensing modalities is not a new approach brought by this paper as it has been discussed [28] and used by other studies making use of sensors to measure various physical variables [15,17,30,47,48]. However, as there is no certainty that in follow-up field data collection activities the acceleration dataloggers will be placed at the same location each time, the developed models need to produce classifications that are invariant to such issues. By fusing the Euclidian Norm data collected on two of the most accessible parts of the tool, this study has facilitated the attempt of making the models invariant to the data sensing location. This is proven by the results obtained on the testing datasets, which returned in all the cases excellent classification results (Tables A1–A3), irrespective of the datalogger placement, operational variability or the individual handling of the tool. Moreover, the developed models were found to deal very well with the intra-class variability of the Euclidian Norm data (Figure 3, events labelled with 2 and 3 for a single sensing modality: engine (TRAIN_E) or transmission shaft (TRAIN_S)), which was mostly generated by the variation of operational behavior. As a fact, intra-class variability may be related to and generated by the same or more individuals performing differently something in a given activity [49]. In this study, intra-class variability was the effect of operational behavior in relation to the crop layout, some portions requiring walking with engine in the idle running, as well as the effect of other issues such as changes in the operational behavior for similar operational conditions. For a comparison, the reader may consult, for instance, Figures A2 and A4. However, a speculation that could be raised here is that the use of vibration data sensed by a direct contact with the tool has more potential in generating more clearly separable events; hence, it could stand for a good approach to eliminate much of the intra-class variability which could be generated by different persons carrying on the same task. Due to the vibration characteristics of the tools equipped with two-stroke engines, the models developed and tested in this study might work well also on data collected by sensor placement on the chainsaws to distinguish between engine working (ON) and non-working states (OFF). For instance, the work of [16] has shown a similar data pattern and vibration magnitudes for engine working states. However, further studies are needed to check if the models would work on tools from other classes that are characterized by contrasting constructive concepts.

Class imbalance [28,49] and inter-class similarity [49] are common issues causing classification problems in various applications of the human activity recognition. On the one hand, class imbalance biases the prediction of conventional models toward the classes holding the majority of data [28]. On the other hand, experiments that are purely observational hold few if no ways to address this challenge [12,14], as the occurrence of the datapoints in given classes is imposed by the operational conditions. Class imbalance was a defining feature of the datasets used in this study, which have shown a data majority attributed to the ON class (Table 2). Given the results of classification performance, however, it seems that this characteristic had small effects on the datasets if compared, for instance, to inter-class similarity (transient events such as turning on and of the engine), which resulted in some misclassifications (Table 4, Figures 5 and 6).

Classification performance of the models was found to be very high, while keeping the error rates at a low level in both the training and testing datasets. For instance, classification accuracy was higher than 99% irrespective of the explored dataset, a value that is frequently termed as being very good [40]. However, there was a tradeoff between achieving high classification performances and keeping the generalization errors low, which in this study was evaluated by a trial-and-error approach which tuned the regularization parameter and lead to a selection of two final best-performing models. The final models (for α set at 0.0001 and 0.1, respectively), which were retained based on their lowest generalization errors, shared similar classification accuracies, excepting that of the TEST_S2 dataset, a case in which the model trained for α = 0.1 performed better. Given the similarity of classification accuracy for the rest of comparisons (Table 3), this outcome could be attributed to the functions and decision boundaries learned by the MLP model.

In addition to the hyperparameters' tuning, classification performance is affected by the architecture of the MLP (particularly the size) but also by the sensing modalities. Most often, the size of the MLP is selected based on rules of thumb [50,51]. However, the work of [16] has shown how an increasing depth (number of hidden layers) and width (number of neurons in a hidden layer) of the MLP may output increasingly accurate classification results for a case study run on triaxial acceleration data collected on a chainsaw. Based on that, as well as on the recommendations of [33], the size of the MLP was set to the maximum allowed by the used software. Sensing by two or more modalities may increase the classification performance. For instance, the work of [15] has found that the use of sound in addition to acceleration and gyroscope-collected data contributed to the performance increment of a Random Forest algorithm by decreasing the classification errors, while the work of [17] has found a better classification performance when fusing the data on acceleration and sound pressure level by horizontal staking before feeding it into MLPs, concluding that the preservation of sensing location may be of high importance in developing more accurate classification models. By comparison, this study removes the problem of sensing location by training the MLPs on dual-sensed signals collected at two locations on the tool. While further studies would be needed to check it, by its specific learning characteristics, the developed MLP could be invariant to the sampling rate of new coming data, because it makes its predictions based on the functions learned and not based on the sampling rate characteristics.

Collecting, processing, analyzing and interpreting large amounts of data is one of the approaches taken today to better understand ecological, social and technical systems, enabling a better decision making based on deeply informed grounds. There are several approaches, techniques and technologies to the problem which have already been described as being opportune for the general forestry [52]. Operational monitoring of SRWCs may benefit from sensor-based data collection approaches coupled with the techniques of artificial intelligence by removing the human error and the effort associated to the traditional observation [53,54]. Moreover, approaches such that described in this study could be used to prevent the safety issues associated with collecting data near dangerous machines and tools or in difficult outdoor conditions [12], being also less intrusive in applications that aim at observing people at work. At the time of purchasing the dataloggers used in this study, they were considered to be very small and useful for operational monitoring of motor-manual operations [47,48]. However, the technology of producing affordable miniaturized sensors in ongoing and had a significant progress, since smaller sensors are already released on the market, facilitating the transition to sensor-based operational monitoring. This progress has been reflected positively in various forestry applications requiring close-range sensing [55], and there is a lot of unexplored potential for such techniques both in forest and in WSRC operations.

There are two main limitations of this study. The first one is related to the MLP's misclassifications which were mainly found for those datapoints characterizing the so-called transient events of turning on and off the tool's engine (Table 4, Figure 5). The second one is related to the fact that in this study the data was segmented only in two classes, therefore the engine idle time was included in the same class as the effective willow felling time. Both of these problems may be easily solved by adding more context by the use of GPS units, an approach that has shown good results in previous studies on the topic [7,10]. In addition, in the phase of data interpretation, one can treat misclassifications as non-felling time having in mind the knowledge gained by this study. However, further studies are needed to evaluate whether adding primary and derived GPS location data into the MLP would help in designing applications able to look more deeply into the underlying process of willow felling operations. For instance, GPS coordinates and speed were used to infer the location and operational behavior of a feller by a traditional human-assisted classification approach [7,10] and they could provide additional context for the design of a multiclass MLP.

5. Conclusions

For a binary classification problem which emulates the most important operational events in SRWCs' motor-manual felling by brush cutters, the developed MLPs have returned high classification accuracies (99.1% to 99.9%) which were invariant to the sensing modality judged by the sensors' location. While the study has addressed hyperparameter tunning by the modification of the regularization parameter term, two final models were retained as being able to (i) provide high classification accuracies, (ii) generalize well on the testing datasets collected by single modalities and (iii) retain low errors in both, the training and testing phases. Given the obtained results, the developed models are assumed to be invariant to the new coming data, making them useful in classification applications and enabling the automation of most of the workflow typically implemented to collect, process, analyze and interpret large amounts of data. Further studies could bring new interesting and valuable insights if focused on evaluating the classification performance by the possibility of adding more context to the developed MLPs. This could be achieved by fusing the triaxial acceleration data with that collected by miniaturized GPS units to be able to classify and describe in more depth the operational tasks.

Author Contributions: The author has read and agreed to the published version of the manuscript.

Funding: This research received no external funding.

Institutional Review Board Statement: Ethical review and approval were waived for this study, due to the fact that the workers observed in the study agreed to participate based on an informed consent.

Informed Consent Statement: Informed consent was obtained from all subjects involved in the study.

Data Availability Statement: Data supporting the study is available on request to the author.

Acknowledgments: The author acknowledges the technical support of the Department of Forest Engineering, Forest Management Planning and Terrestrial Measurements, Faculty of Silviculture and Forest Engineering, Transilvania University of Brasov in designing and conducting this study. The author would like to thank to Eng. Arpad Domokos and to his employees for making this study possible, as well as to Eng. Nicolae Talagai and Eng. Marius Cheța for their help in field data collection activities and for providing the raw data for this study.

Conflicts of Interest: The author declares no conflict of interest.

Appendix A

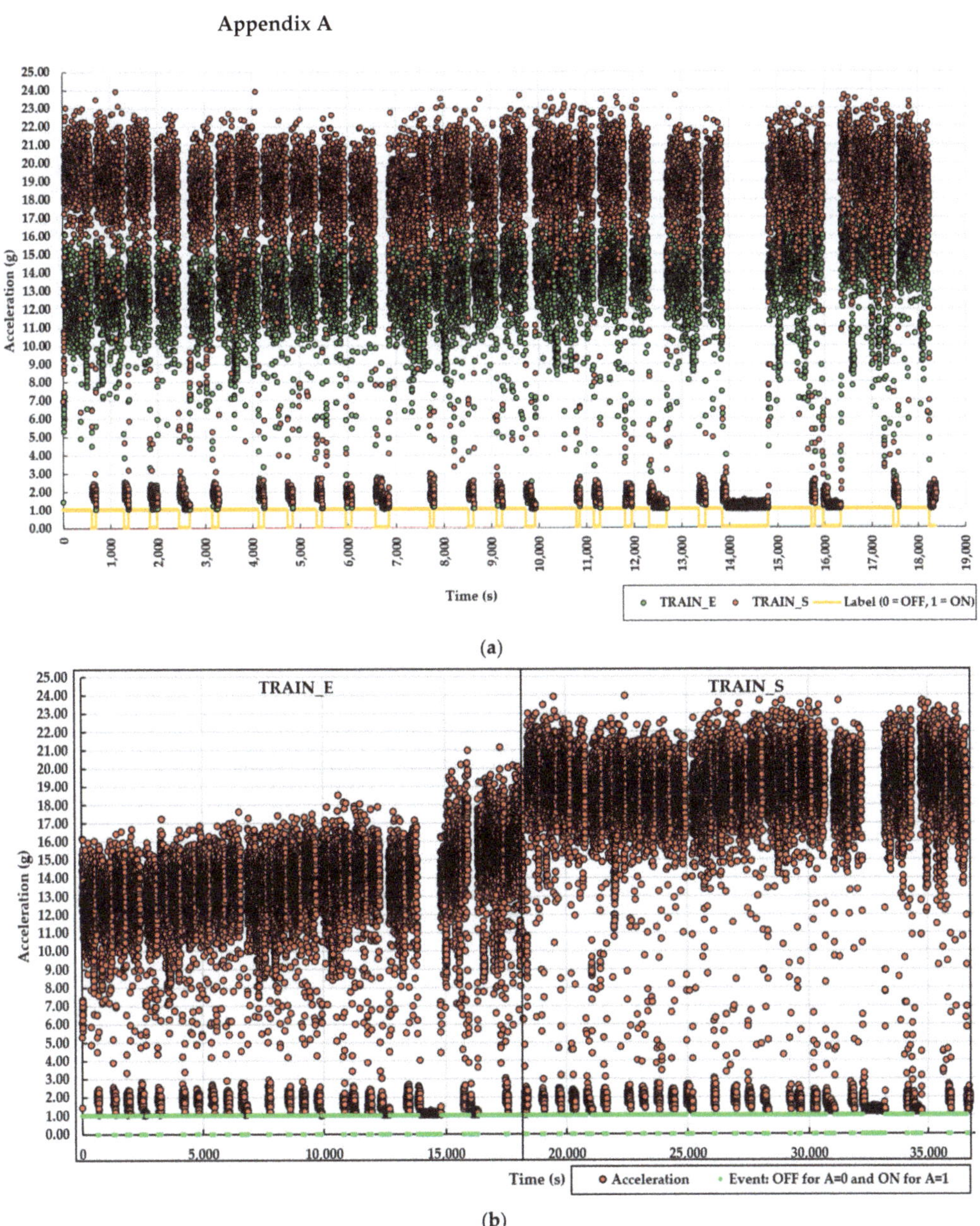

Figure A1. Description of the dataset used for the training phase of the MLPs. Legend: (**a**) training dataset before fusion showing the effect of modality on the acceleration's (A) magnitude; (**b**) training dataset after data fusion: left—data collected on the tool's engine (TRAIN_E), right—data collected on the tool's transmission shaft (TRAIN_S). Note: in both figure panels data is given as the Euclidian Norm of acceleration responses (Equation (1)) in the time domain. Conventionally, a value set at 0 (A = 0) indicates the labels and events documented as OFF, while a value set at 1 (A = 1) indicates the labels and events documented as ON.

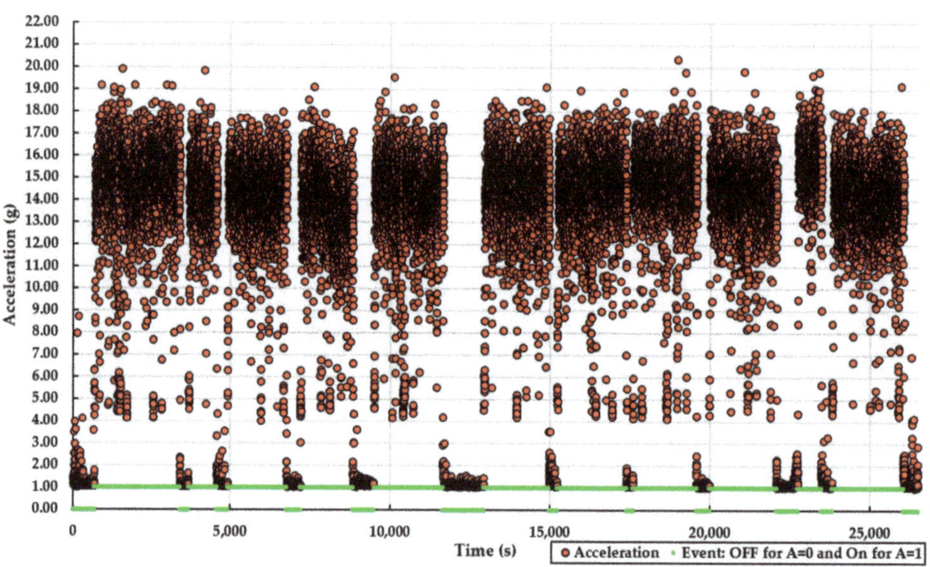

Figure A2. Description of datasets used in the testing phase of the MLPs: TEST_E1. Note: conventionally, a value set at 0 (A = 0) indicates the events documented as OFF, while a value set at 1 (A = 1) indicates the events documented as ON.

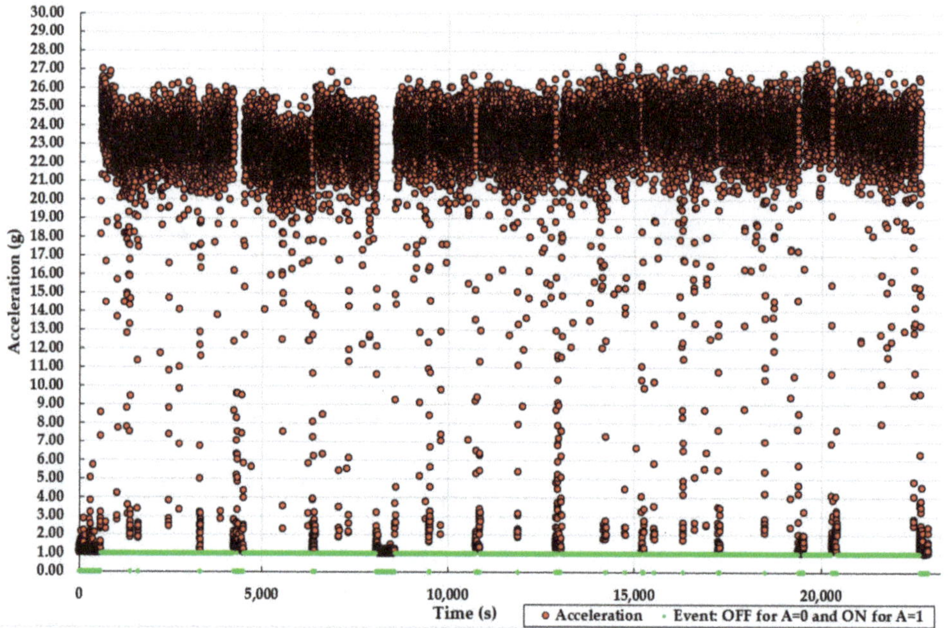

Figure A3. Description of datasets used in the testing phase of the MLPs: TEST_S1. Note: conventionally, a value set at 0 (A = 0) indicates the events documented as OFF, while a value set at 1 (A = 1) indicates the events documented as ON.

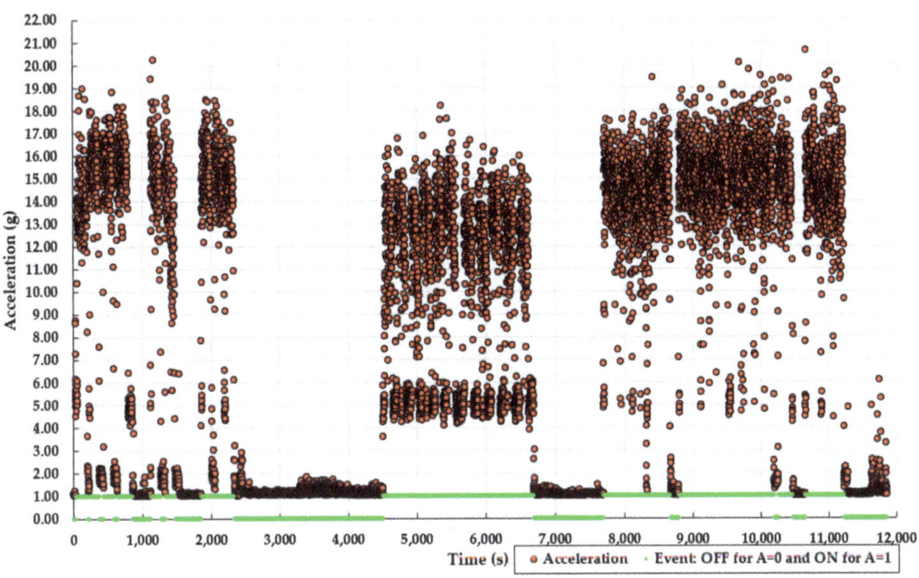

Figure A4. Description of datasets used in the testing phase of the MLPs: TEST_E2. Note: conventionally, a value set at 0 (A = 0) indicates the events documented as OFF, while a value set at 1 (A = 1) indicates the events documented as ON.

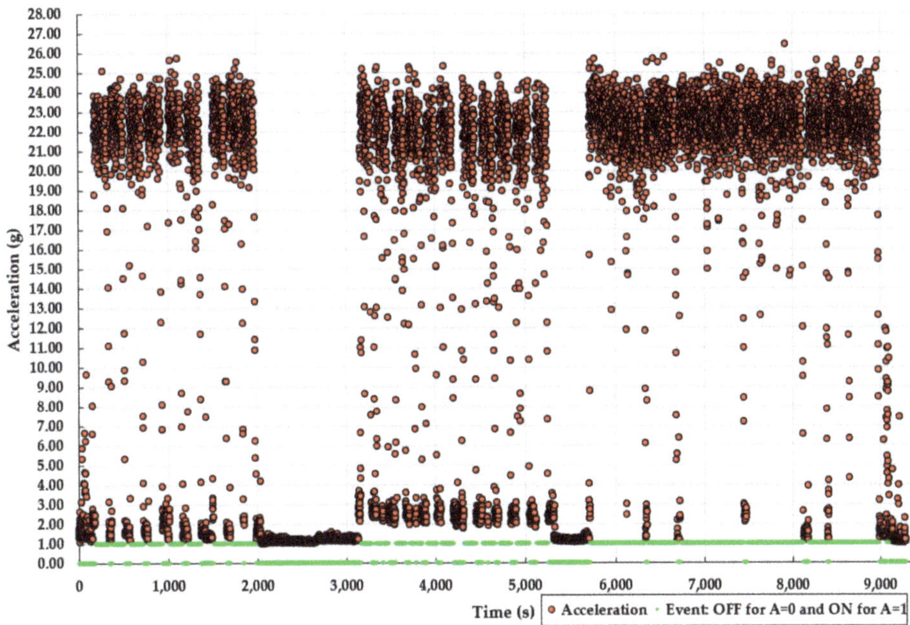

Figure A5. Description of datasets used in the testing phase of the MLPs: TEST_S2. Note: conventionally, a value set at 0 (A = 0) indicates the events documented as OFF, while a value set at 1 (A = 1) indicates the events documented as ON.

Table A1. Classification performance metrics of the training and testing datasets.

Regularization Parameter (Training Time)	Dataset	Area under Curve (AUC)	Classification Accuracy (CA)	F1	Precision (PREC)	Recall (REC)	Specificity (SPEC)
$\alpha = 0.0001$ (261 s) [a]	TRAIN	1.000	0.999	0.999	0.999	0.999	0.997
	TEST_E1	1.000	0.999	0.999	0.999	0.999	0.996
	TEST_S1	1.000	0.996	0.996	0.996	0.996	0.989
	TEST_E2	0.999	0.994	0.994	0.994	0.994	0.995
	TEST_S2	0.999	0.991	0.991	0.991	0.991	0.986
$\alpha = 0.001$ (326 s)	TRAIN	1.000	0.999	0.999	0.999	0.999	0.998
	TEST_E1	1.000	0.999	0.999	0.999	0.999	0.997
	TEST_S1	1.000	0.996	0.996	0.996	0.996	0.992
	TEST_E2	0.999	0.994	0.994	0.994	0.994	0.995
	TEST_S2	0.999	0.995	0.995	0.995	0.995	0.993
$\alpha = 0.01$ (480 s)	TRAIN	1.000	0.999	0.999	0.999	0.999	0.998
	TEST_E1	1.000	0.999	0.999	0.999	0.999	0.997
	TEST_S1	1.000	0.996	0.996	0.996	0.996	0.992
	TEST_E2	0.999	0.994	0.994	0.994	0.994	0.995
	TEST_S2	0.999	0.995	0.995	0.995	0.995	0.994
$\alpha = 0.1$ (443 s) [a]	TRAIN	1.000	0.998	0.998	0.998	0.998	0.998
	TEST_E1	1.000	0.999	0.999	0.999	0.999	0.998
	TEST_S1	1.000	0.995	0.995	0.995	0.995	0.992
	TEST_E2	0.999	0.994	0.994	0.994	0.994	0.995
	TEST_S2	0.999	0.995	0.995	0.995	0.995	0.994
$\alpha = 1$ (482 s)	TRAIN	1.000	0.997	0.997	0.997	0.997	0.998
	TEST_E1	1.000	0.994	0.994	0.994	0.994	0.997
	TEST_S1	1.000	0.994	0.994	0.995	0.994	0.994
	TEST_E2	0.999	0.974	0.974	0.975	0.974	0.980
	TEST_S2	0.999	0.994	0.994	0.994	0.994	0.994
$\alpha = 10$ (269 s)	TRAIN	1.000	0.994	0.994	0.994	0.994	0.998
	TEST_E1	1.000	0.990	0.990	0.991	0.990	0.996
	TEST_S1	1.000	0.994	0.994	0.994	0.994	0.995
	TEST_E2	0.999	0.944	0.944	0.950	0.944	0.958
	TEST_S2	0.999	0.993	0.993	0.993	0.993	0.995

Note: [a] regularization terms which returned the lowest log losses.

Table A2. Classification performance metrics of the training and testing datasets for ON events.

Regularization Parameter (Training Time)	Dataset	Area under Curve (AUC)	Classification Accuracy (CA)	F1	Precision (PREC)	Recall (REC)	Specificity (SPEC)
$\alpha = 0.0001$ (261 s) [a]	TRAIN	1.000	0.999	0.999	0.999	0.999	0.997
	TEST_E1	1.000	0.999	0.999	0.998	1.000	0.994
	TEST_S1	1.000	0.996	0.998	0.999	0.997	0.988
	TEST_E2	0.999	0.994	0.995	0.997	0.992	0.996
	TEST_S2	0.999	0.991	0.993	0.988	0.998	0.979
$\alpha = 0.001$ (326 s)	TRAIN	1.000	0.999	0.999	0.999	0.999	0.997
	TEST_E1	1.000	0.999	0.999	0.999	1.000	0.996
	TEST_S1	1.000	0.996	0.998	0.999	0.996	0.992
	TEST_E2	0.999	0.994	0.995	0.998	0.992	0.998
	TEST_S2	0.999	0.995	0.996	0.995	0.997	0.991
$\alpha = 0.01$ (480 s)	TRAIN	1.000	0.999	0.999	0.999	0.999	0.997
	TEST_E1	1.000	0.999	0.999	0.999	1.000	0.996
	TEST_S1	1.000	0.996	0.997	0.999	0.996	0.992
	TEST_E2	0.999	0.994	0.995	0.998	0.992	0.998
	TEST_S2	0.999	0.995	0.996	0.995	0.996	0.992
$\alpha = 0.1$ (443 s) [a]	TRAIN	1.000	0.998	0.999	0.999	0.998	0.997
	TEST_E1	1.000	0.999	0.999	0.999	1.000	0.997
	TEST_S1	1.000	0.995	0.997	0.999	0.996	0.992
	TEST_E2	0.999	0.994	0.995	0.999	0.991	0.998
	TEST_S2	0.999	0.995	0.996	0.996	0.996	0.994
$\alpha = 1$ (482 s)	TRAIN	1.000	0.997	0.998	1.000	0.996	0.999
	TEST_E1	1.000	0.994	0.996	1.000	0.993	0.998
	TEST_S1	1.000	0.994	0.997	0.999	0.994	0.994
	TEST_E2	0.999	0.974	0.997	0.999	0.956	0.999
	TEST_S2	0.999	0.994	0.995	0.997	0.993	0.995
$\alpha = 10$ (269 s)	TRAIN	1.000	0.994	0.996	1.000	0.993	0.999
	TEST_E1	1.000	0.990	0.994	1.000	0.988	0.999
	TEST_S1	1.000	0.994	0.997	0.999	0.994	0.996
	TEST_E2	0.999	0.944	0.948	0.999	0.903	0.999
	TEST_S2	0.999	0.993	0.995	0.998	0.992	0.996

Note: [a] regularization terms which returned the lowest log losses.

Table A3. Classification performance metrics of the training and testing datasets for OFF events.

Regularization Parameter (Training Time)	Dataset	Area under Curve (AUC)	Classification Accuracy (CA)	F1	Precision (PREC)	Recall (REC)	Specificity (SPEC)
$\alpha = 0.0001$ (261 s) [a]	TRAIN	1.000	0.999	0.998	0.998	0.997	0.999
	TEST_E1	1.000	0.999	0.997	1.000	0.994	1.000
	TEST_S1	1.000	0.996	0.982	0.976	0.988	0.997
	TEST_E2	0.999	0.994	0.993	0.989	0.996	0.992
	TEST_S2	0.999	0.991	0.988	0.997	0.979	0.998
$\alpha = 0.001$ (326 s)	TRAIN	1.000	0.999	0.997	0.997	0.997	0.999
	TEST_E1	1.000	0.999	0.998	1.000	0.996	1.000
	TEST_S1	1.000	0.996	0.980	0.970	0.992	0.996
	TEST_E2	0.999	0.994	0.993	0.989	0.998	0.992
	TEST_S2	0.999	0.995	0.993	0.995	0.991	0.997
$\alpha = 0.01$ (480 s)	TRAIN	1.000	0.999	0.998	0.998	0.997	0.999
	TEST_E1	1.000	0.999	0.998	0.999	0.996	1.000
	TEST_S1	1.000	0.996	0.980	0.968	0.992	0.996
	TEST_E2	0.999	0.994	0.993	0.989	0.998	0.992
	TEST_S2	0.999	0.995	0.993	0.994	0.992	0.996
$\alpha = 0.1$ (443 s) [a]	TRAIN	1.000	0.998	0.996	0.995	0.997	0.998
	TEST_E1	1.000	0.999	0.998	0.999	0.997	1.000
	TEST_S1	1.000	0.995	0.979	0.965	0.992	0.996
	TEST_E2	0.999	0.994	0.993	0.988	0.998	0.991
	TEST_S2	0.999	0.995	0.993	0.993	0.994	0.996
$\alpha = 1$ (482 s)	TRAIN	1.000	0.997	0.994	0.989	0.999	0.996
	TEST_E1	1.000	0.994	0.987	0.976	0.998	0.993
	TEST_S1	1.000	0.994	0.974	0.955	0.994	0.994
	TEST_E2	0.999	0.974	0.970	0.944	0.999	0.956
	TEST_S2	0.999	0.994	0.992	0.988	0.995	0.993
$\alpha = 10$ (269 s)	TRAIN	1.000	0.994	0.988	0.977	0.999	0.993
	TEST_E1	1.000	0.990	0.979	0.960	0.999	0.988
	TEST_S1	1.000	0.994	0.974	0.953	0.996	0.994
	TEST_E2	0.999	0.944	0.938	0.884	0.999	0.903
	TEST_S2	0.999	0.993	0.991	0.986	0.996	0.992

Note: [a] regularization terms which returned the lowest log losses.

References

1. Dickmann, D.I. Silviculture and biology of short-rotation woody crops in temperate regions: Then and now. *Biomass Bioenerg.* **2006**, *30*, 696–705. [CrossRef]
2. Kuzovkina, Y.A.; Volk, T.A. The characterization of willow (Salix L.) varieties for use in ecological engineering applications: Co-ordination of structure, function and autecology. *Ecol. Eng.* **2009**, *35*, 1178–1189. [CrossRef]
3. Berhongaray, G.; El Kasmioui, O.; Ceulemans, R. Comparative analysis of harvesting machines on operational high-density short rotation woody crop (SRWC) culture: One-process versus two-process harvest operation. *Biomass Bioenerg.* **2013**, *58*, 333–342. [CrossRef]
4. Schweier, J.; Becker, G. Harvesting of short rotation coppice—Harvesting trials with a cut and storage system in Germany. *Silva Fenn.* **2012**, *46*, 287–299. [CrossRef]
5. El Kasmioui, O.; Ceulemans, R. Financial analysis of the cultivation of short rotation woody crops for bioenergy in Belgium: Barriers and opportunities. *Bioenerg. Res.* **2013**, *6*, 336–350. [CrossRef]
6. Buchholz, T.; Volk, T.A. Improving the profitability of willow crops—Identifying opportunities with a crop budget model. *Bioenerg. Res.* **2011**, *4*, 85–95. [CrossRef]
7. Borz, S.A.; Talagai, N.; Cheţa, M.; Chiriloiu, D.; Gavilanes Montoya, A.V.; Castillo Vizuete, D.D.; Marcu, M.V. Physical strain, exposure to noise and postural assessment in motor-manual felling of willow short rotation coppice: Results of a preliminary study. *Croat. J. Eng.* **2019**, *40*, 377–388. [CrossRef]
8. Schweier, J.; Becker, G. Motor manual harvest of short rotation coppice in South-West Germany. *Allg. Forst-und Jagdztg.* **2012**, *183*, 159–167.
9. Talagai, N.; Borz, S.A.; Ignea, G. Performance of brush cutters in felling operations of willow short rotation coppice. *Bioresources* **2017**, *12*, 3560–3569. [CrossRef]

10. Borz, S.A.; Talagai, N.; Cheţa, M.; Gavilanes Montoya, A.V.; Castillo Vizuete, D.D. Automating data collection in motor-manual time and motion studies implemented in a willow short rotation coppice. *Bioresources* **2018**, *13*, 3236–3249. [CrossRef]

11. Vanbeveren, S.P.P.; Schweier, J.; Berhongaray, G.; Ceulemans, R. Operational short rotation woody crops plantations: Manual or mechanized harvesting? *Biomass Bioenerg.* **2015**, *72*, 8–18. [CrossRef]

12. Acuna, M.; Bigot, M.; Guerra, S.; Hartsough, B.; Kanzian, C.; Kärhä, K.; Lindroos, O.; Magagnotti, N.; Roux, S.; Spinelli, R.; et al. *Good Practice Guidelines for Biomass Production Studies*; Magagnotti, N., Spinelli, R., Eds.; CNR IVALSA: Sesto Fiorentino, Italy, 2012; Available online: http://www.forestenergy.org/pages/cost-action-fp0902/good-practice-guidelines/ (accessed on 15 April 2018).

13. Björheden, R.; Apel, K.; Shiba, M.; Thompson, M. *IUFRO Forest Work Study Nomenclature*; Swedish University of Agricultural Science, Department of Operational Efficiency: Grapenberg, Sweden, 1995.

14. Borz, S.A. *Evaluarea Eficienţei Echipamentelor şi Sistemelor Tehnice în Operaţii Forestiere*; Lux Libris Publishing House: Braşov, Romania, 2008.

15. Keefe, R.F.; Zimbelman, E.G.; Wempe, A.M. Use of smartphone sensors to quantify the productive cycle elements of hand fallers on industrial cable logging operations. *Int. J. Eng.* **2019**, *30*, 132–143. [CrossRef]

16. Cheţa, M.; Marcu, M.V.; Borz, S.A. Effect of training parameters on the ability of artificial neural networks to learn: A simulation on accelerometer data for task recognition in motor-manual felling and processing. *Bull. Transilv. Univ. Bras. Ser. Wood Ind. Agric. Food Eng.* **2020**, *131*, 19–36. [CrossRef]

17. Cheţa, M.; Marcu, M.V.; Iordache, E.; Borz, S.A. Testing the capability of low-cost tools and artificial intelligence techniques to automatically detect operations done by a small-sized manually driven bandsaw. *Forests* **2020**, *11*, 739. [CrossRef]

18. Borz, S.A.; Păun, M. Integrating offline object tracking, signal processing and artificial intelligence to classify relevant events in sawmilling operations. *Forests* **2020**, *11*, 1333. [CrossRef]

19. Talagai, N.; Borz, S.A. Concepte de automatizare a activităţii de colectare a datelor cu aplicabilitate în monitorizarea performanţei productive în operaţii de gestionare a culturilor de salcie rotaţie scurtă. *Rev. Pădurilor* **2016**, *131*, 78–94.

20. Talagai, N.; Marcu, M.V.; Zimbalatti, G.; Proto, A.R.; Borz, S.A. Productivity in partly mechanized planting operations of willow short rotation coppice. *Biomass Bioenerg.* **2020**, *138*, 105–609. [CrossRef]

21. Borz, S.A.; Niţa, M.D.; Talagai, N.; Scriba, C.; Grigolato, S.; Proto, A.R. Performance of small-scale technology in planting and cutback operations of short-rotation willow crops. *Trans. ASABE* **2019**, *62*, 167–176. [CrossRef]

22. User Guide of the Husqvarna 545 Series (In Romanian). Available online: https://www.husqvarna.com/ro/products/motounelte-motocoase/545rx/966015901/ (accessed on 16 February 2021).

23. Talagai, N.; Cheţa, M.; Gavilanes Montoya, A.V.; Castillo Vizuete, D.D.; Borz, S.A. Predicting time consumption of chipping tasks in a willow short rotation coppice from GPS and acceleration data. In Proceedings of the Biennial International Symposium "Forest and Sustainable Development", Braşov, Romania, 25–27 October 2018; Borz, S.A., Curtu, A.L., Muşat, E.C., Eds.; Transilvania University Press: Braşov, Romania, 2019; pp. 1–12.

24. Boja, N.; Borz, S.A. Energy inputs in motor-manual release cutting of broadleaved forests: Results of twelve options. *Energies* **2020**, *13*, 4597. [CrossRef]

25. Ignea, G.; Ghaffaryian, M.R.; Borz, S.A. Impact of operational factors on fossil energy inputs in motor-manual tree felling and processing operations: Results of two case studies. *Ann. Res.* **2017**, *60*, 161–172. [CrossRef]

26. Technical Specifications of the Extech® VB300 3-Axis G-force USB Datalogger. Available online: http://www.extech.com/products/VB300 (accessed on 17 February 2021).

27. Technical specifications of the Extech® 407760 USB Sound Level Datalogger. Available online: http://www.extech.com/products/407760 (accessed on 17 February 2021).

28. Chen, K.; Zhang, D.; Yao, L.; Guo, B.; Yu, Z.; Liu, Y. Deep learning for sensor-based human activity recognition: Overview, challenges and opportunities. *J. ACM* **2018**, *37*, 111.

29. Van Hees, V.T.; Gorzelniak, L.; Dean Leon, E.C.; Eder, M.; Pias, M.; Taherian, S.; Ekelund, U.; Renström, F.; Franks, P.W.; Horsch, A.; et al. Separating movement and gravity components in acceleration signal and implications for the assessment of human daily physical activity. *PLoS ONE* **2013**, *8*, e61691. [CrossRef] [PubMed]

30. Ha, S.; Yun, J.-M.; Chi, S. Multi-modal convolutional neural networks for activity recognition. In Proceedings of the 2015 IEEE Conference on Systems, Man and Cybernetics, Hong Kong, China, 9–12 October 2015; pp. 3017–3022. [CrossRef]

31. Han, J.; Kamber, M.; Pei, J. *Data Mining: Concepts and Techniques*, 3rd ed.; Morgan Kaufmann Publishers: Burlington, MA, USA, 2006.

32. Demsar, J.; Curk, T.; Erjavec, A.; Gorup, C.; Hocevar, T.; Milutinovic, M.; Mozina, M.; Polajnar, M.; Toplak, M.; Staric, A.; et al. Orange: Data Mining Toolbox in Python. *J. Mach. Learn. Res.* **2013**, *14*, 2349–2353.

33. Goodfellow, J.; Bengio, Y.; Courville, A. *Deep Learning*; MIT Press: Cambridge, MA, USA, 2016; Available online: https://www.deeplearningbook.org/ (accessed on 17 February 2021).

34. Maas, A.L.; Hannun, A.Y.; Ng, A.Y. Rectifier nonlinearities improve neural network acoustic models. In Proceedings of the 30th International Conference on Machine Learning, ICML 2013, Atlanta, GA, USA, 16–21 June 2013.

35. Nair, V.; Hinton, G.E. Rectified linear units improve restricted Boltzmann machines. In Proceedings of the 27th International Conference on Machine Learning (ICML 2010), Haifa, Israel, 21–24 June 2010.

36. Kingma, D.P.; Ba, J.L. ADAM: A method for stochastic optimization. In Proceedings of the 3rd International Conference on Learning Representations, ICLR 2015, San Diego, CA, USA, 7–9 May 2015.
37. Varying Regularization in Multi-Layer Perceptrons. Available online: https://scikit-learn.org/stable/auto_examples/neural_networks/plot_mlp_alpha.html#sphx-glr-auto-examples-neural-networks-plot-mlp-alpha-py (accessed on 17 February 2021).
38. Understanding Binary Cross-Entropy/Log-Loss: A Visual Explanation. By Daniel Godoy. Available online: https://towardsdatascience.com/understanding-binary-cross-entropy-log-loss-a-visual-explanation-a3ac6025181a (accessed on 17 February 2021).
39. Fawcett, T. An introduction to ROC analysis. *Pattern Recogn. Lett.* **2006**, *27*, 861–874. [CrossRef]
40. Kamilaris, A.; Prenafeta-Boldu, F.X. Deep learning in agriculture: A survey. *Comput. Electron. Agric.* **2018**, *147*, 70–90. [CrossRef]
41. Wescoat, E.; Krugh, M.; Henderson, A.; Goodnough, J.; Mears, L. Vibration analysis using unsupervised learning. *Procedia Manuf.* **2019**, *34*, 876–884. [CrossRef]
42. McDonald, T.P.; Fulton, J.P. Automated time study of skidders using global positioning system data. *Comput. Electron. Agric.* **2005**, *48*, 19–37. [CrossRef]
43. Strandgard, M.; Mitchell, R. Automated time study of forwarders using GPS and a vibration sensor. *Croat. J. Eng.* **2015**, *36*, 175–184.
44. Contreras, M.; Freitas, R.; Ribeiro, L.; Stringer, J.; Clark, C. Multi-camera surveillance systems for time and motion studies of timber harvesting equipment. *Comput. Electron. Agric.* **2015**, *135*, 208–215. [CrossRef]
45. McDonald, T.P.; Fulton, J.P.; Darr, M.J.; Gallagher, T.V. Evaluation of a system to spatially monitor hand planting of pine seedlings. *Comput. Electron. Agric.* **2008**, *64*, 173–182. [CrossRef]
46. Borz, S.A.; Marcu, M.V.; Cataldo, M.F. Evaluation of a HSM 208F 14tone HVT-R2 forwarder prototype under conditions of steep-terrain low-access forests. *Croat. J. For. Eng.*. [CrossRef]
47. Marogel-Popa, T.; Cheța, M.; Marcu, M.V.; Duță, I.; Ioraș, F.; Borz, S.A. Manual cultivation operations in poplar stands: A characterization of job difficulty and risks of health impairment. *Int. J. Env. Res. Public Health* **2019**, *16*, 1911. [CrossRef]
48. Cheța, M.; Marcu, M.V.; Borz, S.A. Workload, exposure to noise, and risk of musculoskeletal disorders: A case study of motor-manual tree felling and processing in poplar clear cuts. *Forests* **2018**, *9*, 300. [CrossRef]
49. Bulling, A.; Blanke, U.; Schiele, B. A tutorial of human activity recognition using body-worn inertial sensors. *ACM Comput. Surv.* **2014**, *46*, 33. [CrossRef]
50. Karsoliya, S. Approximating number of hidden layer neurons in multiple hidden layer BPNN architecture. *Int. J. Eng. Technol.* **2012**, *3*, 714–717.
51. Panchal, F.S.; Panchal, M. Review on methods of selecting number of hidden nodes in Artificial Neural Network. *Int. J. Comput. Sci. Mob. Comput.* **2014**, *3*, 455–464.
52. Keefe, R.F.; Wempe, A.M.; Becker, R.M.; Zimbelman, E.G.; Nagler, E.S.; Gilbert, S.L.; Caudill, C.C. Positioning methods and the use of location and activity data in forests. *Forests* **2019**, *10*, 458. [CrossRef]
53. Borz, S.A. Turning a winch skidder into a self-data collection machine. *Bull. Transilv. Univ. Bras. Ser. Wood Ind. Agric. Food Eng.* **2016**, *9*, 1–6.
54. Spinelli, R.; Laina-Relano, R.; Magagnotti, N.; Tolosana, E. Determining observer effect and method effects on the accuracy of elemental time studies in forest operations. *Balt. For.* **2013**, *19*, 301–306.
55. Picchio, R.; Proto, A.R.; Civitarese, V.; Di Marzio, N.; Latterini, F. Recent contributions of some fields of the electronics in development of forest operations technologies. *Electronics* **2019**, *8*, 1465. [CrossRef]

Article

A Forest Fire Detection System Based on Ensemble Learning

Renjie Xu [1], Haifeng Lin [1], Kangjie Lu [1], Lin Cao [2] and Yunfei Liu [1,*]

1 College of Information Science and Technology, Nanjing Forestry University, Nanjing 210037, China; xurenjie@njfu.edu.cn (R.X.); haifeng.lin@njfu.edu.cn (H.L.); lukangjie@njfu.edu.cn (K.L.)
2 Co-Innovation Center for Sustainable Forestry in Southern China, Nanjing Forestry University, Nanjing 210037, China; lincao@njfu.edu.cn
* Correspondence: lyf@njfu.com.cn; Tel.: +86-139-1389-5117

Abstract: Due to the various shapes, textures, and colors of fires, forest fire detection is a challenging task. The traditional image processing method relies heavily on manmade features, which is not universally applicable to all forest scenarios. In order to solve this problem, the deep learning technology is applied to learn and extract features of forest fires adaptively. However, the limited learning and perception ability of individual learners is not sufficient to make them perform well in complex tasks. Furthermore, learners tend to focus too much on local information, namely ground truth, but ignore global information, which may lead to false positives. In this paper, a novel ensemble learning method is proposed to detect forest fires in different scenarios. Firstly, two individual learners Yolov5 and EfficientDet are integrated to accomplish fire detection process. Secondly, another individual learner EfficientNet is responsible for learning global information to avoid false positives. Finally, detection results are made based on the decisions of three learners. Experiments on our dataset show that the proposed method improves detection performance by 2.5% to 10.9%, and decreases false positives by 51.3%, without any extra latency.

Keywords: forest fire detection; deep learning; ensemble learning; Yolov5; EfficientDet; EfficientNet

Citation: Xu, R.; Lin, H.; Lu, K.; Cao, L.; Liu, Y. A Forest Fire Detection System Based on Ensemble Learning. *Forests* **2021**, *12*, 217. https://doi.org/10.3390/f12020217

Academic Editor: Stelian Alexandru Borz

Received: 4 January 2021
Accepted: 12 February 2021
Published: 13 February 2021

1. Introduction

With the change of the earth's climate, forest fires occur frequently all over the world, which not only cause serious economic losses and destroy the ecological environment, but also pose a great threat to the safety of human life.

Forest fires usually spread quickly and are difficult to control in a short time. Therefore, it is imperative to detect the early forest fire before it spreads out, but traditional detection methods have obvious drawbacks in detecting it in open forest areas. Sensors-based [1–3] detection systems have good performance in indoor space, but it is difficult to install them outdoors, considering high coverage cost [4,5]. In addition, they cannot provide important visual information which can help firefighters promptly grasp the situation of the fire scene. Infrared or ultraviolet detectors [6,7] are easy to be interfered by the environment, and considering their short detection distance, they are not suitable for large open areas. Satellite remote sensing [8] is good at detecting large-scale forest fires, but it cannot detect early regional fire.

Impressed by the rising computer vision technology, researchers start to seek an efficient and effective fire detection model based on image processing. Chen et al. [9] proposed an RGB (red, green, blue) model based chromatic and disorder measurement for extracting fire-pixels in the video. The color information is responsible for extracting fire-pixels, and dynamic information is used to verify if it is a real fire. Töreyin et al. [10] used 1D temporal wavelet transform to detect flame flicker, and applied 2D spatial wavelet transform to identify fire moving regions. This method, which integrated color and temporal variation information, reduced false alarms in real-world scenes. Çelik et al. [11] studied diverse video sequences and images, and proposed a fuzzy color model using statistical analysis.

Combined with motion analysis, the model achieves a good discrimination between fire and fire-like objects. Teng et al. [12] analyzed fire characteristics and proposed a real-time fire detection method based on hidden Markov models (HMMs), which extracted candidate fire-pixels using moving pixel detection, fire-color inspection, and pixel clustering. Chino et al. [13] found that most algorithms were designed for video, which had obvious limitations. To solve this problem, a novel fire detection method named BowFire was proposed. The method combined color features with superpixel texture discrimination to detect fire in still images. In conclusion, most traditional fire detection methods based on image processing focused on creating artificial features like color, motion, and texture to detect fires.

However, powerful deep learners begin to replace human intelligence. They are better at learning features than humans, and the features they extract contain much deeper semantic information than manmade ones. Recently, deep learning has outperformed traditional manmade features in many fields, and have been widely used in fire detection. Zhang et al. [14] created a forest fire benchmark, and used Faster R-CNN (region-based convolutional neural network) [15], Yolo (you only look once) [16–19], and SSD (single shot multibox detector) [20] to detect fire. They found that SSD was better regarding efficiency, detection accuracy, and early fire detection ability. Moreover, they proposed an improved tiny-Yolo by adjusting the network architecture. Kim et al. [21] employed faster R-CNN to detect fire and non-fire regions based on their spatial features. In addition, long short-term memory (LSTM) is used to verify the reliability of fire alarm. Lee et al. [22] proposed a video-based fire detection model, which used faster R-CNN to generate a fire candidate region for each frame. Then, structural similarity (SSIM) and mean square error (MSE) were calculated to determine similarity between adjacent frames. Final fire regions were determined based on spatial and temporal features. Pan et al. [23] proposed a camera-based wildfire detection system via transfer learning, in which block-based analysis strategy was used to improve fire detection accuracy. Redundant filters, which had low energy impulse response, were removed to ensure the model's efficiency on edge devices. Wu et al. [24] applied principal component analysis (PCA) to process forest fire images, and then fed them into the training network. The combination of two models was proved to enhance location results. In conclusion, faced with fire detection task, most researchers tend to only assign individual learners to perform object detection tasks, which is considered unreliable, since it may lead to false negatives.

In this paper, a novel method based on ensemble learning for forest fire detection is proposed. First, forest fire detection is a complicated and difficult task, making it highly impractical for individual learners to detect fires in diverse scenarios. Every individual learner has its own expertise, and can extract different features from the image, so integrating different individual learners can significantly improve the robustness of the model and enhance detection performance. Therefore, two individual object detectors Yolov5 [25] and EfficientDet [26] are integrated to detect the fire in parallel. These two learners work synergistically in detecting different types of forest fires, thereby improving the detection accuracy. Second, the object detectors only care about what fire is like, so they do not take the whole image into consideration. In this case, fire-like objects will absolutely affect the detection results. To solve this problem, the EfficientNet image classifier [27] is incorporated into our model, whose role is to enable the model to take full advantage of the global information. Final detection results will be made through the decision strategy according to results of these three learners, which will efficiently increase detection accuracy and decrease the false positives.

2. Materials and Methods

2.1. Datasets

To ensure our learners can handle different kinds of forest fires (ground fire, trunk fire, and canopy fire), we collected images from multiple public fire datasets: BowFire [28], FD-dataset [29], ForestryImages [30], VisiFire [31], etc. After manual filtration, we created a single integrated forest fire dataset containing 10,581 images, with 2976 forest fire images and 7605 non-fire images. Representative samples of our dataset are shown in Figures 1–3.

Figure 1. Representative forest fire images in the fire section of our dataset, including (**a**) ground fire 1, (**b**) ground fire 2, (**c**) trunk fire, and (**d**) canopy fire.

Figure 2. Representative normal forest images in the non-fire section of our dataset, including (**a**) normal forest scene 1, (**b**) normal forest scene 2, (**c**) normal forest scene 3, and (**d**) normal forest scene 4. (**a**–**d**) illustrate normal forest scenes without fire objects.

Figure 3. Representative images in the non-fire section of our dataset, including (**a**) wild scene with sun 1, (**b**) wild scene with sun 2, (**c**) wild scene with sun 3, and (**d**) wild scene with sun 4. (**a**–**d**) illustrate normal wild scenes containing fire-like object (e.g., sun).

2.2. Yolov5

Yolo is a state-of-the-art, real-time object detector, and Yolov5 is based on Yolov1-Yolov4. Continuous improvements have made it achieve top performances on two official object detection datasets: Pascal VOC (visual object classes) [32] and Microsoft COCO (common objects in context) [33].

The network architecture of Yolov5 is shown in Figure 4. There are three reasons why we choose Yolov5 as our first learner. Firstly, Yolov5 incorporated cross stage partial network (CSPNet) [34] into Darknet, creating CSPDarknet as its backbone. CSPNet solves the problems of repeated gradient information in large-scale backbones, and integrates the gradient changes into the feature map, thereby decreasing the parameters and FLOPS (floating-point operations per second) of model, which not only ensures the inference speed and accuracy, but also reduces the model size. In forest fire detection task, detection speed and accuracy is imperative, and compact model size also determines its inference efficiency on resource-poor edge devices. Secondly, the Yolov5 applied path aggregation network (PANet) [35] as its neck to boost information flow. PANet adopts a new feature pyramid network (FPN) structure with enhanced bottom-up path, which improves the propagation of low-level features. At the same time, adaptive feature pooling, which links feature grid and all feature levels, is used to make useful information in each feature level propagate directly to following subnetwork. PANet improves the utilization of accurate localization signals in lower layers, which can obviously enhance the location accuracy of the object. Thirdly, the head of Yolov5, namely the Yolo layer, generates 3 different sizes ($18 \times 18, 36 \times 36, 72 \times 72$) of feature maps to achieve multi-scale [18] prediction, enabling the model to handle small, medium, and big objects. A forest fire usually develops from small-scale fire (ground fire) to medium-scale fire (trunk fire), then to big-scale fire (canopy fire). Multi-scale detection ensures that the model can follow size changes in the process of fire evolution.

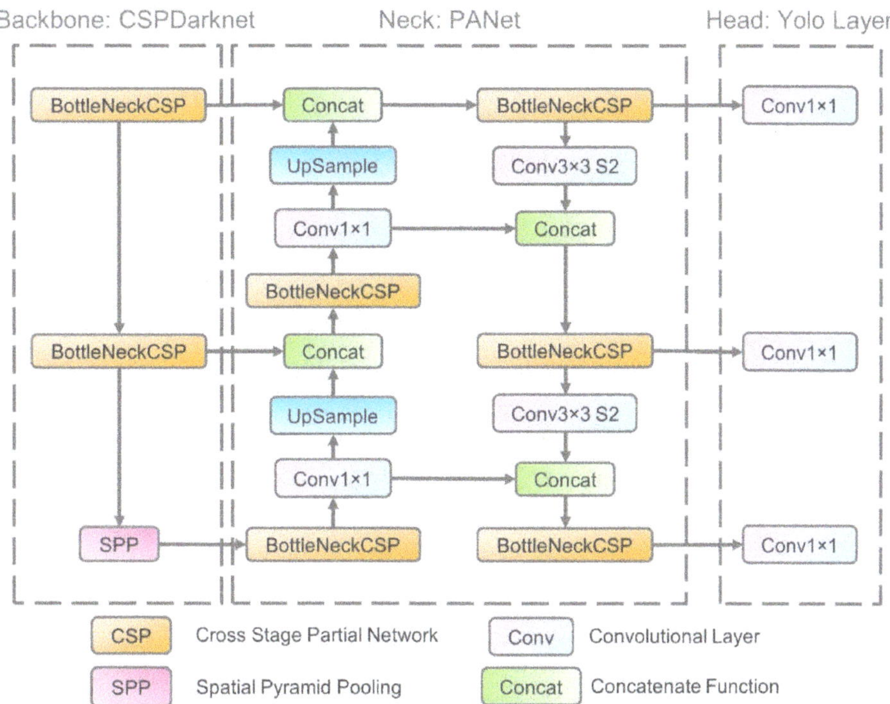

Figure 4. The network architecture of Yolov5. It consists of three parts: (1) Backbone: CSPDarknet, (2) Neck: PANet, and (3) Head: Yolo Layer. The data are first input to CSPDarknet for feature extraction, and then fed to PANet for feature fusion. Finally, Yolo Layer outputs detection results (class, score, location, size).

2.3. EfficientDet

EfficientDet is a new family of object detectors developed by Google, and it consistently achieves better efficiency than prior art across a wide spectrum of resource constraints. Similar to Yolov5, EfficientDet has also achieved remarkable performances in Pascal VOC and Microsoft COCO tasks, and is widely used in real-world applications.

The network architecture of EfficientDet is shown in Figure 5. There are three reasons why we choose EfficientDet as our second learner. Firstly, EfficientDet employed state-of-the-art network EfficientNet [27] as its backbone, making that the model has sufficient ability to learn the complex feature of diverse forest fires. Secondly, it applied an improved PANet, named bi-directional feature pyramid network (Bi-FPN) as its neck, to allow easy and fast multi-scale feature fusion. Bi-FPN introduces learnable weights, enabling network to learn the importance of different input features, and repeatedly applies top-down and bottom-up multi-scale feature fusion. Compared with Yolov5's neck PANet, Bi-FPN has better performances with less parameters and FLOPS. Meanwhile, different feature fusion strategy brings different semantic information, thereby bringing different detection results. Thirdly, similar to EfficientNet, it integrates a compound scaling method that uniformly scales the resolution, depth, and width for all backbone, feature network, and box/class prediction networks at the same time, which ensures the maximum accuracy and efficiency under the limited computing resources. With more available resources, accuracy will be consistently improved. Our second learner, EfficientDet, with different backbone, neck, and head, can learn different information that Yolov5 cannot.

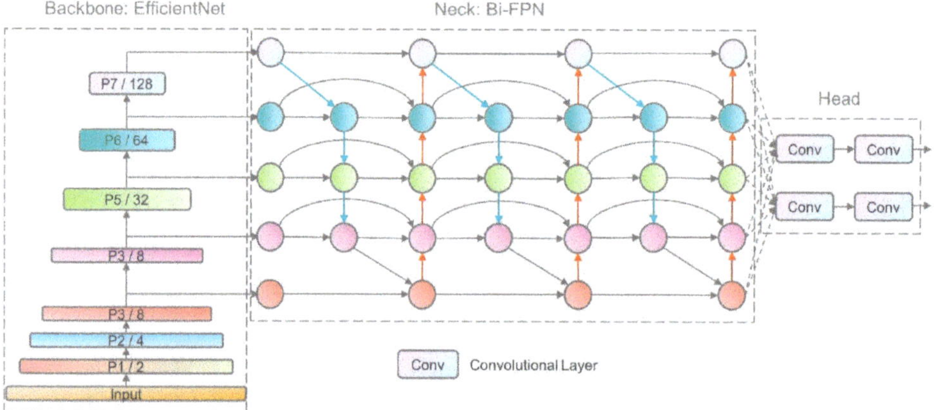

Figure 5. The network architecture of EfficientDet. It consists of three parts: (1) Backbone: EfficientNet, (2) Neck: Bi-FPN, (3) Head. Similar to Yolov5, the data are first input to EfficientNet for feature extraction, and then fed to Bi-FPN for feature fusion. Finally, head outputs detection results (class, score, location, size).

2.4. EfficientNet

EfficientNet is a new efficient network proposed by Google. It applied a novel model scaling strategy, namely compound scaling method, to balance network depth, network width, and image resolution for better accuracy at a fixed resource budget. With this, EfficientNet outperformed other hot networks like ResNet [36], DenseNet [37], ResNeXt [38] with the highest Top-1 accuracy in ImageNet image classification task.

The network architecture of EfficientNet is shown in Figure 6. The reason why we choose EfficientNet as our third learner is that it achieves a superior trade-off between accuracy and efficiency. In our model, the third learner plays the most important role. It is responsible for learning the whole image to guide the detection, meaning that its decisions directly determine the final results. Meanwhile, it must be highly efficient, otherwise it will slow down the speed of the entire model.

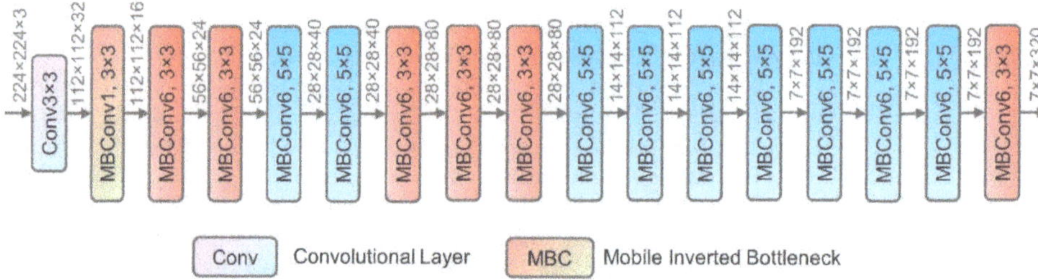

Figure 6. The network architecture of EfficientNet. It can output a feature map with deep semantic information after the input data flows through the multi-layer network.

2.5. Our Model

In real-world forest fire detection task, we need to handle different types of forest fires like ground fire, trunk fire, canopy fire. These fires, influenced by the environment, are diverse in shape, texture, or even color, bringing great difficulty for individual learner to extract effective features. By careful observations, we find that Yolov5 is better at learning long-area fires (Figure 7), but it sometimes misses objects (Figure 8). Meanwhile, even though EfficientDet is not sensitive to long-area fires (Figure 7), it is more careful than Yolov5, meaning that EfficientDet can make a complementary detection (Figure 8). Therefore, we consider that integrating these two efficient learners with different specialties to make detection together can improve detection accuracy.

Figure 7. Yolov5 is better at detecting long-area fires than EfficientDet. (**a**) True positive of Yolov5; (**b**) true positive of Yolov5; (**c**) false negative of EfficientDet; (**d**) false negative of EfficientDet. (**a,b**) illustrate that Yolov5 detect long-area fires successfully, while (**c,d**) show that EfficientDet fails to detect them.

Figure 8. EfficientDet is a more careful object detector than Yolov5, meaning that it seldom losses potential objects easily. (**a**) Yolov5 fails to cover all fire areas; (**b**) Yolov5 misses two fire objects; (**c**) EfficientDet covers all fire areas; (**d**) EfficientDet detects four fire objects.

Another issue is that the ability of the object detector is limited. It only learns the fire region, which is just a local pattern of the whole image, but ignores the other information like background. As a result, the object detector may treat fire-like objects (e.g., sun) as fires (Figure 9), thereby making false alarms. Therefore, a good leader EfficientNet that has a full understanding of the whole image is needed to guide the detection process.

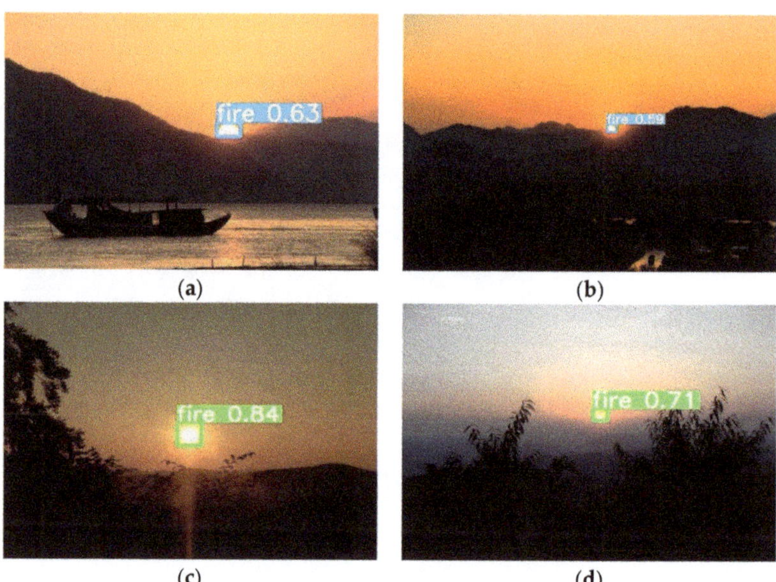

Figure 9. Object detectors Yolov5 and EfficientDet are easy to be deceived by fire-like objects (e.g., sun). (**a**) False positive of Yolov5 (confidence score: 0.63); (**b**) false positive of Yolov5 (confidence score: 0.59); (**c**) false positive of EfficientDet (confidence score: 0.84); (**d**) false positive of EfficientDet (confidence score: 0.71).

To address the above two issues and make sure our model is robust to diverse scenarios, three deep learners are integrated to make decisions together (Figure 10). The first and second learners Yolov5 and EfficientDet act as object detectors, to detect fire locations in images by generating candidate boxes, respectively. Then, the non-maximum suppression algorithm [39] (Algorithm 1) is employed to eliminate redundant boxes, preserving boxes with top confidence. The third learner EfficientNet acts as a binary classifier, responsible for learning the whole image to determine whether the image contains fire objects. Finally, the object detection results, and image classification results are sent into a decision strategy module, in which if the image is considered to contain fire objects, retaining object detection results, otherwise ignoring them.

In addition, integrating multiple learners will not affect the overall efficiency of model, because the three learners are structurally independent, and the whole model is executed by multi processes, meaning that each learner has a separate process responsible for it.

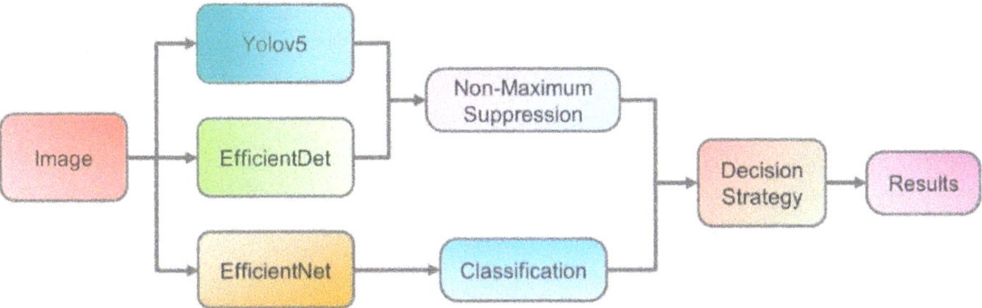

Figure 10. Structure of the proposed model in this paper. Three deep learners are ensembled in parallel. Two object detectors Yolov5 and EfficientDet are integrated to perform object detection task, and the classifier EfficientNet is in charge of discriminating whether the image contains fire objects. Final detection results are made based on the decisions of three learners.

Algorithm 1. Non-Maximum Suppression (NMS)

INPUT: $B = \{b_1, \ldots, b_N\}, S = \{s_1, \ldots, s_N\}, N_t$
B is the list of initial detection boxes
S contains corresponding detection scores
N_t is the NMS threshold
Begin:
$D \leftarrow \{\}$
while $B \neq$ empty **do**

$\quad m \leftarrow \text{argmax } S$
$\quad M \leftarrow b_m$
$\quad D \leftarrow D \cup M; B \leftarrow B - M$

for b_i in B **do**
if $\text{iou}(M, b_i) \geq N_t$ **then**
$B \leftarrow B - b_i; S \leftarrow S - s_i$
end
end

$\quad$ **end**
$\quad$ **Return** D, S
End

2.6. Model Evaluation

We evaluate models using Microsoft COCO criteria (Table 1), which is widely used in object detection tasks. However, fire is a special object, which is diverse in shape, texture, and color. Bounding box generated by object detectors may slightly differ from ground truth (Figure 11), thereby influencing the calculation of average precision, but detectors do identify the fire areas successfully. Therefore, to evaluate models more comprehensively, we introduce two additional evaluation metrics, namely frame accuracy (FA) and false positive rate (FPR). For one image, if the detector misses any fire object, we call it is a frame false (FF), otherwise frame true (FT). If the detector treats any fire-like object as fire, we call it is a false positive (FP), otherwise true positive (TP). Note that FA is calculated on the test set containing 476 forest images, and FPR is calculated on our challenging non-fire dataset containing 641 images with fire-like objects (e.g., sun). The FA and FPR can be calculated as Equation (1) and Equation (2), respectively:

$$FA = \frac{FT}{FT + FF} \times 100, \tag{1}$$

$$FPR = \frac{FP}{FP + TP} \times 100. \tag{2}$$

Table 1. Microsoft COCO criteria—commonly used in object detection task for evaluating the model precision and recall across multiple scales.

Average Precision (AP)	
$AP_{0.5}$	AP at IoU = 0.5
AP Across Scales:	
AP_S	$AP_{0.5}$ for small objects: area < 32^2
AP_M	$AP_{0.5}$ for medium objects: 32^2 < area < 96^2
AP_L	$AP_{0.5}$ for big objects: area > 96^2
Average Recall (AR)	
$AR_{0.5}$	AR at IoU = 0.5
AR Across Scales:	
AR_S	$AR_{0.5}$ for small objects: area < 32^2
AR_M	$AR_{0.5}$ for medium objects: 32^2 < area < 96^2
AR_L	$AR_{0.5}$ for big objects: area > 96^2

Figure 11. Bounding boxes generated by (**a**) Yolov5, (**b**) EfficientDet, and (**c**) our model (3 learners) are different from (**d**) ground truth, but still has good detection performance.

3. Results

3.1. Training

We applied different strategies to train our three learners: Yolov5, EfficientDet, and EfficientNet. Object detectors, namely Yolov5 and EfficientDet, are trained with 2381 forest fire images, and tested with 476 forest fire images. The image classifier, namely EfficientNet, is trained with 2381 forest fire images and 5804 non-fire images, and tested with 476 forest fire images and 1160 non-fire images. Note that non-fire images contain normal images, and images with fire-like objects (e.g., sun). Each model is built up by Pytorch [40] and trained on NVIDIA GTX 2080TI. The details of our training strategy are shown in Table 2.

Table 2. Detailed training strategies of models.

Model	Train	Test	Optimizer	LR	Batch Size	Epoch
Yolov5	2381	476	SGD [41,42]	1×10^{-2}	8	300
EfficientDet	2381	476	AdamW [43]	1×10^{-4}	4	300
EfficientNet	8185	1636	SGD	1×10^{-2}	8	300

LR: learning rate, SGD: stochastic gradient descent, AdamW: Adam with decoupled weight decay.

3.2. Comparison

We compare our model with typical one-stage object detectors. As is shown in Table 3, even though Yolov5 and EfficientDet are the most powerful detectors in this task, the high false positive rate and missing detections cannot be ignored. By integrating them (2 learners), all evaluation metrics are significantly improved, but the false positive rate is increased to 51.6%, since the false positives come from both Yolov5 and EfficientDet. Under the guide of our third learner EfficientNet, the false positive rate is reduced to 0.3%. What is also worth mentioning is that, after introducing the third learner, some metrics are slightly decreased. It is because that EfficientNet wrongly treats some fire images as non-fire ones, and then ignores the object detection results, but we consider it is worthwhile to sacrifice a tiny decrease in average precision and recall for substantial improvement in the false positive rate. To sum up, our model (3 learners) is superior in $AP_{0.5}$, AP_S, AP_M, AP_L, $AR_{0.5}$, AR_S, AR_M, AR_L, FPR, and FA compared with other typical object detectors. Comprehensive improvements make the model have better performance in detecting different types of forest fires: small-scale fires, medium-scale fires, big-scale fires, ground fires, trunk fires, canopy fires, and fires at night (Figures 12 and 13). Faced with fire-like objects (e.g., sun), our model will not be interfered. (Figure 14).

Table 3. Experiments on our dataset—evaluating models using Microsoft COCO criteria, FPR, FA, and latency.

Model	$AP_{0.5}$	AP_S	AP_M	AP_L	$AR_{0.5}$	AR_S	AR_M	AR_L	FPR	FA	Latency (ms)
SSD	66.8	37.8	42.4	78.6	70.1	39.1	45.7	82.7	45.6	92.6	88.8
Yolov3	66.4	26.0	44.6	78.1	71.1	26.1	52.5	82.5	22.9	88.0	**15.6**
Yolov3-SPP	68.3	56.3	49.9	76.7	73.9	60.9	56.6	81.9	30.7	93.3	**15.6**
Yolov4	69.6	53.7	48.9	78.4	75.5	60.9	57.5	83.9	61.9	94.1	20.5
Yolov5	70.5	51.9	53.7	79.2	75.6	56.5	61.2	83.0	22.6	94.7	28.0
EfficientDet	75.7	63.7	58.5	83.0	79.2	65.2	63.9	86.5	41.8	95.5	65.6
Ours (2 learners)	**79.7**	**72.2**	**65.6**	**85.5**	**84.1**	**76.1**	**73.1**	**89.3**	51.6	**99.4**	66.8
Ours (3 learners)	79.0	**72.2**	64.9	84.7	83.8	**76.1**	72.6	88.9	**0.3**	98.9	66.8

Note that $AP_{0.5}$, AP_S, AP_M, AP_L, $AR_{0.5}$, AR_S, AR_M, AR_L, FPR, and FA are all percentages. The best figure of each metric are highlighted in bold.

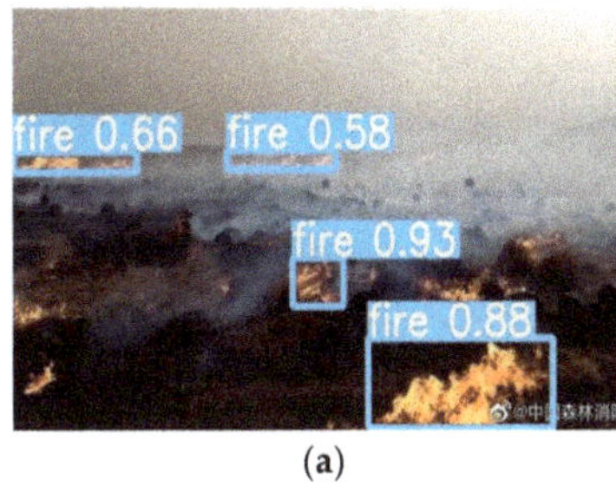

(a)

(b)

(c)

Figure 12. *Cont.*

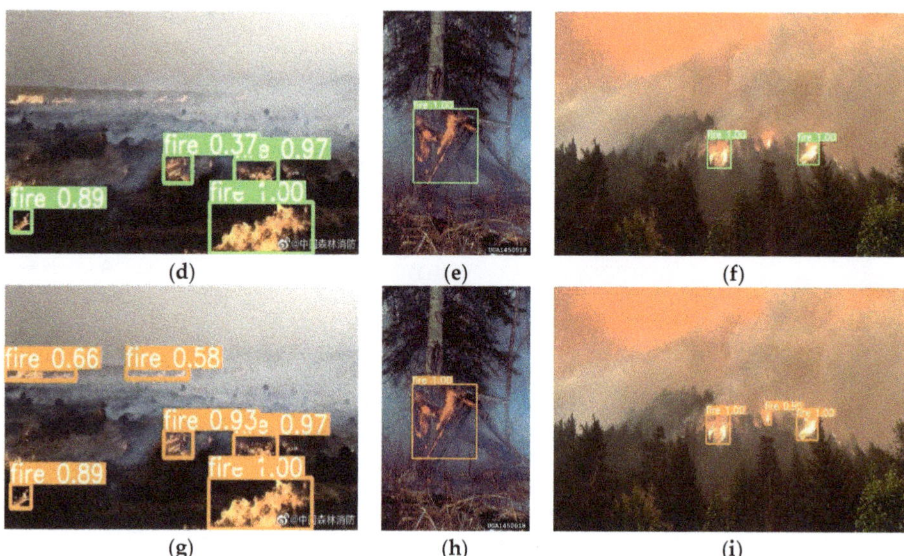

Figure 12. Our ensemble model (3 learners) has better performance on ground fires, trunk fires, and canopy fires. (**a**) Four ground fires detected by Yolov5; (**b**) Yolov5 fails to detect the trunk fire; (**c**) three canopy fires detected by Yolov5; (**d**) four ground fires detected by EfficientDet; (**e**) the trunk fire detected by EfficientDet; (**f**) two canopy fires detected by EfficientDet; (**g**) six ground fires detected by our model; (**h**) the trunk fire detected by our model; (**i**) three canopy fires detected by our model.

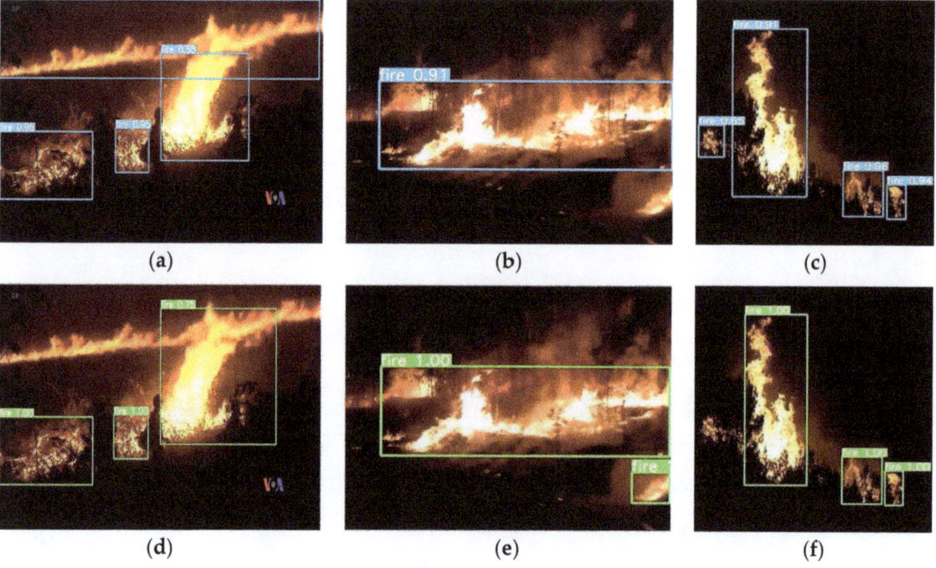

Figure 13. *Cont.*

(g)   (h)   (i)

Figure 13. Our improved model has better performance on small-scale, medium-scale, and big-scale fires at night. (**a**) Medium-scale and big-scale fires detected by Yolov5; (**b**) medium-scale and big scale fires detected by Yolov5; (**c**) small-scale, medium-scale and big-scale fires detected by Yolov5; (**d**) medium-scale and big-scale fires detected by EfficientDet; (**e**) medium-scale and big scale fires detected by EfficientDet; (**f**) small-scale, medium-scale, and big-scale fires detected by EfficientDet; (**g**) medium-scale and big-scale fires detected by our model; (**h**) medium-scale and big scale fires detected by our model; (**i**) small-scale, medium-scale, and big-scale fires detected by our model.

(a)   (b)

(c)   (d)

(e)   (f)

Figure 14. Under the guide of EfficientNet, our ensemble model has a good discriminability between fire and fire-like objects (e.g., sun). (**a**) True negative of Yolov5; (**b**) false positive of Yolov5 (confidence score: 0.59); (**c**) false positive of EfficientDet (confidence score 0.71); (**d**) true negative of EfficientDet; (**e**) true negative of our model; (**f**) true negative of our model.

4. Discussion

Compared with other common objects that have fixed form, forest fire is a dynamic object [44]. In the real-world scenario, a forest fire usually starts from small-scale fire, develops to medium-scale fire, and then to big-scale fire [45]. In terms of types, it starts from ground fire, then spreads to the trunk, and finally to the canopy [46]. The various shapes, sizes, textures, and colors of forest fires make the fire evolution a complex process, and bring great difficulty in fire detection.

Therefore, it is highly imperative for detectors to be sensitive to different types of fires. Through careful experimental comparisons, we find that no single detector that can handle all kinds of fires. They have respective advantages and disadvantages. Yolov5 is excellent at detecting long-area fires (Figure 7), but it frequently misses objects (Figure 8). EfficientDet is a more careful detector, compared to Yolov5; even though it has a bad performance on long-area fires (Figure 7), it can detect fires that Yolov5 cannot (Figure 8), meaning that it is a good partner for Yolov5. Our model, which efficiently integrates decisions of these two powerful learners, boost detection performance by 2.5–10.9%, in terms of $AP_{0.5}$, AP_S, AP_M, AP_L, $AR_{0.5}$, AR_S, AR_M, AR_L. The significant improvements of average precision and average recall for small, medium, and big objects make the model more sensitive to the size changes of fires, thereby enhancing detection performance on different types of forest fires: ground fire, trunk fire, canopy fire, and fires at night in the fire evolution (Figures 12 and 13).

Another problem is that the false positive rate of the improved model (2 learners) becomes higher: 22.6% to 51.6% since the model also integrates wrong detection results from both learners. To address this issue, we use 8185 images containing 2381 forest fire images and 5804 non-fire images (containing fire-like images and normal forest images) to train our third learner EfficientNet. Sufficient training sets enabled EfficientNet to show a good discriminability between fire objects and fire-like objects, with 99.6% accuracy on 476 fire images, and 99.7% accuracy on 676 fire-like images. With the help of the leader learner EfficientNet, wrong detection results are eliminated, and the false positive rate is significantly decreased to 0.3% (Figure 14). Noticeably, the join of EfficientNet reduces $AP_{0.5}$, AP_M, AP_L, $AR_{0.5}$, AR_M, AR_L by roughly 1%, which is because that EfficientNet wrongly ignores 2 fire images containing medium-scale and big-scale fire objects.

In terms of latency, the Yolo family is superior compared to EfficientDet and SSD. Excellent inference speed makes Yolo family widely used in real-world applications, but experimental results show that they are not able to have a satisfactory performance on forest fire detection tasks. The latency of EfficientDet is 65.6 ms, which is over twice that of Yolov5 (28.0 ms), but EfficientDet outperforms Yolov5 by over 5% regarding detection performance. We ensemble these three learners Yolov5 (28.0 ms), EfficientDet (65.6 ms), EfficientNet (31.3 ms) in parallel to make sure that our model can achieve the best performance without any extra latency. The final latency of our model (3 learners) is 66.8 ms, which shows that an excellent trade-off between detection performance and efficiency has been achieved, and the model is applicable for real-time detection task.

For further improvement, we plan to study the labeling strategy for forest fires, since the quality of training data directly determines the detection performance. Another interesting extension is to investigate the network architecture of backbones, and modify them to make sure that they are specially designed for forest fire detection task. Additionally, we will work on developing a forest fire tracking system, which can classify different types of forest fires: ground fire, trunk fire and canopy fire, to track the evolution and spread of forest fires.

5. Conclusions

The successful application of convolutional neural networks significantly improves the performance of object detection. However, forest fire is a dynamic object with no fixed form, which the individual object detector cannot handle. In addition, object detectors are easy to be deceived by fire-like objects and generate false positives due to their limited visual field. To address these problems, a novel ensemble learning method for real-time forest fire detection is proposed in this paper. Two powerful object detectors (Yolov5 and EfficientDet) with different expertise are integrated to make the whole model more robust to diverse forest fire scenarios. Then, a leader (EfficientNet) is introduced to guide the detection process to reduce false positives. Experimental results show that, compared with other popular object detectors, our model achieves a superior trade-off among average precision, average recall, false positive rate, frame accuracy, and latency. The significant improvements make it possible for the model to perform well in real-world forestry applications.

Author Contributions: R.X. devised the programs and drafted the initial manuscript. H.L. and K.L. helped with data collection, data analysis, and figures and tables. L.C. contributed to fund acquisition and writing embellishment. Y.L. designed the project and revised the manuscript. All authors have read and agreed to the published version of the manuscript.

Funding: This research was funded by the National Key R&D Program of China (grant number 2017YFD0600904) and the Priority Academic Program Development of Jiangsu Higher Education Institutions (PAPD).

Data Availability Statement: Publicly available datasets were analyzed in this study. The data can be found here: BowFire [28], FD-dataset [29], ForestryImages [30], VisiFire [31].

Conflicts of Interest: The authors declare no conflict of interest.

References

1. Zhang, J.; Li, W.; Yin, Z.; Liu, S.; Guo, X. Forest fire detection system based on wireless sensor network. In Proceedings of the 4th IEEE Conference on Industrial Electronics and Applications (ICIEA 2009), Xi'an, China, 25–27 May 2009; pp. 520–523.
2. Yu, L.; Wang, N.; Meng, X. Real-time forest fire detection with wireless sensor networks. In Proceedings of the International Conference on Wireless Communications, Networking and Mobile Computing (WiCOM 2005), Wuhan, China, 26 September 2005; pp. 1214–1217.
3. Chen, S.J.; Hovde, D.C.; Peterson, K.A.; Marshall, A.W. Fire detection using smoke and gas sensors. *Fire Saf. J.* **2007**, *42*, 507–515. [CrossRef]
4. Zhang, F.; Zhao, P.; Xu, S.; Wu, Y.; Yang, X.; Zhang, Y. Integrating multiple factors to optimize watchtower deployment for wildfire detection. *Sci. Total Environ.* **2020**, *737*, 139561. [CrossRef] [PubMed]
5. Zhang, F.; Zhao, P.; Thiyagalingam, J.; Kirubarajan, T. Terrain-influenced incremental watchtower expansion for wildfire detection. *Sci. Total Environ.* **2018**, *654*, 164–176. [CrossRef] [PubMed]
6. Lee, B.; Kwon, O.; Jung, C.; Park, S. The development of UV/IR combination flame detector. *J. KIIS* **2001**, *16*, 1–8.
7. Kang, D.; Kim, E.; Moon, P.; Sin, W.; Kang, M. Design and analysis of flame signal detection with the combination of UV/IR sensors. *J. Korean Soc. Int. Inf.* **2013**, *14*, 45–51. [CrossRef]
8. Fernandes, A.M.; Utkin, A.B.; Lavrov, A.V.; Vilar, R.M. Development of neural network committee machines for automatic forest fire detection using lidar. *Pattern Recognit.* **2004**, *37*, 2039–2047. [CrossRef]
9. Chen, T.H.; Wu, P.H.; Chiou, Y.C. An early fire-detection method based on image processing. In Proceedings of the IEEE International Conference on Image Processing (ICIP 2004), Singapore, 24–27 October 2004; pp. 1707–1710.
10. Töreyin, B.U.; Dedeoğlu, Y.; Güdükbay, U.; Cetin, A.E. Computer vision based method for real-time fire and flame detection. *Pattern Recognit. Lett.* **2006**, *27*, 49–58. [CrossRef]
11. Çelik, T.; Özkaramanlı, H.; Demirel, H. Fire and smoke detection without sensors: Image processing based approach. In Proceedings of the IEEE 15th European Signal Processing Conference (EUSIPCO 2007), Poznan, Poland, 3–7 September 2007; pp. 1794–1798.
12. Teng, Z.; Kim, J.H.; Kang, D.J. Fire detection based on hidden Markov models. *Int. J. Control Autom. Syst.* **2010**, *8*, 822–830. [CrossRef]
13. Chino, D.Y.; Avalhais, L.P.; Rodrigues, J.F.; Traina, A.J. Bowfire: Detection of fire in still images by integrating pixel color and texture analysis. In Proceedings of the 28th SIBGRAPI Conference on Graphics, Patterns and Images, Salvador, Brazil, 26–29 August 2015; pp. 95–102.
14. Wu, S.; Zhang, L. Using popular object detection methods for real time forest fire detection. In Proceedings of the 11th International Symposium on Computational Intelligence and Design (ISCID 2018), Hangzhou, China, 8–9 December 2018; pp. 280–284.
15. Ren, S.; He, K.; Girshick, R.; Sun, J. Faster r-cnn: Towards real-time object detection with region proposal networks. *IEEE Trans. Pattern Anal. Mach. Intell.* **2016**, *39*, 1137–1149. [CrossRef] [PubMed]
16. Redmon, J.; Divvala, S.; Girshick, R.; Farhadi, A. You only look once: Unified, real-time object detection. In Proceedings of the IEEE Conference on Computer Vision and Pattern Recognition (CVPR 2016), Las Vegas, NV, USA, 26 June–1 July 2016; pp. 779–788.

17. Redmon, J.; Farhadi, A. YOLO9000: Better, faster, stronger. In Proceedings of the IEEE Conference on Computer Vision and Pattern Recognition (CVPR 2017), Honolulu, Hawaii, 21–26 July 2017; pp. 7263–7271.
18. Redmon, J.; Farhadi, A. Yolov3: An incremental improvement. *arXiv* **2018**, arXiv:1804.02767.
19. Bochkovskiy, A.; Wang, C.Y.; Liao, H.Y.M. YOLOv4: Optimal Speed and Accuracy of Object Detection. *arXiv* **2020**, arXiv:2004.10934.
20. Liu, W.; Anguelov, D.; Erhan, D.; Szegedy, C.; Reed, S.; Fu, C.Y.; Berg, A.C. Ssd: Single shot multibox detector. In Proceedings of the European Conference on Computer Vsion (ECCV 2016), Amsterdam, The Netherlands, 8–16 October 2016; pp. 21–37.
21. Kim, B.; Lee, J. A video-based fire detection using deep learning models. *Appl. Sci.* **2019**, *9*, 2862. [CrossRef]
22. Lee, Y.; Shim, J. False Positive Decremented Research for Fire and Smoke Detection in Surveillance Camera using Spatial and Temporal Features Based on Deep Learning. *Electronics* **2019**, *8*, 1167. [CrossRef]
23. Pan, H.; Badawi, D.; Cetin, A.E. Computationally Efficient Wildfire Detection Method Using a Deep Convolutional Network Pruned via Fourier Analysis. *Sensors* **2020**, *20*, 2891. [CrossRef] [PubMed]
24. Wu, S.; Guo, C.; Yang, J. Using PCAand one-stage detectors for real-time forest fire detection. *J. Eng.* **2020**, *2020*, 383–387. [CrossRef]
25. Ultralytics-Yolov5. Available online: https://github.com/ultralytics/yolov5 (accessed on 1 January 2021).
26. Tan, M.; Pang, R.; Le, Q.V. Efficientdet: Scalable and efficient object detection. In Proceedings of the IEEE Conference on Computer Vision and Pattern Recognition (CVPR 2020), Washington, DC, USA, 14–19 June 2020; pp. 10781–10790.
27. Tan, M.; Le, Q. EfficientNet: Rethinking Model Scaling for Convolutional Neural Networks. In Proceedings of the International Conference on Machine Learning (ICML 2019), Long Beach, CA, USA, 9–15 June 2019; pp. 6105–6114.
28. BoWFire Dataset. Available online: https://bitbucket.org/gbdi/bowfire-dataset/downloads/ (accessed on 1 January 2021).
29. FD-Dataset. Available online: http://www.nnmtl.cn/EFDNet/ (accessed on 1 January 2021).
30. ForestryImages. Available online: https://www.forestryimages.org/browse/subthumb.cfm?sub=740 (accessed on 1 January 2021).
31. VisiFire. Available online: http://signal.ee.bilkent.edu.tr/VisiFire/ (accessed on 1 January 2021).
32. Everingham, M.; Eslami, S.A.; Van Gool, L.; Williams, C.K.; Winn, J.; Zisserman, A. The pascal visual object classes challenge: A retrospective. *Int. J. Comput. Vis.* **2015**, *111*, 98–136. [CrossRef]
33. Lin, T.Y.; Maire, M.; Belongie, S.; Hays, J.; Perona, P.; Ramanan, D.; Dollár, P.; Zitnick, C.L. Microsoft coco: Common objects in context. In Proceedings of the 13th European Conference on Computer Cision (ECCV 2014), Zurich, Switzerland, 6–12 September 2014; pp. 740–755.
34. Wang, C.Y.; Mark Liao, H.Y.; Wu, Y.H.; Chen, P.Y.; Hsieh, J.W.; Yeh, I.H. CSPNet: A new backbone that can enhance learning capability of cnn. In Proceedings of the IEEE Conference on Computer Vision and Pattern Recognition (CVPR 2020), Washington, DC, USA, 14–19 June 2020; pp. 390–391.
35. Wang, K.; Liew, J.H.; Zou, Y.; Zhou, D.; Feng, J. Panet: Few-shot image semantic segmentation with prototype alignment. In Proceedings of the IEEE International Conference on Computer Vision (ICCV 2019), Seoul, Korea, 20–26 October 2019; pp. 9197–9206.
36. He, K.; Zhang, X.; Ren, S.; Sun, J. Deep residual learning for image recognition. In Proceedings of the IEEE Conference on Computer Vision and Pattern Recognition (CVPR 2016), Las Vegas, NV, USA, 26 June–1 July 2016; pp. 770–778.
37. Huang, G.; Liu, Z.; Van Der Maaten, L.; Weinberger, K.Q. Densely connected convolutional networks. In Proceedings of the IEEE Conference on Computer Vision and Pattern Recognition (CVPR 2017), Honolulu, Hawaii, 21–26 July 2017; pp. 4700–4708.
38. Xie, S.; Girshick, R.; Dollár, P.; Tu, Z.; He, K. Aggregated residual transformations for deep neural networks. In Proceedings of the IEEE Conference on Computer Vision and Pattern Recognition (CVPR 2016), Las Vegas, NV, USA, 26 June–1 July 2016; pp. 1492–1500.
39. Neubeck, A.; Van Gool, L. Efficient non-maximum suppression. In Proceedings of the 18th International Conference on Pattern Recognition (ICPR 2006), Hong Kong, China, 20–24 August 2006; pp. 850–855.
40. Paszke, A.; Gross, S.; Massa, F.; Lerer, A.; Bradbury, J.; Chanan, G.; Killeen, T.; Lin, Z.; Gimelshein, N.; Antiga, L. Pytorch: An imperative style, high-performance deep learning library. In Proceedings of the Neural Information Processing Systems (NIPS 2019), Vancouver, BC, Canada, 8–14 December 2019; pp. 8026–8037.
41. Bottou, L. Large-scale machine learning with stochastic gradient descent. In Proceedings of the 19th International Conference on Computational Statistics (COMPSTAT 2010), Paris, France, 22–27 August 2010; pp. 177–186.
42. Zinkevich, M.; Weimer, M.; Li, L.; Smola, A.J. Parallelized stochastic gradient descent. In Proceedings of the Neural Information Processing Systems (NIPS 2010), Vancouver, BC, Canada, 6–11 December 2010; pp. 2595–2603.
43. Loshchilov, I.; Hutter, F. Decoupled weight decay regularization. *arXiv* **2017**, arXiv:1711.05101.
44. Merino, L.; Caballero, F.; Martínez-de-Dios, J.R.; Maza, I.; Ollero, A. An unmanned aircraft system for automatic forest fire monitoring and measurement. *J. Intell. Robot. Syst.* **2012**, *65*, 533–548. [CrossRef]
45. Serón, F.J.; Gutiérrez, D.; Magallón, J.; Ferragut, L.; Asensio, M.I. The evolution of a wildland forest fire front. *Vis. Comput.* **2005**, *21*, 152–169. [CrossRef]
46. Pimont, F.; Dupuy, J.L.; Linn, R.R.; Dupont, S. Impacts of tree canopy structure on wind flows and fire propagation simulated with FIRETEC. *Ann. For. Sci.* **2011**, *68*, 523–530. [CrossRef]

MDPI
St. Alban-Anlage 66
4052 Basel
Switzerland
Tel. +41 61 683 77 34
Fax +41 61 302 89 18
www.mdpi.com

Forests Editorial Office
E-mail: forests@mdpi.com
www.mdpi.com/journal/forests